SOLID WASTE MANAGEMENT
IN THE WORLD'S CITIES

SOLID WASTE MANAGEMENT IN THE WORLD'S CITIES

WATER AND SANITATION IN THE WORLD'S CITIES 2010

United Nations Human Settlements Programme

Routledge
Taylor & Francis Group

LONDON AND NEW YORK

First published 2010
by Earthscan

Published 2014 by Routledge
2 Park Square, Milton Park, Abingdon, Oxon OX14 4RN
711 Third Avenue, New York, NY, 10017, USA

Routledge is an imprint of the Taylor & Francis Group, an informa business

DISCLAIMER

The designations employed and the presentation of the material in this publication do not imply the expression of any opinion whatsoever on the part of the Secretariat of the United Nations concerning the legal status of any country, territory, city or area, or of its authorities, or concerning delimitation of its frontiers or boundaries, or regarding its economic system or degree of development. The analysis, conclusions and recommendations of the report do not necessarily reflect the views of the United Nations Human Settlements Programme (UN-HABITAT), the Governing Council of UN-HABITAT or its Member States.

Earthscan publishes in association with the International Institute for Environment and Development

Typeset by MapSet Ltd, Gateshead, UK
Cover design by Susanne Harris

A catalogue record for this book is available from the British Library

Library of Congress Cataloging-in-Publication Data has been applied for

ISBN 13: 978-1-849-71169-2 (hbk)
ISBN 13: 978-1-849-71170-8 (pbk)

At Earthscan we strive to minimize our environmental impacts and carbon footprint through reducing waste, recycling and offsetting our CO_2 emissions, including those created through publication of this book.

Printed and bound by CPI Group (UK) Ltd, Croydon, CR0 4YY

FOREWORD

Regardless of the context, managing solid waste is one of biggest challenges of the urban areas of all sizes, from mega-cities to the small towns and large villages, which are home to the majority of humankind. It is almost always in the top five of the most challenging problems for city managers. It is somewhat strange that it receives so little attention compared to other urban management issues. The quality of waste management services is a good indicator of a city's governance. The way in which waste is produced and discarded gives us a key insight into how people live. In fact if a city is dirty, the local administration may be considered ineffective or its residents may be accused of littering. Available data show that cities spend a substantial proportion of their available recurrent budget on solid waste management, yet waste collection rates for cities in low- and middle-income countries range from a low of 10 per cent in peri-urban areas to a high of 90 per cent in commercial city centres.

Many developing and transitional country cities have active informal sector recycling, reuse and repair systems, which are achieving recycling rates comparable to those in the West, at no cost to the formal waste management sector. Not only does the informal recycling sector provide livelihoods to huge numbers of the urban poor, but they may save the city as much as 15 to 20 per cent of its waste management budget by reducing the amount of waste that would otherwise have to be collected and disposed of by the city. This form of inclusion in solid waste management shows how spectacular results can be achieved where the involvement of the informal sector is promoted.

The struggle for achieving the Millennium Development Goal and related targets for water and sanitation is being waged in our cities, towns and villages where solid wastes are generated. It is at this level that policy initiatives on solid waste management become operational reality and an eminently political affair: conflicts have to be resolved and consensus found among competing interests and parties.

This publication, *Solid Waste Management in the World Cities*, is the third edition in UN-HABITAT's State of Water and Sanitation in the World Cities series. It aims to capture the world's current waste management trends and draw attention to the importance of waste management, especially regarding its role in reaching the UN Millennium Development Goals. The publication acknowledges the escalating challenges in solid waste management across the globe. It seeks to showcase the good work that is being done on solid waste by cities around the world, large and small, rich and poor. It achieves this by looking at what drives change in solid waste management, how cities find local solutions and what seems to work best under different circumstances. The publication endeavours to help decision-makers, practitioners and ordinary citizens understand how a solid waste management system works and to inspire people everywhere to make their own decisions on the next steps in developing a solution appropriate to their own city's particular circumstances and needs. Most readers will never travel to all the 20 cities featured in this report, but through this publication they will have access to real experiences of people working on the ground. We hope it will provide a reference point for managing solid waste in the world's cities and towns, and that many will follow in the footsteps of our authors, and we can move to an improved set of global reference data.

Anna Tibaijuka
Under-Secretary General, United Nations
Executive Director, UN-Habitat

ACKNOWLEDGEMENTS

The Water and Sanitation in the World's Cities series, published every three years, was mandated by UN-HABITAT Governing Council Resolution 19/6, adopted on 9 May 2003, following the publication of the first edition, *Water and Sanitation in the World's Cities: Local Action for Global Goals*, in March 2003. This was followed by the production of the second edition, *Meeting Development Goals in Small Urban Centres: Water and Sanitation in the World's Cities 2006*.

This current production of *Solid Waste Management in the World Cities* was funded by the Water and Sanitation Trust Fund of UN-HABITAT, currently supported by the Governments of Spain, the Netherlands and Norway.

The report was prepared under the overall substantive guidance of Bert Diphoorn, Ag. Director, Human Settlements Financing Division. The conceptualization of the report and the management of its production was undertaken by Graham Alabaster who was supported by Paul Onyango.

The substantive preparation of the report was undertaken by WASTE, Advisers on Urban Environment and Development. Anne Scheinberg, (WASTE and Wageningen University and Research Centre), David C. Wilson (Imperial College) and Ljiljana Rodic-Wiersma (Wageningen University and Research Centre) were the principal authors/editors. They were supported by a team who made substantial contributions, comprising Lilia G. C. Casanova (CAPS); Bharati Chaturvedi (Chintan Environmental Research and Action Group); Manus Coffey (Manus Coffey Associates); Sanjay K. Gupta and Jeroen IJgosse (Independent Consultants); Reka Soos (Green Partners Romania); and Andrew Whiteman (Wasteaware). The report would not have been possible without the dedication, commitment, professionalism, and passion which the contributing authors undertook their work.

The report benefited from a number of consultations attended by eminent researchers and solid waste management experts. In early 2009, an outline of the report was prepared and in May 2009 the first expert group meeting took place in Gouda, the Netherlands to review the annotated outline. Participants in the expert group included Manus Coffey (Manus Coffey Associates), Edward Stentiford (Leeds University), Sonia Dias, Oscar Espinoza, Sanjay Gupta, Kossara Kisheva, Michael Simpson, Portia Sinnott, Reka Soos, Ljiljana Rodic, Anne Scheinberg, Andy Whiteman, David Wilson, Verele de Vreede, Lilliana Abarca, Ivo Haenen, Valentin Post and Alodia Ishengoma. Bert Diphoorn, Graham Alabaster and Paul Onyango oversaw the process for UN-HABITAT. A number of contributors and researchers were identified at the consultations for preparing the 20 city inserts and the key sheets for the report.

The contributors and writers for the city profiles were: Adelaide, Australia by Andrew Whiteman (Wasteaware) and Rebecca Cain (Hyder Consulting); Bamako, Mali by Modibo Keita (CEK, Cabinet d'Etudes Kala Saba), Erica Trauba (WASTE Intern), Mandiou Gassama, Bakary Diallo, and Mamadou Traoré, (CEK); Bengaluru, India by Sanjay K. Gupta (Senior Advisor and Consultant Water, Sanitation and Livelihood), Smt. Hemalatha (KBE, MTech.), BMP Environmental Engineer, Bruhat Bangalore Mahangara Palike, and Anselm Rosario (WasteWise Resource Centre); Belo Horizonte, Brazil by Sônia Maria Dias, Jeroen IIgosse, Raphael T. V. Barros (UFMG), and the team of the Planning Department of SLU; Cañete, Peru by Oscar Espinoza, Humberto Villaverde (IPES), Jorge Canales and Cecilia Guillen; Curepipe, Mauritius by Professor Edward Stentiford (University of Leeds, UK) and Professor Romeela Mohee (The University of Mauritius, Mauritius); Delhi, India by Malati Gadgil, Anupama Pandey, Bharati Chaturvedi and Prakash Shukla (Chintan Environmental Research and Action Group), Irmanda Handayani (WASTE and Chintan Evironmental), Jai Prakash Choudhury/Santu and Safai Sena; Dhaka, Bangladesh by Andrew Whiteman (Wasteaware), Monir Chowdhury (Commitment

Consultants), Shafiul Azam Ahmed, Dr Tariq bin Yusuf (DCC), Prof. Ghulam Murtaza (Khulna University) and Dr Ljiljana Rodic (Wageningen University); Ghorahi, Nepal by Bhushan Tuladhar (ENPHO, Environment and Public Health Organization); Kunming, China by Ljiljana Rodic and Yang Yuelong (Wageningen University); Lusaka, Zambia by Michael Kaleke Kabungo (Lusaka City Council – Waste Management Unit) and Rueben Lupupa Lifuka (Riverine Development Associates); Managua, Nicaragua by Jane Olley (Technical Advisor for UN-HABITAT Improving Capacity for Solid Waste Management in Managua Programme), Jeroen IJgosse (Facilitator and International Consultant for UN-HABITAT Programme "Improving Capacity for Solid Waste Management in Managua"), Victoria Rudin (Director of the Central American NGO ACEPESA, partner organization for UN-HABITAT Improving Capacity for Solid Waste Management in Managua Programme), Mabel Espinoza, Wilmer Aranda, Juana Toruño and Tamara Yuchenko (Technical Committee from Municipality of Managua for UN-HABITAT Improving Capacity for Solid Waste Management in Managua Programme); Moshi, Tanzania by Mrs Alodia Ishengoma (Independent Consultant), Ms. Bernadette Kinabo (Municipal Director Moshi), Dr Christopher Mtamakaya (Head of Health and Cleansing Department), Ms Viane Kombe (Head of Cleansing Section), Miss Fidelista Irongo (Health Secretary), Mr Lawrence Mlay (Environmental Health Officer); Nairobi, Kenya by Misheck Kirimi (NETWAS), Leah Oyake-Ombis, Director of Environment, City Council of Nairobi / Wageningen University) and Ljiljana Rodic (Wageningen University); Quezon City, the Philippines by Ms Lizette Cardenas (SWAPP), Ms Lilia Casanova (CAPS), Hon Mayor Feliciano Belmonte, Hon Vice Mayor Herbert Bautista, Quezon City Councilors, Quezon City Environmental Protection and Waste Management Department headed by Ms Frederika Rentoy, Payatas Operations Group, Quezon City Planning and Development Office, Ms Andrea Andres-Po and Mr Paul Andrew M. Tatlonghari of QC-EPWMD; Rotterdam, The Netherlands by Frits Fransen (ROTEB retired), Joost van Maaren, Roelof te Velde (ROTEB), Anne Scheinberg and Ivo Haenen (WASTE); San Francisco, USA by Portia M. Sinnott (Reuse, Recycling and Zero Waste Consultant) and Kevin Drew (City of San Francisco Special Projects and Residential Zero Waste Coordinator); Sousse, Tunisia by Verele de Vreede (WASTE), Tarek Mehri, and Khaled Ben Adesslem (Muncipality of Sousse); Tompkins County, USA by Barbara Eckstrom (Tompkins County Solid Waste Manager), Kat McCarthy (Tompkins County Waste Reduction and Recycling Specialist), Portia M. Sinnott (Reuse, Recycling and Zero Waste Consultant) and Anne Scheinberg (WASTE); and Varna, Bulgaria by Kossara Bozhilova-Kisheva and Lyudmil Ikonomov (CCSD Geopont-Intercom).

The key sheet authors were Ellen Gunsilius and Sandra Spies (GTZ), Lucia Fernandez and Chris Bonner (WIEGO), Jarrod Ball (Golder Associates), Nadine Dulac (World Bank Institute), Alodia Ishengoma (ILO), A. H. M. Maqsood Sinha and Iftekhar Enayetullah (Waste Concern), Jeroen IJgosse, Manus Coffey, Lilia Casanova, Ivo Haenen, Michael Simpson, David C. Wilson, Valentin Post (WASTE), Choudhury R. C. Mohanty (UNCRD), Bharati Chaturvedi, Anne Scheinberg, Andy Whiteman (WasteAware), Valentina Popovska (IFC), Frits Fransen, Arun Purandhare, Sanjay K. Gupta and Ljiljana Rodic-Wiersma.

The second Expert Group Meeting was jointly organized by UN-HABITAT and WASTE, with a great deal of support from Dr Laila Iskandar and the staff of CID Consulting, Zemalek, in Cairo, Egypt in October 2009. The expert group meeting reviewed the draft and recommended a reorganization and further expansion of report.

The production management for the team was handled by Verele de Vreede (WASTE). Ivo Haenen (WASTE), Michael Simpson (Antioch New England) and Portia M. Sinnott (Lite) were in charge of data management while the writing and research team members were Sônia M. Dias, Oscar Espinoza, Irmanda Handayani, Kossara Kisheva-Bozhilovska and Erica Trauba. Photo and graphic editors were Verele de Vreede and Jeroen IJgosse.

The report was widely circulated among UN-HABITAT professional staff members and consultants, and benefited from their review, comments and inputs. Valuable contributions were made by Graham Alabaster, Andre Dzikus, Paul Onyango, Daniel Adom and Debashish Bhattacharjee.

Antoine King and his team of the Programme Support Division, Marcellus Chege, Veronica Njuguna, Grace Wanjiru and Helda Wandera of Water, Sanitation and Infrastructure Branch of UN-HABITAT and Margaret Mathenge of UNON provided valuable administrative support.

Special thanks are due to our publishers, Earthscan, led by Jonathan Sinclair Wilson, with Hamish Ironside and Claire Lamont. Their support in working to extreme deadlines is gratefully acknowledged. Special recognition goes to Arnold van de Klundert, Founder of WASTE, who, with support from Justine Anschutz, distilled the ISWM framework in the course of the Urban Waste Expertise Programme.

This publication is dedicated to the memory of Brian Williams, who passed away during its production. Brian was heavily involved in previous editions of this report.

CONTENTS

LIST OF FIGURES, TABLES AND BOXES

FIGURES

TABLES

BOXES

LIST OF ACRONYMS AND ABBREVIATIONS

3-R	reduce, reuse, recycle
3-R Forum	A programme of UNCRD
AAU	assigned amount unit
ACEPESA	Asociación Centroamericana para la Economía, la Salud y el Ambiente (Central American Association for Economy, Health and Environment)
ACR+	Association of Cities and Regions for Recycling and Sustainable Resource Management
ADB	Asian Development Bank
AECID	Spanish Agency for International Cooperation for Development
AIE	accredited independent entity
AIT	Asian Institute of Technology
APFED	Asia-Pacific Forum for Environment and Development
APO	Asian Productivity Organization
ASEAN	Association of Southeast Asian Nations
BATF	Bangalore Agenda Task Force
BCRC	Basel Convention Regional Coordinating Centre for Asia and the Pacific
BMP	Bengaluru Municipal Corporation
BMW	biomedical waste
BoI	Board of Investment
BOO	build–own–operate
3Cs	confine, compact, cover
C&D	construction and demolition
C&I	commercial and industrial
CAN	Coupe d'Afrique de Nations
CBE	community-based enterprise
CBO	community-based organization
CCX	Chicago Climate Exchange
CDIA	Cities Development Initiative for Asia
CDM	Clean Development Mechanism
CER	certified emissions reduction
CIDA	Canadian International Development Agency
CIWMB	California Integrated Waste Management Board
CO_2	carbon dioxide
CO_2e	carbon dioxide equivalent
COFESFA	Coopérative des Femmes pour l'Éducation, la Santé Familiale et l'Assainissement
CONPAM	Ceará State Council for Environment Policy and Management
CSD	Commission on Sustainable Development
CTRS	Centro de Tratamento de Resíduos Sólidos (Brazil)

CWG	Collaborative Working Group on Solid Waste Management in Low- and Middle-Income Countries
DANIDA	Danish International Development Agency
DBOO	design–build–own–operate
DCC	Dhaka City Corporation
DCCl	Dar es Salaam City Council
DCCn	Dar es Salaam City Commission
Defra	UK Department for Environment, Food and Rural Affairs
DOE	designated operational entity
PDUD	Projet de Développement Urbain et Décentralisation
DFID	UK Department for International Development
DIY	do it yourself
DNA	Designated National Authority
DSD	Division for Sustainable Development
EIA	environmental impact assessment
EMC	Environmental Municipal Commission
EnTA	Environmental Technology Assessment
EPA	US Environmental Protection Agency
EPR	extended producer responsibility
ERM	Environmental Resources Management
EU	European Union
EU ETS	European Union Emissions Trading Scheme
FEAM	Federal State Environmental Authority (Brazil)
GDP	gross domestic product
GEF	Global Environment Facility
GHG	greenhouse gas
GIE	*Groupement d'Intérêt Économique*
GIS	green investment schemes
GNP	gross national product
GS	Gold Standard
GTZ	Deutsche Gesellschaft für Technische Zusammenarbeit (German Technical Cooperation)
HDI	Human Development Index
HDPE	high-density polyethylene
HHW	household hazardous waste
HLC	High-Level Consultation (Group)
ICLEI	International Council for Local Environmental Initiatives
ID	identification
IETC	International Environmental Technology Centre (UNEP)
IFC	International Finance Corporation
IFI	international financial institution
IFP	ILO Programme on Boosting Employment through Small Enterprise Development
IGES	Institute for Global Environmental Strategies
IGNOU	Indira Gandhi National Open University
ILO	International Labour Organization
IPC	intermediate processing centre
IPCC	Intergovernmental Panel on Climate Change
IPF	intermediate processing facility
ISHWM	Indian Society of Hospital Waste Management

ISP	informal service provider
ISWM	integrated sustainable waste management
ISWMP	Integrated Solid Waste Management Programme
IUF	International Union of Food, Agricultural, Hotel, Restaurant, Catering, Tobacco and Allied Workers
IULA	International Union of Local Authorities
IWB	itinerant waste buyer
JI	joint implantation
JICA	Japan International Cooperation Agency
JV	joint venture
KKPKP	Trade Union of Waste-Pickers in Pune, India
KSTP	Keppel Seghers Tuas Waste-to-Energy Plant
LDPE	low-density polyethylene
LEI	Dutch Agricultural Economics Institute
LF	landfill
LFG	landfill gas capture/extraction
LGA	local government authority
LTS	large transfer station
MCD	Municipal Corporation of Delhi
MDG	Millennium Development Goal
MEIP	Metropolitan Environmental Improvement Programme
METAP	Mediterranean Environmental Technical Assistance Programme
MoEF	Ministry of Environment and Forests
MRF	materials recovery facility
MSE	micro- and small enterprise
MSW	municipal solid waste
MW	megawatt
NDMC	New Delhi Municipal Council
NGO	non-governmental organization
NIMBY	not in my backyard
NOC	No-Objection Certificate
NO_x	nitrogen oxide
OECD	Organisation for Economic Co-operation and Development
5-Ps	pro-poor public–private partnerships
PAH	polycyclic aromatic hydrocarbon
PAHO	Pan-American Health Organization
PBDE	polybrominated diphenyl ether
PCB	polychlorinated biphenyl
PET	polyethylene terephthalate
PFD	process flow diagram
PGAP	Multi-Annual Municipal Action Plan (Brazil)
PIL	public interest litigation
PP	polypropylene
PPP	public–private partnership
PPP-SD	public–private partnership for sustainable development
PS	polystyrene
PSP	private-sector participation
3Rs	reduce, reuse, recycle

R&D	research and development
RLP	Recycling Linkages Programme
SBC	Secretariat of the Basel Convention
SCP	Global Sustainable Cities Programme
SDP	Sustainable Dar es Salaam Project
SEALSWIP	South-East Asia Local Solid Waste Improvement Project
SEAM	Support for Environmental Assessment and Management project
SEIA	strategic environmental impact assessment
SEWA	Self-Employed Women's Association (India)
Sida	Swedish International Development Cooperation Agency
SLU	Superintendência de Limpeza Urbana (Brazil)
SME	small- and medium-sized enterprise
SO$_2$	sulphur dioxide
SPG	*Strategic Planning Guide for Municipal Solid Waste Management*
STS	small transfer station
SWAPP	Solid Waste Management Association of the Philippines
SWM	solid waste management
TPD	(metric) tonnes per day
TPY	(metric) tonnes per year
UBC	used beverage container
UCLG	United Cities and Local Governments
UNCHS	United Nations Centre for Human Settlements (Habitat) (now UN-Habitat)
UK	United Kingdom
UNCRD	United Nations Centre for Regional Development
UNDESA	United Nations Department of Economic and Social Affairs
UNDP	United Nations Development Programme
UNEP	United Nations Environment Programme
UN ESCAP	United Nations Economic and Social Commission for Asia and the Pacific
UNESCO	United Nations Educational, Scientific and Cultural Organization
UNFCCC	United Nations Framework Convention on Climate Change
UN-Habitat	United Nations Human Settlements Programme (formerly UNCHS (Habitat))
UNIDO	United Nations Industrial Development Organization
US	United States
UWEP	Urban Waste Expertise Programme
VCS	Voluntary Carbon Standard
VOC	volatile organic compound
VOS	Voluntary Offset Standard
WEEE	waste electrical and electronic equipment
WHO	World Health Organization
WHO SEARO	World Health Organization Regional Office for South East Asia
WIEGO	Women in Informal Employment: Globalizing and Organizing
WRAP	UK Waste and Resources Action Programme
WREP	Waste and Resources Evidence Programme
WTE	waste-to-energy
ZW	zero waste
ZWSA	Zero Waste South Australia

A NOTE TO DECISION-MAKERS

A good solid waste management system is like good health: if you are lucky to have it, you don't notice it; it is just how things are, and you take it for granted. On the other hand, if things go wrong, it is a big and urgent problem and everything else seems less important.

Managing solid waste well and affordably is one of the key challenges of the 21st century, and one of the key responsibilities of a city government. It may not be the biggest vote-winner, but it has the capacity to become a full-scale crisis, and a definite vote-loser, if things go wrong.

This note to decision-makers introduces UN-Habitat's Third Global Report on Water and Sanitation in the World's Cities: *Solid Waste Management in the World's Cities*.

A unique feature of the book is that it is based on new information, collected in a standardized format, from 20 reference cities around the world. The cities demonstrate a range of urban solid waste and recycling systems across six continents and illustrate how solid waste management works in practice in tropical and temperate zones, in small and large cities, in rich and poor countries, and at a variety of scales.

The book shows that cities everywhere are making progress in solid waste management – even relatively small cities with very limited resources – but also that there is plenty of room for improvement. The authors are interested in understanding and sharing insights on what drives change in solid waste management, how things work in cities and what seems to work better under which circumstances.

If you take just one message from this book, it should be that there are no perfect solutions, but also no absolute failures: the specific technical and economic approaches that work in, say, Denmark or Canada or Japan may not work in your country. As in most other human endeavours, 'the best is the enemy of the good'.

There is only one sure winning strategy, and that is to understand and build upon the strengths of your own city – to identify, capitalize on, nurture and improve the indigenous processes that are already working well. These may well be outside the 'formal' waste management system provided by the city – the research for his book shows that the informal and micro-enterprise sectors in many developing country cities are often achieving recycling rates, comparable to those reached in Europe and North America only after years of high investment by the city. For example, the research for this book shows that informal recyclers handle 27 per cent of the waste generated in Delhi; if they were to disappear, the city would have to pay its contractors to collect and dispose of an additional 1800 tonnes of waste every day.

The overall aim of the book is to facilitate actors in cities everywhere – the mayor, other politicians, officials, citizens, non-governmental

Urban cities continue to expand to areas with difficult accessibility, posing a challenge for collecting waste from these neighbourhoods. The example from Caracas, Venezuela is representative for many Latin American cities.

© Jeroen IJgosse

organisations, the formal and informal private sector, and indeed the national government – to make their own decisions on the next steps in developing a solution appropriate to their own city's particular circumstances and needs.

We hope that this book will inspire you to be both creative and critical: to design your own models, to pick and mix, adopt and adapt the components and strategies that work in your particular circumstances. You and your citizens and stakeholders deserve the best system, and nothing less. If this book can contribute to that, we will have done our work well.

THE ISWM FRAMEWORK

This book is built around the concept of integrated and sustainable (solid) waste management, known as ISWM. We have divided an ISWM system for convenience into two 'triangles', the physical elements and the governance features. The first triangle comprises the three key physical elements that *all* need to be addressed for an ISWM system to work well and to work sustainably over the long term:

Solid waste is a vital municipal responsibility. You need to be able to put all three elements in place – collection, disposal and materials recovery.

1   **public health:** maintaining healthy conditions in cities, particularly through a good waste collection service;

2   **environment:** protection of the environment throughout the waste chain, especially during treatment and disposal; and

3   **resource management:** 'closing the loop' by returning both materials and nutrients to beneficial use, through preventing waste and striving for high rates of organics recovery, reuse and recycling.

Triangle 2 focuses on ISWM 'software': the governance strategies to deliver a well functioning system. Until the 1990s, this would probably have been framed primarily around technology; but there is consensus today on the need for a much broader approach. Three interrelated requirements for delivering ISWM are distinguished here under the framework of 'good waste governance'. There is a need for the system to:

1   be **inclusive**, providing transparent spaces for stakeholders to contribute as users, providers and enablers;

2   be **financially sustainable**, which means cost-effective and affordable; and

3   rest on a base of **sound institutions and pro-active policies**.

THREE KEY SYSTEM ELEMENTS IN ISWM

Public health (collection)

The safe removal and subsequent management of solid waste sits alongside the management of human excreta (sanitation) in representing two of the most vital urban environmental services. Other essential utilities and infrastructures, such as water supply, energy, transport and housing, often get more attention (and much more budget); however, failing to manage properly the 'back end' of the materials cycle has direct impacts on health, length of life, and the human and natural environment.

Uncollected solid waste blocks drains, and causes flooding and subsequent spread of water-borne diseases. This was the cause of a major flood in Surat in India in 1994, which resulted in an outbreak of a plague-like disease, affecting 1000 people and killing 56. Annual floods in East and West African, and Indian cities are blamed, at least in part, on plastic bags blocking drains.

The responsibility of municipalities to provide solid waste collection services dates back to the mid-19th century, when infectious diseases were linked for the first time to poor sanitation and uncollected solid waste. There are major cities in all continents that have had collection services in place for a century or more.

The data collected for this book, and other UN-Habitat data, show waste collection coverage for cities in low- and middle-income countries ranging from a low of 10 per cent in peri-urban areas to a high of 90 per cent or more in commercial city centres. This means that many households in many cities receive no services at all, with the result that far too much waste ends up in the environment. UN-Habitat health data also show that rates of diarrhoea and acute respiratory infections are significantly higher for children living in households where solid waste is dumped, or burned in the yard, compared to households in the same cities that receive a regular waste collection service.

Perhaps surprisingly, even in Europe and North America uncollected waste can still hit the headlines, as in the 2008 example of Naples, Italy, where mountains of solid waste lined the streets for months; collectors stopped picking up the waste because all of the region's landfills were full, and residents protested fiercely.

The 20 reference cities in this report provide many examples of different approaches that have been successful in providing collections services across the city. For example, both Bengaluru (Bangalore) in India and Quezon City in the Philippines have collection coverage rates over 90 per cent. One key message is to adopt and adapt technology that is appropriate, and can easily be maintained locally. Just as it is amusing to picture a cycle rickshaw collecting waste in Adelaide, it is ridiculous to send a giant compactor truck designed for Australian roads into the lanes of the old city in Dhaka, or even onto the main roads which have not been designed for such high axle loading rates. Another key message is to 'mix and match' the methods of service delivery. New Delhi is an example of a city where primary collection is done by authorized informal sector collectors/recyclers, who deliver the waste by hand cart to a large private sector operator who provides secondary collection from communal bins.

Environmental protection (waste treatment and disposal)

Until the environmental movement emerged in the 1960s, most wastes were disposed of with

Sorted and crushed ferrous and non-ferrous metals being loaded on a truck for transportation to the recycling industry (Caínde, Brazil)

© Jeroen IJgosse

little or no control: to land, as open dumping; to air, by burning or evaporation of volatile compounds; or to water, by discharging solids and liquids to surface, groundwater or the ocean. There was little regard for the effects on drinking water resources and health of those living nearby – the philosophy was 'out of sight, out of mind'.

Over the last 30 to 40 years, countries and cities seeking to take control of growing quantities of waste and to maintain a clean environment have built up experience about what works. Moving towards modern disposal has generally followed a step-by-step process: first phasing out uncontrolled disposal, then introducing, and gradually increasing, environmental standards for a disposal facility. In the process, controlling water pollution and methane emissions from sanitary landfills, and air pollution from incinerators, receive increasing attention.

Attention in high-income countries may now be moving on to other aspects, but many cities in low- and middle-income countries are still working on phasing out open dumps and establishing controlled disposal. This is a necessary first step towards good waste management; a properly controlled landfill site is an essential part of any modern waste management system.

Whatever technologies and equipment are used, they should be appropriate for and adapted to the local conditions. The small and relatively remote city of Ghorahi in Nepal shows what can be achieved with limited local resources: their well-sited and managed facility includes waste sorting and recycling, sanitary landfilling,

Phasing out and upgrading open dumps and controlling the disposal of waste is a necessary first step.

leachate collection and treatment, and a buffer zone with forests, gardens and a bee farm that shields the site from the surrounding area.

Many 'new' technologies are being developed to treat solid wastes, and salesmen target both developed and developing country cities. In principle, this is fine, but it is important that decision-makers have the information they need to make informed choices. Unfortunately, experience shows that there are no magic solutions: technologies developed for relatively dry wastes with high calorific value in the 'North' may not work when confronted with wet and mainly organic wastes with low calorific value in the 'South'. If a solution seems 'too good to be true', it's probably not true.

Resource management (valorization of recyclables and organic materials)

Prior to the industrial revolution, most cities had few material resources, money was scarce and households had more needs than they could meet. Wastage was minimized, products were repaired and reused, materials were recycled and organic matter was returned to the soil.[1] Extensive informal recycling systems flourished, but began to be displaced by emerging formal municipal waste collection systems in the late 19th century. Recycling and materials recovery became large, but almost invisible, private industrial activities.

During the past 10–20 years, high-income countries have been rediscovering the value of recycling as an integral part of their waste (and resource) management systems, and have invested heavily in both physical infrastructure and communication strategies to increase recycling rates. Their motivation is not primarily the commodity value of the recovered materials, which was the only motivation of the earlier, informal or private sector, systems. Rather, the principal driver is that the recycling market offers a competitive 'sink', as an alternative to increasingly expensive landfill, incineration of other treatment options.

Many developing and transitional country cities still have an active informal sector and micro-enterprise recycling, reuse and repair

systems, which often achieve recycling and recovery rates comparable to those in the West; the average recovery rate across the 20 reference cities is 29 per cent. Moreover, by handling such large quantities of waste, which would otherwise have to be collected and disposed of by the city, the informal recycling sector has been shown to save the city 20 per cent or more of its waste management budget. In effect, the poor are subsidizing the rest of the city.

There is a major opportunity for the city to build on these existing recycling systems, to increase further the existing recycling rates, to protect and develop people's livelihoods, and to reduce still further the costs to the city of managing the residual wastes. The formal and informal sectors need to work together, for the benefit of both.

The priorities of good resource management are expressed by the '3Rs' – reduce, reuse, recycle. The last can be further split between 'dry' recyclables and bio-solids or organic wastes:

1. **Reduce** the quantities of waste being generated. This is the new focus of modernization in developed countries; but it is important also for rapidly growing cities in middle- and low-income countries to bring their waste growth rates under control.
2. **Reuse** products that can be reused, repaired, refurbished, or remanufactured to have longer useful lives.
3. **Recycle** materials that can be extracted, recovered and returned to industrial value chains, where they strengthen local, regional and global production.
4. **Return nutrients to the soil**, by composting or digesting organic wastes ('bio-solids') – plant and animal wastes from kitchen, garden and agricultural production, together with safely managed and treated human excreta. These are sources of key nutrients for the agricultural value chain, and their proper utilization is important to food security and sustainable development.

Beware the 'magic solution' salesman, whose technology will solve your problem at little or no cost.

Waste is a resource, and the entire waste system should be designed to maximize the benefits from the discarded materials.

THREE ISWM GOVERNANCE FEATURES

Inclusivity

The municipal government is responsible for solid waste management in a city, but cannot deliver on that responsibility by prescribing or undertaking measures in isolation, entirely on their own. The best-functioning solid waste systems involve *all* the stakeholders in planning, implementing, and monitoring the changes.

A solid waste system consists of three main groups of stakeholders: the providers, including the local authority, who actually offer the service; the users, who are the clients; and the external agents in the enabling environment, including both national and local government, who organize the boundary conditions and make change possible.[2]

Users, or waste generators, are key stakeholders in waste management, as are the NGOs, women's unions, and other organizations that represent them in the policy and governance processes. The reference cities demonstrate a range of good practices, in areas such as:

- consultation, communication, and involvement of users;
- participatory and inclusive planning;
- inclusivity in siting facilities; and
- institutionalizing inclusivity – the solid waste 'platform'.

Service providers include the formal municipal waste organization, in partnership with a variety of private, informal and/or community actors of widely varying sizes and capabilities. They can supplement the knowledge and capacity of the local authority to implement recycling, manage organic waste and serve households with waste collection. In urban waste systems in most low- and middle-income countries, the informal and micro-enterprise collection and recycling sector is particularly important, providing a livelihood for an average of 0.5 per cent of the urban population across 10 of reference cities, and of more than 1 per cent in both Delhi and Dhaka. The numbers are much lower, but the informal sector does operate also in the US, Canada, Europe and Japan.

Financial sustainability

Financial sustainability in solid waste management is a major issue for cities all over the world. In developing and transitional country cities, solid waste management represents a significant proportion of the total recurrent budget of the city, with figures of 3 to 15 per cent being reported by the reference cities. When the solid waste budgets are divided by the population, and this per capita figure is expressed as a percentage of per capita GDP, most of the cities are in the range of 0.1–0.7 per cent, with two greater than 1 per cent. Yet in spite of relatively high costs, collection service coverage is often low and disposal standards remain poor.

Costs in high-income country cities are continuing to increase as wastes are collected in several separate streams to facilitate recycling, wastes are diverted from landfill to higher cost facilities, and the costs of environmental protection at treatment and disposal sites have increased.

For most cities in low- and middle-income countries, the coming years will see increased waste, more people, more vehicles, more labour needed for collection, more transfer stations, more separated waste types of collection and more administration. As the city spreads and

The private sector includes both formal and informal enterprises, of widely varying sizes and capabilities. They can supplement the knowledge and capacity of the local authority to implement recycling, manage organic waste, and serve households with waste collection.

An appropriate time schedule of collection routes needs to coincide with local circumstances: in Managua, Nicaragua, it is early morning when citizens are at home

© UN-Habitat, Jeroen Ijgosse

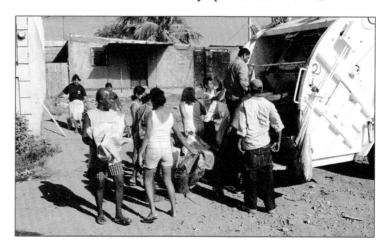

Poster encouraging the people to take a subscription to primary waste collection service provided by local NGOs in Benin

© WASTE, Justine Anschütz

standards improve, suitable sites for landfills will be scarcer, further from the city centre, and (much) more expensive. Making service delivery more efficient should free up some resources, but many cities can expect to see costs rise substantially. It will therefore be imperative to find both regular sources of revenue and significant amounts of investment finance.

Where international donors, or other investors, are involved in providing finance to cities for new waste management vehicles, equipment or infrastructure, one precondition is often that the city can demonstrate that they are able to pay both the recurrent costs and to repay any capital that has been borrowed. This usually involves discussion both on establishing the full current costs of providing the service, which is commonly underestimated by up to 50 per cent, and on the introduction of user fees, which in turn raises the issues of equity, affordability and willingness to pay.

Discussions with international donors are often complicated by their internal rules, which may restrict them to funding waste facilities that meet the latest international environmental standards, which may make them unaffordable to the city. This conflicts with one of the basic recommendations of this book: that each city should select next steps in the development of their waste management system that are appropriate, and thus affordable, in their own particular circumstances.

Experience has shown that service users

An essential first step is to establish your current costs, to serve as a baseline for comparing the costs of any proposed improvements to the waste management system.

are prepared to pay for their waste to be removed when they agree with the service levels, when the charging system is transparent and when services are provided for locally acceptable prices. Even in slum areas, people are generally willing to pay for appropriate primary collection services. Moving from a position where solid waste management is paid for through general revenues, to one where it is paid for entirely from user charges, is likely to be a gradual transition, particularly if the overall costs are rising at the same time. So, at least in the medium term, a significant proportion of the total cost will still have to be paid for by the municipality or the national government from general revenues, as part of its public health and environmental protection responsibilities.

Sound institutions and proactive policies

A strong and transparent institutional framework is essential to good governance in solid waste. Without such a framework, the system will not work well over the long term. Indeed, it was suggested at the 2001 UN-Habitat World Urban Forum that the cleanliness of a city and the effectiveness of its solid waste management system could be useful as proxy indicators of good governance. The adequacy of services to lower-income communities also reflects how successfully a city is addressing issues of urban poverty and equity.

If waste services are to be effective, a city must have the capacity and the organizational structure to manage finances and services in an efficient and transparent manner, streamline management responsibilities with communities, and listen to users. For waste management to work well, the city needs to address underlying issues relating to management structures, contracting procedures, labour practices, accounting, cost recovery and corruption. Clear budgets and lines of accountability are essential.

Private sector involvement in service delivery is an option for improving cost-effectiveness, quality and coverage. However, private sector involvement in waste management is not simple 'privatization'. The municipal authorities remain responsible and, as the contracting body, need to have sufficient understanding and capacity to carry out their 'client' function. The necessary conditions that must be met for successful private sector involvement include competition, transparency and accountability, all of which help to ensure that the contracting process is free from corruption and that citizens receive the services as contracted. The concept of pro-poor public–private partnerships (5-Ps) develops this more explicitly, by addressing the need to engage users, the rights of small and micro-enterprises and the informal sector to hold on to their livelihoods, and the obligation to serve poor communities fairly and effectively.

 ## CONCLUSIONS

The stories from our 20 reference cities – rich and poor, and in all parts of the world – show that it is possible to make progress in tackling solid waste management. There is no 'one size fits all': any successful solution must address all three physical elements of ISWM and all three features of good governance. But a reliable approach is to be critical and creative; to start from the existing strengths of the city and to build upon them; to involve all the stakeholders to design their own models; and to 'pick and mix', adopting and adapting the solutions that will work in any particular situation.

Notice of municipal ordinance published in two languages in Quezon City, Philippines advising of the legal consequences of not keeping your premises clean

There is no single 'right answer' to solid waste management; rather, a mix of measures and approaches. The solutions that will work in your city have to fit your own circumstances.

NOTES

1   Strasser, 1999.
2   Spaargaren and van Vliet, 2000; Scheinberg, in press.

KEY SHEET 1
MODERNIZING SOLID WASTE IN THE ASIAN TIGERS

Lilia Casanova (Center for Advanced Philippine Studies – CAPS)

The sanitary landfill of San Fernando City, La Union was developed using the so-called design–build–own–operate DBOO scheme for private-sector involvement

© Quezon City

Rapid urbanization, increasing industrialization, rising incomes and a more sophisticated form of consumerism are leading to an increase in the amount and toxicity of waste in middle-income Asian countries, especially in the cities. According to the World Bank, urban areas in Asia generate about 760,000 tonnes of municipal solid waste (MSW), or approximately 2.7 million cubic metres, per day. In 2025, this figure will increase to 1.8 million tonnes of waste per day, or 5.2 million cubic metres of waste.[1] Countries report a rapid increase in hazardous materials in the waste stream, as well as in recyclable paper, plastic and metal. Densely populated cities in Singapore, Japan, Thailand, Malaysia, South Korea, Indonesia, China and the Philippines are thus under pressure to modernize their solid waste systems, bring their waste streams under control, and shift from pure disposal to recovery of both energy and materials.

Singapore and South Korea are responding to this challenge by testing the usefulness of public–private partnership (PPP) schemes for waste-to-energy plants. While being clear that government has to play the dominant role to ensure efficient waste management, regional interest in private-sector involvement is growing. In 2008, Singapore announced the Keppel Seghers Tuas Waste-to-Energy Plant (KSTP) to be operated under a design–build–own–operate (DBOO) scheme.[2] In South Korea, a similar PPP with Aquentium Inc will implement and operate a 1000-tonne-per-day waste-to-energy processing plant.

Malaysia, the Philippines, Thailand and Sri Lanka, with somewhat lower gross domestic product (GDP) and a later entry into the modernization process, are making progress in developing regulatory frameworks to institutionalize ISWM, in part to cope with increasing waste challenges due to tourism. Alerted by the collapse of the Payatas dumpsite in 2000, which killed more than 200 waste-pickers, the Philippines passed the Ecological Solid Waste Management Act in 2001 and created a National Solid Waste Management Commission to oversee policy implementation nationwide.

In the case of Malaysia, the desire to federalize solid waste management (SWM) and move the country to the status of a developed nation by 2020 has led it to introduce a new 2007 Solid Waste Management and Public Cleansing Act.[3]

In Thailand, the growing tourism industry is triggering infrastructure improvements in transportation, hotels and ISWM in Bangkok, Chiang Mai, Phuket and other tourist destinations. Sanitary landfills and incineration plants have replaced open dumpsites. New recycling centres, such as, the Pobsuk Recycling Centre, are being operated on a pilot scale.

Starting in the 1990s, Asia has been host to a number of national and regional initiatives in SWM. The World Bank's Metropolitan Environmental Improvement Programme (MEIP)[4] is credited for solid waste management improvements in large cities in Asia, such as Beijing, Bombay, Colombo, Jakarta, Metro Manila and, later, Kathmandu. Between 1994 and 1998, the South-East Asia Local Solid Waste Improvement Project (SEALSWIP),[5] a Canadian International Development Agency (CIDA) assistance programme, successfully assisted communities in the Philippines, Thailand and Indonesia in various aspects of SWM, including organizing waste-pickers and junk shops; setting up a 'waste bank' for recyclables; siting landfills; and providing training on hazardous waste management.

From 1996 to 2001, the Sustainable Cities Programme, a joint programme of UN-Habitat and the United Nations Environment Programme (UNEP), assisted cities in China, India, Sri Lanka, the Philippines and Indonesia to set up solid waste management systems. In 2002, UNEP and its International Environmental Technology Centre (IETC) organized an Association of Southeast Asian Nations (ASEAN) High-Level Consultation (HLC) Group to establish a policy-level forum to deal with solid waste issues. The HLC heightened awareness on the urgency of the problem in the region and encouraged the pursuit of national initiatives in SWM. The Kitakyushu Initiative, which came out during the Ministerial Conference on Environment and Development in 2002 in Kitakyushu, Japan, has been a continuing source of technical and financial assistance by many countries in Asia for SWM.

The latest initiative in ISWM in Asia is the Regional 3Rs Forum financed by the Japanese

Waste-pickers on the Payatas landfill site in Quezon City in 2007. A major landslide at this site in 2000 killed some 200 people.

© SWAPP

Environmental Ministry, and organized in 2008 by the Institute for Global Environmental Strategies (IGES), a research and development (R&D) institution in Japan.[6] It involves governments, donor agencies and scientific institutes in 12 Asian countries. With support from the Japanese government, it aims to promote policy development and projects on the 3Rs (reduce, reuse, recycle). Another new regional initiative is the Asian Development Bank's (ADB's) Cities Development Initiative for Asia (CDIA),[7] established in 2008, which is investing in public transport, solid waste, methane capture and other urban management priorities. During the same year, the European Union (EU) reintroduced its Asia Pro Eco Programme[8] on environmental management, which includes solid waste management. AusAid, on the other hand, directs more of its support to the Pacific Island countries than to Asia.[9]

Donor contributions have come in different forms (i.e. grants, loans, and technical and expert assistance) and have influenced Asian people's habits and perceptions about solid waste in different ways. The World Bank is, especially, perceived to have been effective in introducing the concept of the 3Rs, materials recovery and sanitary landfills. CIDA was found to be effective in community organizing for waste collection and recycling, while the Deutsche Gesellschaft für Technische Zusammenarbeit (GTZ) gets some credit for introducing sanitary landfills to some countries in South-East Asia. But Japan, through the Kitakyushu Initiative, has been the champion for establishing composting, integrated solid waste management systems and recycling.

Not all donor assistance is equally effective in providing lasting positive impacts and many activities ceased or failed after external support was removed. Various technical, financial, institutional, economic and social constraints faced by both the recipient countries/cities and external support agencies form constraints, combined with limited resources available to resolve the problems. During the early 1990s, for instance, many incineration plants installed in some cities in the Philippines and Indonesia with assistance from the World Bank ended up as white elephants, never to be used because the high organic content of the waste streams meant that the waste was not incinerable.

But, on the whole, in Asia regional initiatives appear to be particularly good investments. The main funders of development in the region, the World Bank and the ADB, have allocated substantial amounts for solid waste management over the last 30 years. From US$151.9 million invested between 1974 and 1988,[10] the World Bank's investment in SWM increased to US$1.538 billion in 2004,[11] or almost ten times in a period of 16 years. The ADB, on the other hand, just recently signed (in 2009) a US$200 million loan package for China Everbright International Limited to develop a clean waste-to-energy project in the People's Republic of China.[12] China has received a total of US$21 billion in loans financing since joining the ADB in 1986, making it the second largest ADB borrower and client for private-sector financing. These investments made during the last 20 years have served as 'seed' financing for Asian countries to effectively develop a framework for SWM – in particular, to establish and modernize their institutions and institutional arrangements that will embed in the minds of their nationals the importance of waste segregation and the basic principles of the 3Rs. In the process, the foreign-assisted projects or programmes have also supported national champions in SWM who will ensure that the work will be sustained.

NOTES

1   World Bank (1999) *What a Waste: Solid Waste Management in Asia*, May, www.worldbank.org/urban/solid_wm/erm/CWG%20folder/uwp1.pdf.

2   'NEA raises efforts to achieve recycling target of 70% by 2030', 12 August 2009, www.nea.gov.sg.

3   National Seminar on Sustainable Environment through the New Solid Waste Management and Public Cleansing Act 2007, 31 July 2008, www.wmam.org/index.php?option=com_content&task=view&id=91&Itemid=2.

4   Metropolitan Environment Improvement Programme in Asia, World Bank, 1998, www-wds.worldbank.org/.

5   SEALSWIP, www.icsc.ca/content/history/sealswip/lessons.html.

6   Press release: *Towards the Establishment of the Regional 3R Forum*, www.uncrd.or.jp/env/spc/.

7   ADB's Cities Development Initiatives in Asia: www.adb.org.

8   Sustainable Waste Management Cycle (SWMC) brochure, www.iclei-europe.org/fileadmin/user_upload/newsbits/EU-Asia_Waste_Management_Conference.pdf.

9   Hisashi Ogawa (undated) 'Sustainable solid waste management in developing countries', Paper presented at the Seventh ISWA International Congress and Exhibition, www.gdrc.org/uem/waste/swm-fogawa1.htm.

10 Bartone et al, 1990.

11 Annamari Paimela-Wheler, 2004, *World Bank Financed Waste Management Projects: Overview of the Waste Management Markets and Trends in Developing Countries and Prospects for Finnish Companies*, 5 March, http://akseli.tekes.fi/opencms/opencms/OhjelmaPortaali/ohjelmat/Streams/fi/Dokumenttiarkisto/Viestinta_ja_aktivointi/Julkaisut/Kansainvaliset_selvitykset/WB_Waste_management_report.pdf.

12 'ADB supports $200 M waste-to-energy project', ADB, 2009, www.greeneconomyinitiative.com/news/183/ARTICLE/1688/2009-09-03.html.

CHAPTER 1

EXECUTIVE SUMMARY

Bharati Chaturvedi

Of the many common threads that bind cities across the world, waste handling, is possibly one of the strongest. Regardless of the context, waste, directly and indirectly, is one of biggest challenges of the urban world. It's also a city's calling card. If a city is dirty, the local administration is written off as ineffective. If not, governance is presumed in the public eye to be effective.

This Third Global Report acknowledges escalating challenges without boundaries. Yet, it is not prescriptive – that would go against its fundamental premise in highlighting the value of local innovation and knowledge. What it seeks to do instead is to follow another one of the beliefs that it lays out: to build capacity through networking. The contributors have dug out vast amounts of knowledge and experience, and distilled it in this volume. Most readers might never travel to all of the 22 diverse cities upon which this Global Report is based. Yet, they will have access to real experiences of people working on the ground. Indeed, this is an entirely new kind of networking: that of ideas. Perhaps reading about what one city has been able to do will light up an idea in another.

If you, as a city planner, are hoping that reading this report will be like popping a wisdom pill, be warned. There is only one mantra here: use what you have and build on it with an army of partners. If anything, the report warns against imagining as ideal the systems, technologies and solutions of the developed world and trying to copy them as a means of cleaning the city. It might not work if it lacks local relevance and local buy-in. Just as it is amusing to picture a cycle rickshaw collecting waste in Adelaide, it's ridiculous to send a giant compactor into the lanes of the old city in Dhaka, Bangladesh. Clearly, modernization is not necessarily motorization. Delve into this idea and you'll find a few more strands of thought to build from.

All over the world, municipalities and counties have shown how inclusion can achieve spectacular results. There are two kinds of inclusion identified here: service users and service providers. In Varna, Bulgaria, it took a consultant to inform the community that the municipality was picking up half-empty bins. Changing the pick-up frequency came as a relief because it reduced costs in a none-too-wealthy system. In 2007, Quezon City's waste collection services in the Philippines received a 100 per cent satisfied report card from its households, in large part because it was guided by their choices.

When many developed world cities began solid waste modernization processes, their informal sectors had ceased to be robust. As a result, they had to 'reinvent' recycling, almost from scratch. Today's developing world cities aspiring for modernization aren't in the same situation. They are already serviced by numerous private players – individuals or micro-enterprises (often informal-sector players) – offering waste collection services, or picking waste from streets and dumps, and trading in it. Their contribution is substantial. In Bamako, Mali, over 120 self-employed micro-enterprises collect approximately 300,000 tonnes of waste annually, while in Lusaka, Zambia, informal service providers reach out to 30 per cent of the city. In Bengaluru and Delhi, India, micro-enterprises function similarly, covering as much as 25 per cent of Delhi across income groups, apart from picking and recycling the valuable waste. The informal sector is clearly any city's key ally. These human resources can be best deployed in the public interest through appropriate legal and institutional spaces. But that doesn't imply that the informal recycling sector is a distinctly developing world phenomenon. The research here shows that it exists even in San Francisco, California, and Tompkins County, Ithaca, New York, and in some people's opinion plays a positive role.

.Globally, the thinking is shifting from merely removing waste before it becomes a health hazard to creatively minimizing its environmental impact. Waste reduction is desirable; but, typically, it is not monitored anywhere. Recycling, this Global Report emphasizes, has universal buy-in and a range of approaches are applied. Yet, it must be seen with new eyes. While the commodity value of materials is taken for granted, the service aspect of recycling is relatively new everywhere. Besides, the greatest value of recycling is, literally, as a sink. It absorbs the various costs otherwise incurred were the waste treated using other options, such as landfills or incinerators. This opportunity cost is recognized by enlightened planners. In Rotterdam, The Netherlands, reuse enterprises are given diversion credits from the waste management budget, while in Kunming, China, resource management is so important, it is an institutionally separate set of activities. The cities here suggest that recycling grows as the modernization process expands and begins to control disposal and its costs. Moreover, as both Quezon City and San Francisco demonstrate, strong policies and systems adaptation also give recycling an important push, coupled with change in user behaviour. Meanwhile, Dhaka, Bangladesh, has met global standards to receive carbon credits from composting.

Teasing out these trends requires data. Often, such raw data was not easily available, forcing the question of institutionalization of information generation and storage. Without proper data collection and management, it is difficult to be accountable, transparent and even; to make effective strategies; and to budget for them. The absence of all of this, in turn, creates barriers for modern waste management systems.

As the linkages between valorizing and climate change become clear, proper waste handling has become an important tool to mitigate greenhouse gases. This Third Global Report expresses the hope that it can offer optimism that this is a battle to win, regardless of what kind of city decides to join in this fight. In fact, the report expresses the hope that it will persuade everyone to enlist because our urban future will only be the maturing of our urban present.

2

INTRODUCTION AND KEY CONCEPTS

INTRODUCING THIS BOOK

This publication is UN-Habitat's Third Global Report on Water and Sanitation in the World's Cities. It focuses on the state of solid waste management, which is an important challenge facing all of the world's cities. Previous volumes focused on water supply and sanitation. The book has four main aims:

1   to showcase the good work that is being done on solid waste by cities around the world, large and small, rich and poor;

2   to look at what drives change in solid waste management, how things work in cities and what seems to work better under which circumstances;

3   to help decision-makers, practitioners and ordinary citizens understand how a solid waste management system works; and

4   to inspire people everywhere, in good communication with their neighbours, constituents and leaders, to make their own decisions on the next steps in developing a solution appropriate to their own city's particular circumstances and needs.

This book is designed both to fill a gap in the literature and knowledge base about solid waste management in low-, middle- and high-income countries, and to provide a fresh perspective and new data. The book distinguishes itself in a number of ways:

- First and foremost, it is based on the framework of integrated sustainable waste management (ISWM), especially the concepts of sustainability and inclusive good practice that have broadened and enriched the field.

- The 20 real city examples provide up-to-date data and are used to inform questions of waste policy, good and bad practice, management, governance, financing and many other issues. The focus is on processes rather than technologies, and the goal is to encourage a different kind of thinking.

- It uncovers the rich diversity of waste management systems that are in place around the world. This book brings out common elements and develops a lens for 'viewing' a solid waste management system, while at the same time encouraging every city to develop its own individual solution, appropriate to its specific history, economy, demography and culture, and to its human, environmental and financial resources.

- A central tenet of the book is that there is no one right answer that can be applied to all cities and all situations. In this, the book challenges the notion that all a developing country city needs to do is to copy a system that works in a particular developed country city.

- This is neither a 'how-to' book nor a 'let's fix it' book, although the discerning reader will find elements of both, but more of a 'how do they do it now and what do they need to do more or less of' kind of discussion.

A typical view of a canal in Amsterdam, The Netherlands, in the 1960s

© Stadsarchief Amsterdam, used with permission

Waste thrown into a watercourse alongside a neighbourhood in Nairobi, Kenya, 2008

© UN-Habitat

This book's ambition is to look at solid waste and the world's cities in a fresh new way; to observe what works and what does not; and to let this inform the policy process and contribute to rethinking the whole waste management concept. The authors see an urgent need for this in transitional, low- and middle-income countries; but it may well be that looking from another viewpoint gives new insights to developed countries as well. The goal is to provide an honest look at how cities – large and small, complex and simple, coastal and inland, in rich, poor and transitional countries – do and do not succeed to

make reasonable choices that serve their citizens and protect their environment at acceptable financial cost.

Looking beyond what is happening to what could be improved, the book seeks to make the principles and elements of sound practice in waste management clear and accessible. The book explores both expensive 'best practice' technologies, as used in high-income countries, and moderate-cost creative alternatives that improve the environment.

Most books on solid waste treat the solid waste systems in developing and transitional countries as imperfect or incomplete copies of an ideal system that operates in developed countries such as Canada, Denmark or Japan. Many, if not most, waste interventions seek to perfect or improve the copying process and spread the ideal.

What is frequently overlooked is that the higher-income countries in Europe and North America have been busy with solid waste for the last 40 years or so. The systems and technologies in use there were not developed overnight, and they fit the climates, social conditions and economies of Northern European society. What is not always clear to the visitor to Denmark or Germany is that even these 'clean giants' did not move from open dumping to current best practice in one step. They and their citizens debated and struggled and agreed to disagree. Their engineers took risks, made innovations and made their share of mistakes. Some things that were designed 20 years ago – such as the Dutch producer-responsibility agreement for packaging – have never worked, while other innovations such as dual collection of organic waste and residuals have made contributions to both economy and environment.

This book responds to a growing global consensus that cities in low-income, middle-income and transitional countries need to take charge of the modernization process and to develop their own models for waste management that are more than simply 'imperfect copies'. Citizens of the world need to have solid waste and recycling systems that serve their needs and match their wishes and what they can and want to afford. This calls for a larger variety of models and approaches tailored to fit specific local conditions.

A good baseline analysis and a transparent stakeholder process will reveal one or more logical 'next steps' that each city can take to improve what they have and move the whole system towards effective, affordable performance. Because modern waste management is about much more than a 'technical fix', such next steps can relate to making the institutional framework stronger, sending waste system employees to training, shifting the recycling strategy to be easier for citizens, or phasing out energy-intensive approaches to collection. Technologies are visible evidence of humanity's best intentions to transform solid waste into a safe, inert substance. They carry the system, but they are not the system. And if they work at all, they do so because of the far less visible institutional, governance, policy and participative frameworks that are highly varied and complex, and directly related to local conditions.

The book combines experience and case studies with analysis of some 20 city profiles, each of which has been created for the book itself. The method combines data collection, analysis, modelling, reflection and comparison. Cities have been asked unusual questions, which are designed to stimulate city officials and the readers of this book to look differently at management of waste in cities, and to dare to think outside the box – or in this case outside the waste bin or trash barrel. A range of specialists on the writing team are using new and existing information about waste and recycling in cities to look at, analyse and reflect upon solid waste in cities worldwide.

If we take a step back and look with a fresh perspective at urban waste management in the 21st century, we might dare to ask the question: how much progress have we really made, even in the best European solid waste systems, when we still generate ever-increasing amounts of waste, and still rely on burying our discarded products under the ground as a legitimate 'method' of waste management?

Boys transferring waste from collection point to truck in Bengaluru, India

©WASTE, Jeroen IJgosse

About the authors

This book has been a joint effort of UN-Habitat and a group of international solid waste and recycling specialists:

- Around 35 people from 15 countries have worked actively on the book.
- The team has worked on solid waste management in six continents and in many different kinds of cities, towns, villages, counties, provinces, countries and non-national zones.
- Most of the team are experienced solid waste professionals ('garbologists'), but come from a variety of backgrounds and disciplines, including an industrial designer; a journalist; several architects; environmental specialists; university professors; diplomats; students; politicians; scientists; environmental, human rights and sustainability activists; sociologists; engineers; planners; economists; artists; and teachers.
- The collective experience of the team has led them to be critical of business as usual in solid waste, management of organic waste and recycling because they know from first-hand experience what doesn't work well – or at all – when transferred from a high-income to a middle- or low-income country.

About the organization of this book

The book is written around three key *physical elements* of an integrated sustainable (solid) waste management (ISWM) system, and three ISWM *governance approaches* that are used to manage it. Even before the book begins there is a decision-makers' guide (pages xix–xxviii), a short introduction for decision-makers to read and decide whether they have time to read the rest. Chapter 1 is an executive summary, prepared by Bharati Chaturvedi of Chintan-Environmental in Delhi.

Chapter 2, the current chapter, begins with a short introduction explaining what this book is designed to do, and why and how it has been written. The reader can skip this section – or come back to it later; but its purpose is to introduce the book's lightly unconventional way of looking at waste management. The second part of Chapter 2 explores the nature of the solid waste problem and presents key concepts used throughout the book.

Chapter 3 introduces the cities, giving the reader a flavour for location, size, material and waste intensity of the society, economic situation, climate, and a range of other factors. Chapter 3 also includes an appendix of specific two-page summaries of each city (City Inserts section, pages 41–85), focusing the reader on unique features, but also enabling comparison through presenting a common set of seven indicators drawn from information provided by each city.

Chapters 4 and 5 present the main new information that the book hopes to contribute to the global solid waste and recycling discussion. These two chapters present the information from the reference cities, focusing the discussion of successes and problems on real things happening in real places. Chapter 4 looks at the 'first triangle' of the ISWM physical systems: collection, disposal and resource management. Chapter 5 focuses on the 'second supporting triangle' of governance aspects in ISWM: inclusivity, financial sustainability, and sound institutions and proactive policies.

Chapter 6 closes the book with some reflections about ISWM, structured around emerging issues, followed by a glossary and references.

THE SCALE OF THE SOLID WASTE PROBLEM

What is municipal solid waste (MSW)?

Definitions of municipal solid waste (MSW) vary between countries, so it is important to establish at the outset just what is being discussed in this book. A working definition is 'wastes generated by households, and wastes *of a similar nature* generated by commercial and industrial premises, by institutions such as schools, hospitals, care homes and prisons, and from public spaces such as streets, markets, slaughter houses, public toilets, bus stops, parks, and gardens'. This working definition includes most commercial and business wastes as municipal solid waste, with the exception of industrial process and other hazardous wastes. Different countries define municipal solid waste rather differently – for example, depending on which sector does the collecting – so it is important to ask in each city what the definition is and not assume that they are all the same.[1] Some experts suggest that all industrial and construction and demolition (C&D) wastes should be included in the definition of municipal solid waste.[2]

Still, the picture is not always so clear and some *waste generators* produce both municipal and non-municipal wastes:

- Manufacturing industries generate municipal solid waste from offices and canteens, and industrial wastes from manufacturing processes. Some industrial wastes are hazardous and this part of the waste stream requires special management, separate from other wastes.
- Small workshops in urban areas generate both municipal and process wastes, some of which may be hazardous.

- Hospitals and healthcare establishments/services generate municipal solid waste fractions that include food waste, newspapers and packaging, alongside specialized healthcare hazardous wastes that are often mixed with body fluids, chemicals and sharp objects.
- Construction sites generate some municipal solid waste, including packaging and food and office wastes, together with C&D wastes containing materials such as concrete, bricks, wood, windows and roofing materials.
- Construction and demolition wastes from household repairs and refurbishment, particularly 'do-it-yourself' wastes, are most likely to enter the municipal solid waste stream.

In most cities, municipal solid waste includes 'household hazardous wastes' (HNWs; e.g. pesticides, paints and coatings, batteries, light bulbs and medicines). Similar hazardous wastes may come from small businesses. Cities in developed countries have systems that are designed to collect and handle these separately, or to prevent their generation and reduce their toxicity; but there are few cities in which this works completely and most MSW streams include some of these hazardous components when they reach disposal. Parallel collection systems sometimes exist for end-of-life vehicles and for waste electrical and electronic equipment (WEEE), some parts of which may again be classified as hazardous waste.

There are also largely non-municipal waste streams, such as agricultural wastes and mining and quarrying waste.

The working definition implies that parallel waste management systems will exist within an urban area, one for municipal solid waste run by, or on behalf of, the municipality, and others for industrial, C&D, healthcare, end-of-life vehicles and other hazardous wastes.

Definitions also change over time. Prior to rapid modernization, when a city depends on 'open access' to uncontrolled dumping, such sites normally receive all kinds of wastes, including

hazardous, industrial and healthcare wastes.

During development and modernization of an ISWM system, there is both a tendency and a need to refine definitions, analyse and classify types of generators and types of wastes, and gradually increase the precision with which separate streams are directed to appropriate and specific management subsystems. Good governance for solid waste management advances this process, which is also influenced by questions of cost and affordability.

While this book acknowledges the importance of good management of specific hazardous, industrial and healthcare wastes, it addresses them only by specifically excluding them from its main areas of focus. These streams are best managed when they are clearly and effectively segregated from municipal solid waste in management, policy and financing. Substantial guidance on managing hazardous wastes is available, for example, from the Basel Convention[3] and the United Nations Environment Programme (UNEP)[4], and on managing healthcare hazardous wastes from the World Health Organization (WHO)[5].

Construction waste, tree trunks and green trimmings mixed with domestic waste requiring special equipment for cleaning up in Managua, Nicaragua

© UN-Habitat
Jeroen IJgosse

Bulky household waste set out for collection in Paramaribo, Suriname

© Jeroen IJgosse

KEY SHEET 2
SPECIAL WASTE STREAMS

Valentin Post (WASTE)

Healthcare waste has become a serious health hazard in many countries. Careless and indiscriminate disposal of this waste by healthcare institutions can contribute to the spread of serious diseases such as hepatitis and AIDS (HIV) among those who handle it and also among the general public

Infectious and non-infectious wastes are dumped together within the hospital premises, resulting in a mixing of the two, rendering these both hazardous. Collection by the municipality and disposing these at the dumping sites in the city makes it freely accessible to rag-pickers who become exposed to serious health hazards due to injuries from sharp needles and other types of material.

Healthcare waste has infectious and hazardous characteristics. It can result in:

- development of resistant strains of micro-organisms;
- trade in waste materials and disposed of or expired drugs that are recovered and repacked to be sold as new;
- spread of disease through contact with people or animals who pick or eat waste;
- increased risk of infections and sharp injuries to hospital staff, municipal waste workers and waste-pickers;
- organic pollution.

Most hospitals do not have any treatment facility for infectious waste. The laboratory waste materials and liquid wastes are disposed of directly into the municipal sewer without proper disinfection of pathogens and, ultimately, reach nearby water streams.

Unsafe healthcare settings can contribute significantly to some diseases. Legionellosis is a risk associated with healthcare facilities. On average, nearly 10 per cent of infections are acquired in hospital.

The quantity of healthcare waste varies in accordance with income levels (see Table K2.1).

It is estimated that in low-income countries about 5 to 10 per cent of healthcare waste consists of hazardous/infectious waste (Indian Society of Hospital Waste Management). The World Health Organization (WHO)[1] has made a similar estimate for all countries that, generally, only 7 per cent of healthcare waste is infectious and requires red-bag handling.

Sharp waste, although produced in small quantities, is highly infectious. Contaminated needles and syringes represent a particular threat because they are sometimes scavenged from waste areas and dumpsites and then reused. Poorly managed, they expose healthcare work-

Table K2.1

Healthcare waste generation by income level

© Halbanch (1994); Commission of the European Union (1995); Durand (1995)

National income level	Type of waste	Annual waste generation (kg per capita)
High-income countries	All healthcare waste	1.1kg–12.0kg
	Hazardous healthcare waste	0.4kg–5.5kg
Middle-income countries	All healthcare waste	0.8kg–6.0kg
	Hazardous healthcare waste	0.3kg–0.4kg
Low-income counties	All healthcare waste	0.5kg–3.0kg

ers, waste handlers and the community to infections. The WHO estimates that, in 2000, injections with contaminated syringes caused 21 million hepatitis B virus (HBV) infections (32 per cent of all new infections); 2 million hepatitis C virus (HCV) infections (40 per cent of all new infections); and 260,000 HIV infections (5 per cent of all new infections).

THE WORLD HEALTH ORGANIZATION: THE UNITED NATIONS AGENCY FOR HEALTH AND HEALTHCARE WASTE

In 2002, the results of a WHO assessment conducted in 22 developing countries showed that the proportion of healthcare facilities that do not use proper waste disposal ranges from 18 to 64 per cent.

The WHO focuses on healthcare waste. It supports information collection and exchange, development of national policies and training. National agencies focus on implementation of national policies, guidelines on safe practices, training and promotion of effective messages.

Effective healthcare waste management will decrease infections and also benefit visitors, and will be reflected in communities through good practices in safe water, sanitation and hygiene

GOOD PRACTICE: INDIA IS A WORLD LEADER

India is a world leader in working on preventing, reducing and managing healthcare waste. Biomedical waste management and handling rules established in 1998 are in force as part of the Environment Protection Act. The legislation is still in the process of development and promulgation in another ten countries of the region. Although India has advanced in having legisla-

tion, informal sources reveal compliance to the legislation may not be more than 15 per cent. A critical area is its compliance and enforcement.

In general, Indian experience suggests that a multipronged approach focusing on enforcement of legislation, training and education, development of common treatment facilities, and research into unexplored areas are the common denominators in achieving good practice and a stable, effective integrated sustainable waste management (ISWM) system for healthcare waste.

One of the first cases of Action Research in India was in Bengaluru at the Ramaiah Medical College. An internal group, called the Malleshwaram Healthcare Waste Management Cell, created a system for source separation and management of all wastes in this large medical college. Today, in more locations in India, segregation at source helps to avoid the mixing of infectious waste with general solid waste. Based on the Malleshwaram experiment, it has been

Nurse carrying the separated hospital waste to the incinerator, Sri Lanka

© WASTE, Valentin Post

Burning healthcare waste in a barrel in Sri Lanka. Unfortunately, the temperature does not become high enough to destroy all pathogens

© WASTE, Valentin Post

found very effective if nurses, with the support of doctors and hospital management, are trained and equipped to be responsible for healthcare waste management on the ward level.

Common healthcare waste treatment facilities have developed or are planned in many Asian cities. Two operational facilities in Begaluru, India, are run by private entrepreneurs and cater to 250 healthcare institutions each. This has resulted in a parallel transport system for healthcare waste – a safe practice that inspires people to segregate waste at source.

Common treatment facilities will reduce the use of small-scale incinerators or open burning of waste. These facilities have incinerators that use high-temperature operations, where the production of dioxins and furans is negligible.

INSTITUTIONALIZATION

In 2001, the Indian Society of Hospital Waste Management (ISHWM) came into existence. A technical journal, the *Journal of ISHWM*, is published every year and annual conferences promote the exchange of information and foster collaboration with different stakeholders.

The Healthcare Waste Management Cell of the Department of Community Medicine of Ramaiah Medical College also offers national and international training programmes on healthcare waste management and infection control, as well as consultancy services.

The Indira Gandhi National Open University (IGNOU) and the WHO South-East Asia Regional Office (SEARO) have designed a six-month certificate course on healthcare waste management for the capacity-building of healthcare personnel.

NOTE

1   WHO (2002) *Basic Steps in the Preparation of Health Care Waste Management Plans for Health Care Establishments*, Health Care Waste Practical Information Series no. 2, WHO-EM/CEH/100/E/L, World Health Organization, Geneva

KEY SHEET SOURCE

Dr S. Pruthvish, Biomedical Waste Management in India, Department of Community Medicine, Begaluru, and *Indian Journal of Health Care Waste Management*

Taking the measure of MSW

Solid waste data in many cities is largely unreliable and seldom captures informal activities or system losses. Some developed countries, such as The Netherlands, support their city administrations to generate regular and reliable statistics on municipal solid waste, based on weighbridge records and regular monitoring; but many do not. Even cities such as San Francisco in California lack information on the specifics of what is in some waste streams. And when waste data exists, it is difficult to compare even within a city due to inconsistencies in data recording, collection methods and seasonal variations in the quantities of waste generated. Systems for weighing or measuring wastes disposed of are rare in low- and middle-income countries, and small cities such as Moshi, Tanzania, have to estimate waste generation. They generally do this either by estimating based on the design capacity of the vehicles used in collection, or by extrapolating back to the household using imperfect information on what is disposed of.

Table 2.1 shows data from the reference cities on the quantity of municipal solid waste generated per capita per year. For cities that

don't have their generation figures, the amount handled or disposed of is the basis for extrapolating to waste generated.

Data on waste volumes as well as quantities are important in planning waste collection. In a low gross domestic product (GDP) city, waste density can be as high as 400kg per cubic metre due to high fractions of wet organics. In some Organisation for Economic Co-operation and Development (OECD) cities, densities may be less than 100kg per cubic metre because the large volumes of packaging waste don't weigh much.[6]

Weighing wastes entering disposal sites is critical to obtaining accurate data on waste quantities – if you don't measure it, you can't manage it. But care is needed, e.g. to avoid additional people being weighed and weighing the collection vehicles both full and empty

© Jeroen IJgosse

City	Population	Kilograms per capita		Kilograms per household	
		Year	Day	Year	Day
Adelaide, Australia	1,089,728	490	1.3	1176	3.2
Bamako, Mali	1,809,106	256	0.7	1712	4.7
Belo Horizonte, Brazil	2,452,617	529	1.4	1639	4.5
Bengaluru, India	7,800,000	269	0.7	942	2.6
Canete, Peru	48,892	246	0.7	1083	3.0
Curepipe, Republic of Mauritius	83,750	284	0.8	1135	3.1
Delhi, India	13,850,507	184	0.5	938	2.6
Dhaka, Bangladesh	7,000,000	167	0.5	761	2.1
Ghorahi, Nepal	59,156	167	0.5	805	2.2
Kunming, China	3,500,000	286	0.8	903	2.5
Lusaka, Zambia	1,500,000	201	0.6	1107	3.0
Managua , Nicaragua	1,002,882	420	1.1	2182	6.0
Moshi, Tanzania	183,520	338	0.9	1386	3.8
Nairobi, Kenya	4,000,000	219	0.6	1314	3.6
Quezon City, Philippines	2,861,091	257	0.7	1286	3.5
Rotterdam, Netherlands	582,949	528	1.4	1030	2.8
San Francisco, USA	835,364	609	1.7	1400	3.8
Sousse, Tunisia	173,047	394	1.1	1586	4.3
Tompkins County, USA	101,136	577	1.6	1340	3.7
Varna, Bulgaria	313,983	435	1.2	1131	3.1
Average	2,462,386	343	0.9	1243	3.4
Median	1,046,305	285	0.8	1155	3.2

Table 2.1

Municipal solid waste generation in the reference cities.

This table has few surprises. Developed countries have higher generation than developing and transitional ones. The generation for both Belo Horizonte and Managua seems rather high, which may be a general characteristic in the Americas.

Figure 2.1

A world of variation in classifying MSW composition in the reference cities

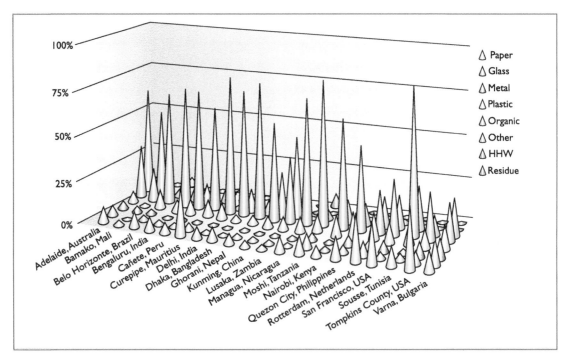

Untertaking a waste characterization (composition) study in Bengaluru, India

©WASTE, Jeroen IJgosse

When there is information, it suggests that while *composition* of municipal solid waste varies widely, both within and between countries, and between different seasons of the year, there are common patterns. Figure 2.1 shows how different the waste stream is in the reference cities, located as they are in high-, middle- and low-income countries. The fact that they classify their waste differently makes the comparison a challenge.

Table 2.2

Waste composition in the reference cities (percentage).

This table says a lot about globalization of products and packages, and also indicates what is not really globalized. For example, the large percentages of sand and grit in Bamako, and 'other' in Lusaka, make these cities exceptions to the generally high percentage of organics. The differences in the way composition is analysed are also quite noticeable.

City	Paper (%)	Glass (%)	Metal (%)	Plastic (%)	Organic (%)	Other (%)	Household hazardous waste (HHW) (%)	Residue (%)	Total (%)
Adelaide, Australia	7	5	5	5	26	52	0	0	100
Bamako, Mali	4	1	4	2	21	52	0	0	83
Belo Horizonte, Brazil	10	3	2	11	66	4	0	5	100
Bengaluru, India	8	2	0	7	72	9	1	0	100
Canete, Peru	6	2	2	9	70	11	0	0	100
Curepipe, Republic of Mauritius	23	2	4	16	48	7	0	0	100
Delhi, India	7	1	0	10	81	0	0	0	100
Dhaka, Bangladesh	9	0	0	4	74	13	0	0	99
Ghorahi, Nepal	6	2	0	5	79	7	0	0	99
Kunming, China	4	2	1	7	58	26	0	0	98
Lusaka, Zambia	3	2	1	7	39	48	0	0	100
Managua, Nicaragua	9	1	1	8	74	1	0	5	100
Moshi, Tanzania	9	3	2	9	65	5	0	7	100
Nairobi, Kenya	6	2	1	12	65	15	0	0	100
Quezon City, Philippines	13	4	4	16	50	12	0	0	100
Rotterdam, Netherlands	27	8	3	17	26	19	0	0	100
San Francisco, USA	24	3	4	11	34	21	3	0	100
Sousse, Tunisia	9	3	2	9	65	11	0	1	100
Tompkins County, USA	36	6	8	11	29	11	0	0	100
Varna, Bulgaria	13	15	10	15	24	23	0	1	100
Average	12	3	3	10	53	17	0	1	
Median	9	2	2	9	61	12	0	0	

The immediate impression from looking at Figure 2.1 is that organic waste is a very large part of all cities' waste streams. Perhaps the most neglected part of the modernization project, closing nutrient cycles by capturing organic waste, is a topic that has profound impacts upon both the city and upon global cycling of carbon and nitrogen.

Managing more and more waste

In spite of the data issues, it is clear that in most cities, waste quantities are increasing rapidly for a number of reasons:

- The number of people living and working in the city is increasing.
- The amount of waste generated per person is rising, together with increases in wealth (as shown by GDP).
- The amount of waste from businesses is increasing.
- The substances in waste are increasing in complexity and variety.

Waste quantities have been growing steadily in high-income countries over the last few decades, often at a rate of 3 per cent per year. Cities in low- and middle-income countries are experiencing even higher growth rates due to a combination of the factors listed above. Although there are some indications that growth may now be slowing down in some developed countries, in most of the world substantial growth rates are likely to continue for some time to come.

Estimating global waste generation figures is difficult given the unreliability of the data, particularly for low- and middle-income countries. One estimate puts municipal solid waste generation worldwide in 2006 at 2 billion tonnes, with a 37 per cent increase forecast by 2011.[7] The world population in 2006 was around 6.5 billion,[8] giving an average per capita generation rate of just over 300kg per year.

Table 2.3 presents current and projected estimates of world municipal solid waste genera-

Waste-pickers working at the dump site in Managua, Nicaragua

© UN-Habitat

tion, using the current average OECD and the high-income OECD kilogram per capita rates. If, in 2025, everyone in the world generated waste at the current average OECD per capita rate, with no allowance for growth in that rate, the total world generation would become 5.9 billion tonnes – nearly three times the current estimate. This could equate to around 59 billion cubic metres of MSW each year, enough to cover a country the size of Costa Rica or Ireland to a depth of one metre.

Calls for waste reduction are often mistakenly confused with a plea to slow economic growth. Rather, strategies are needed to decouple waste growth from economic growth, as, for example, in The Netherlands.[9] However, even in countries with active waste prevention initiatives, waste quantities have generally continued to grow and only recently have begun to level off in some countries.[10] Developing country cities are still experiencing rapid population growth, so one element of an integrated solid waste management solution has to be how to tackle exponential growth in waste quantities.

Table 2.3

Estimates of current world MSW generation and projections of what it might become

Source: developed from public sources by David C. Wilson

World municipal solid waste	Kilograms per capita per year	Billion tonnes per year	
		2006	2025
Current estimates	310	2.0	2.4
At simple average for the reference cities	343	2.2	2.7
At average current rate for OECD	580	3.8	4.6
At current OECD maximum rate	760	4.9	5.9

KEY SHEET 3
HEALTH RISKS RELATED TO SOLID WASTE MANAGEMENT

Sandra Spies (GTZ)

BACKGROUND

How healthy is work in solid waste management? Health risks are always present when handling wastes, but the exposure intensity and the incidence frequency vary significantly in industrialized and in developing countries. The differences have a wide spectrum, beginning with the quantity and composition of generated wastes, going through the collection systems and ending with the disposal methods. Typically, countries with higher incomes produce more packaging materials and recyclable wastes; lower-income countries have less commercial and industrial activity and therefore lower waste generation rates, more organics and higher water content. Municipal wastes end up together with faecal matter, infectious medical wastes and other hazardous materials. Inadequate collecting recipients and irregular or no formal collection service expose the population to wastes and the associated risks.

The following is an overview of possible health risks linked to the activities around wastes, beginning with the dangers that affect the general population when hygienic conditions are poor, analysing the related occupational risks to which workers of this sector are exposed, and finally emphasizing the elevated dangers amidst which informal waste-pickers work.

RISKS FOR THE POPULATION

Most non-industrialized countries have low levels of formal collection rates, going from 30 to 60 per cent in low-income countries to 50 to 80 per cent in middle-income countries.

The accumulation of wastes in the street increases contact possibilities and offers very good conditions for the propagation of germs, insects, rats and other disease vectors. Where sanitation infrastructure is insufficient, human excreta or toilet paper is mixed with the municipal wastes and increases its critical character. Sometimes, the wastes are burned in order to lower critical hygienic problems; but this causes the emission of toxic substances to the air, such as dioxins and furans.

Uncollected wastes often clog drains and cause the stagnation of water, the breeding of mosquitoes or the contamination of water bodies from which the population normally takes water for consumption, cooking and cleaning. In tropical countries, the high temperatures and humid conditions accelerate degradation, increase the amount of leachate and directly affect the surrounding ecosystems by penetrating the soil and contaminating groundwater.

At the same time, animals look for food among wastes, becoming vectors of different diseases and increasing the spread of litter. Direct or indirect contact with these vectors or being bitten by certain insects can cause dangerous bacterial, viral and parasitic diseases.[1] The consumption of animals fed with wastes can cause parasitic infestations of the central nervous system and infections of the digestive tract.[2] Alimentary intoxication or food poisoning occurs when eating contaminated groceries.

Children are especially vulnerable to the risks associated to wastes because of both their behaviour and physiological characteristics. They often play outside and might pick up dangerous materials which adults would know to avoid. Moreover, children have a faster rate of breathing than adults and thinner layers of skin, which makes them more susceptible to airborne hazards, chemical absorption and burns.

The exposure to polluting compounds is more critical because they ingest more water, food and air per unit of body weight. In addition, their metabolic pathways to detoxify and excrete toxins are not fully developed. In the same way, the disorders during childhood can be mirrored in the adult years in terms of diseases, malformations or malfunction of some organs and systems.

The local problems of waste management in poor countries are additionally influenced by the illegal export of toxic wastes from industrialized countries. One example was the illicit dumping of 528 tonnes of poisonous liquid wastes in Abidjan during autumn 2006. According to the government of the Côte d'Ivoire, several people died and around 9000 individuals reported severe health problems such as respiratory difficulties, vomiting and irritations.

RISKS FOR THE SECTOR WORKERS

Workers involved in waste management are constantly exposed to specific occupational risks and the injury rate is higher than in industrial work. While standards and norms for handling municipal solid wastes in industrialized countries have reduced occupational and environmental impacts significantly, the risk levels are still very

Water streaming down the hill, polluted by the people living uphill, threatens to enter a woman's house

©WASTE, I. Haenen

high in most developing countries because of poor financial resources and inadequate understanding of the magnitude of the problem. Several scientific studies have shown that the relative risk of infections and parasites is three to six times higher for solid waste workers than for the control baseline populations, while acute diarrhoea occurs ten times more often. Pulmonary problems have an incidence 1.4 to 2.6 times higher. Respiratory disorders may result from inhaling particulate matter, bio-aerosols and volatile organic compounds (VOCs) during collection and disposal. Additional emissions of methane, carbon dioxide and carbon monoxide cause headache, nausea and vomiting. If hazardous wastes are present in the garbage, contact with critical compounds[3] may occur. The exposure to these can cause cancer, birth defects, metabolic problems and failure of organs, among other effects.

Additional particular occupational risks are related to the handling of wastes and containers, transport activities, the operation of equipment and stress factors, all of which are influenced by the infrastructure and use of personal safety equipment such as gloves, protective cloths, masks or belts. Punctures caused by pieces of glass, needles or other objects are very common. This can lead to infections, tetanus, hepatitis or HIV, especially if the wastes contain hazardous and medical materials. Other injuries occur when being hit by heavy objects or being wounded by

Box K3.1 Health risks for waste workers and recyclers

Investigations of by Lund University in Sweden have shown that informal waste-pickers at the dumpsite of Managua, Nicaragua, are exposed to very high concentrations of pollutants: the blood analysis of children between 11 and 15 years of age evidenced the presence of high traces of polybrominated diphenyl ethers (PBDEs) (a chemical flame retardant), heavy metals, pesticides and polychlorinated biphenyls (PCBs). The presence of these compounds can be attributed to the direct contact with the contaminants, the inhalation of contaminated particulate material and dust, and the consumption of polluted food. The health risks of those children who have always lived and worked in the dumpsite were especially high due to prenatal and post-natal accumulation of toxic substances.

machines and trucks when unloading, sorting or disposing of the litter. Possible surface subsidence, fires and slides in unstable fields are also a threat.

The intense physical activity related to lifting containers and using heavy equipment, as well as the risk of vibrations and constantly going down and up the collection trucks causes back and joint injuries, especially if the equipment is not ergonomically designed. Common stress factors are the exposure to noise generated by plants and machinery, heavy traffic, and night-time and shift work. In extreme weather conditions, weakness, dehydration, sunstroke and disorientation may occur, increasing the risk of accidents. In countries with severe rainfalls, the danger of slides is greater, which was the case in Manila in July 2007, when more than 200 people died. Musculoskeletal complaints occur almost twice as frequently and accidents happen up to ten times more often to sector workers.

Some sources report general relative risks

Burning tyres, creating terrible smoke

© GTZ

of mortality up to 30 per cent higher for waste workers. In Mexico, for example, the average life expectancy of waste workers is only 39 years, while the rest of the population reaches an average age of 69.

RISKS FOR INFORMAL WASTE-PICKERS

In many developing countries, informal waste-pickers are part of the collection and recycling chain of wastes. In Lima, Peru, only 0.3 per cent of the wastes are recycled by the formal sector and 20 per cent by waste-pickers. In Pune, India, only informal enterprises are in charge of the recycling activities. Informal-sector workers operate independently and normally lack the minimum protective equipment. Although many waste-pickers say that they do not like doing this work and think that it is dangerous, many of them still prefer to sell their gloves, shoes and special clothes in order to receive money and buy food.

The working environment of waste-pickers is very critical because it combines unhygienic conditions and risks of accidents. Additional dangers for the informal sector are also greater because their living and working environments usually overlap. Sometimes children and adults even look for food among the wastes because they cannot afford to buy it. Some studies detected that important morbidities among street sweepers are chronic bronchitis, asthma, anaemia and conjunctivitis, probably because of exposure to dust and particulate material, undernourishment and contact with infectious media.

Young waste-pickers are especially susceptible because specific risks are added to the general vulnerability of children. They have to carry heavy loads that affect their soft bones, they may eat food wastes, are normally undernourished, and have no sanitation facilities at work and often none at home. The percentage of prevalence of illness is always higher for child waste-pickers than for non-waste-pickers.

RECOMMENDATIONS

These described health risks emphasize the importance of sound waste management that contributes to the healthiness of both the population and the individuals involved in the waste sector. In order to minimize the health risks of the population, it is necessary to ensure the effective disposal of the wastes of all inhabitants. At the same time, in order to optimize the working conditions of the staff involved in waste management, it is necessary to improve the infrastructure and to consider occupational health aspects.

In the short term, waste workers need to be provided with adequate protective equipment such as gloves, footwear and tools to sort waste. First, sanitation facilities have to be near the working places. Workers need training about the risks and the importance of using these tools and facilities correctly. In the case of formal workers, supervisors should control the usage of this equipment. Waste-pickers also have to be motivated – for example, with a method that combines a refund system for protective equipment and food or health checks and medicine bonuses. Vaccination campaigns are fundamental for the prevention of several diseases and should include family members who also have contact with wastes or very unhygienic places. Micro-insurance might be an interesting solution for specific health and pension systems. In order to have an overview and control of these activities, initiating an identification system is recommended for which formal and informal workers receive an ID card and their information is registered.

The separation of hazardous materials from municipal wastes has to be encouraged within the population, not only for environmental reasons but also to lower the risks for waste workers. Child labour has to be at least significantly reduced because of children's health vulnerability, and alternative income options have to be offered to their families. The working situation has to be improved, especially in terms of hygienic conditions and, if possible, taking into account ergonomic characteristics that reduce occupational risks.

The recognition of waste-picking as dignified work is a first step in preventing harassment, violations and personal attacks, and in facilitating access to health services. The integration of waste-pickers in the formal waste management chain has to be accompanied by training that helps to increase the efficiency of collection, sorting and recycling activities. Some successful examples were observed in the projects of the Deutsche Gesellschaft für Technische Zusammenarbeit (GTZ). The training of informal e-waste recyclers in India increased work efficiency and usage of protective equipment. The work of waste-pickers in pre-sorting activities in treatment plants in Thailand and Brazil helped to improve sorting efficiency and to minimize risky picking and separation activities in dumpsites. The establishment of adequate infrastructure that minimizes environmental contamination and the adoption of modified technology that fits local circumstances would be a sustainable solution for developing countries. By improving these aspects, waste management can become much healthier work.

NOTES

1   For example, plague, murine typhus, leptospirosis, Haverhill fever, Rickettsialpox, diarrhoea, dysentery, rabies, typhoid fever, salmonellosis, cholera, amoebiasis, giardiasis, leprosy, toxoplasmosis, malaria, leishmaniasis, yellow fever, dengue fever, filariasis, viral encephalitis, onchocerciasis, Chagas's disease, sleeping sickness, filariasis, fascioliasis, tularaemia, bartonellosis and Oroya fever, among others.
2   For example, cysticercosis, toxoplasmosis, trichinellosis and tapeworm infections.
3   For instance, VOCs, heavy metals, polycyclic aromatic hydrocarbons (PAHs), PCBs, chlorinated hydrocarbons, pesticides, dioxins, asbestos and pharmaceuticals.

REFERENCES

Athanasiadou, M., Cuadra, S. N., Marsh, G., Bergman, A. and Jakobsson, K. (2008) 'Polybrominated diphenyl ethers (PBDEs) and bioaccumulative hydroxylated PBDE metabolites in young humans from Managua, Nicaragua', *Environmental Health Perspectives*, vol 116, no 3

Coad, A. (2007) *The Economics of Informal Sector Recycling*, GTZ, Eschborn, Germany

Cointreau, S. (2006) *Occupational and Environmental Health Issues of Solid Waste Management: Special Emphasis on Middle- and Lower-Income Countries*, World Bank Group, Washington, DC

DESA/UPMG, FEMA/MG (1995) *Manual de saneamento e proteção ambiental para os municípios*, Departamento de Engenharia Sanitária e Ambiental, DESA/UPMG, Fundação Estadual do Meio Ambiente, FEMA/MG

Hunt, C. (1996) 'Child waste pickers in India: The occupation and its health risks', *Environment and Urbanization*, vol 8, no 2, October

Landrigan, P. J., et al (1998) 'Vulnerable populations', in Herzstein, J. A. (ed) *International Occupational and Environmental. Medicine*, Mosby, MO

Medina, M. (2006) *Informal Recycling Around the World: Waste Collectors*, www.wiego.org/occupational_groups/waste_collectors/Medina_Informal_Recycling.php

Mitis, F. et al (2007) 'Waste and health in southern Italy', *Epidemiology*, September, vol 18, no 5, pS134

OPS/OMS (2008) *Guías Técnicas Roedores en casos de desastres: Control de vectores en situaciones de desastres*, Desastres y Asistencia Humanitaria – Guías Técnicas

Sabde, Y. et al (2008) 'A study of morbidity pattern in street sweepers: A cross-sectional study', *Indian Journal of Community Medicine*, vol 33, no 4, October

WHO (World Health Organization) *Waste Collection*, World Health Organization, Regional Office for Europe

LEARNING FROM HISTORY

The role of development drivers in solid waste modernization[11]

What have been the main driving forces for development? In parallel with industrialization and urbanization, the specific drivers for the development and modernization of waste management have related to improvement of public health, protection of the environment and (first and last) the resource value of the waste.

■ Driver 1: Public health

Starting in the middle of the 19th century, as cholera and other infectious diseases reached the cities of Europe and North America, legislation was gradually introduced to address the problem of poor sanitation conditions. This legislation both established strong municipal authorities and charged them with increasing responsibility for removing solid waste and keeping streets clean and litter free.

■ Driver 2: Environment

The focus of solid waste management remained on waste collection, getting waste out of the city, for a century – right up to the emergence of the environmental movement during the 1960s and 1970s. New laws were introduced, first, on water pollution, and from the 1970s on solid waste management, prompted by crises of contamination of water, air and land and their impacts upon the health of those living close to abandoned hazardous waste dumps. The initial response focused on phasing out uncontrolled disposal, both on land and by burning. Subsequent legislation gradually tightened environmental standards – for example, to minimize the formation of contaminated water ('leachate') and to prevent its release into groundwater and surface water from 'sanitary landfills'; and to reduce still further urban air pollution related to the incineration of solid waste in cities.

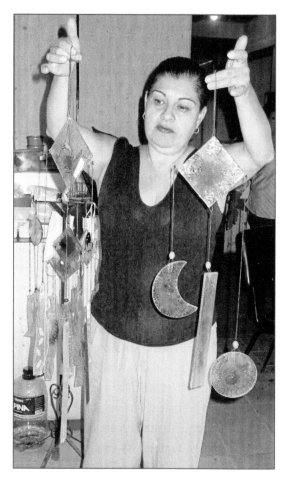

Recycled glass, Costa Rica

© ACEPESA

■ Driver 3: The resource value of the waste

In pre-industrial times, resources were relatively scarce, so household goods were repaired and reused.[12] Food and garden waste entered the agricultural supply chain as animal feed or fertilizer. As cities grew from the 19th century with industrialization, large numbers of people found an economic niche as 'rag-pickers' or 'street buyers', collecting and using or selling materials recovered from waste; in many cases, this activity was done by peddlers who collected rags and bones from the people to whom they sold.[13] This activity continues today – virtually unchanged – in many developing and transitional country cities, where informal-sector activities in solid waste management and recycling secure the livelihood of millions of people.

■ Emerging driver 4: Climate change[14]

Since the early 1990s, climate change has directed attention in the West on the need to keep biodegradable municipal waste, such as

20

Box 2.1 Waste management and climate change

Data shows that municipal solid waste management and wastewater contribute about 3 per cent to current global anthropogenic greenhouse gas emissions, about half of which is methane from landfills. One forecast suggests that without mitigation, this could double by 2020 and quadruple by 2050. It is ironic that these forecast increases are largely due to improved disposal in low- and middle-income countries – open dumps decompose partly aerobically and therefore generate less methane than an anaerobic sanitary landfill.

Mitigation needs to be a mix of the 'technical fix' approach, such as landfill gas collection and utilization, and upstream measures, particularly reduction, reuse, recycling and composting. Reduction is especially beneficial, as it also reduces the amount of 'embedded' carbon used to make the products that are being thrown away as waste.

kitchen and garden wastes and paper, out of landfills in order to reduce emissions of methane (a powerful greenhouse gas). Methane forms when organic materials decompose in the absence of air, a process called *anaerobic decomposition*. This provides a new reason for city officials to focus on diverting biodegradable municipal waste from landfills. Partly as a result, recycling and organic diversion rates, which had declined to single figure percentages as municipal authorities focused on waste collection, began to rise in cities modernizing their waste systems, in some cases dramatically. Policy measures – including laws with targets for diversion from landfill, extended producer responsibility, landfill bans for recyclable waste materials, and recycling and composting goals – pushed the recovery rates up to 50 per cent and beyond, as exemplified by three of the reference cities: Adelaide, San Francisco and Tompkins County. One could argue that history has come 'full circle' now that waste management is beginning to evolve into a mixed system for sustainable resource management.

Modernization of solid waste management systems in developed countries

For most 'developed'[15] countries, the most recent wave of what is termed here as 'modernization' of solid waste management began around the 1970s, when there was a crisis of contamination from waste, either in the city, at the disposal site, or in groundwater or surfacewater. More important than the crisis itself, the political and media discussion around it has usually provided the immediate stimulus for change.

Modernization usually begins with climbing onto the disposal-upgrading ladder – that is, with the phasing out of open dumps. Driver 2 usually results in the closing of town dumps and a plan, often not realized for many years, to develop and operate a 'state-of-the-art' regional landfill. The relatively high costs for building and operating environmental controls means that economies of scale are substantial, which favours large regional landfills, serving a number of cities and towns. Public opposition to new sites, based at least in part on bad experiences with previous uncontrolled sites (not in my backyard, or NIMBY) is a compounding factor, so that the regional landfills tend to be relatively distant from the main population centres. The geographical, logistical and institutional *regionalization* associated with upgrading disposal sets in motion a series of rapid changes in how the waste system functions and how much it costs. The combination of higher technology, more management and longer distance to the new landfill creates a rapid upward spiral in costs for cities and their contractors:

- The newly introduced landfill gate fees, based on weighing the waste, are much higher than the costs of local (largely uncontrolled) disposal.
- Collection and transport costs are much higher, as the longer distances imply increased time on the road and increased fuel consumption, and possibly the need for local transfer stations.
- There are also increased (and often unbudgeted) administration costs involved in organizing 3, 15 or even 50 separate cities and towns together to agree on where the landfill should be, which community should host it, and how the laws, regulations and administration should work.
- Political NIMBY opposition to siting introduces legal battles that cost the local

authority time and money to answer challenges in court – and in the political arena.

It is in part to illustrate this process that the reference 'cities' actually include two multi-municipality regions: Adelaide, Australia, is a regional municipality with 19 cities or towns; and Tompkins County, in New York state, is a typical North American unit of government that combines one city, Ithaca, with ten other towns in a relatively rural area.

In many developed countries, this upward spiral of costs triggered a search for less expensive ways to be modern and environmentally responsible. Some part of the strong interest in recycling and composting came about because, when compared to regional disposal, these activities began to appear to be less expensive, as well as environmentally preferable. During the period of active modernization in the US, for example, recycling goals in many states increased from 15 per cent of total waste to more than 50 per cent in a relatively short period of time at the end of the 1980s.[16]

Modern *municipal recycling*, as it has been reintroduced in Europe and North America since the 1970s, depends on households segregating materials *at the source*. This means that waste system users, the households, need to change their habitual behaviour and to separate their waste into several categories, which they store separately, rather than mixing it all together in one basket, bag or bin. Collecting several source-separated waste streams without greatly increasing collection costs is a similar challenge to the waste collection providers and operators: they also have to change the way in which they think and behave. This has led, in some instances, to a reduction in collection frequency for the residual waste.

The solid waste challenge in developing and transitional country cities

Experience in low- and middle-income countries can also be related to the same drivers. The plague epidemic in Surat is one example of a public health crisis that stimulated new initiatives to collect the waste and clean up the city, now known as one of the cleanest in India.

The landslide at the Payatas dumpsite in Quezon City, the Philippines in July 2000 killed 200 people – a terrible tragedy – but it also catalysed the political process that resulted in the passage of Republic Act 9003, the Ecological Waste Management Act, one of the most complete and progressive solid waste management laws in Asia. Reawakening interest in resource management has inspired a public–private partnership in Dhaka, Bangladesh, that was one of the first to be issued climate credits.

Solid waste management is a major challenge for many cities in developing and transitional countries. The urban areas of Asia were estimated to spend about US$25 billion on solid waste management each year in 1998.[17] Solid waste management represents 3 to 15 per cent of the city budget in our reference cities, with 80 to 90 per cent of that spent on waste collection before modernization.[18] Collection coverage in the reference cities, as in urban areas in general, varies widely, ranging from 25 to 75 per cent in cities where the norm for waste disposal is still open dumping.

Why should the authorities choose to invest in a waste system when such investment is likely to raise costs and offer competition for scarce financial resources to other critical municipal systems, such as schools and hospitals?

Box 2.2 Plague-like epidemic in Surat, India[19]

Uncollected solid waste blocking drains caused a major flood, leading to an outbreak of a plague-like disease in Surat, India, in 1994. The disease caused panic countrywide, and while the citizens blamed the municipality, the public authorities, in turn, blamed the citizens for their lack of civic sense.

Over 1000 plague-suspected patients were reported, with the final death toll of 56 people. The city incurred a daily loss of 516 million Indian rupees during the plague period and a total loss amounting to 12 billion rupees. This was a high price to pay for negligence in the area of solid waste management.

Alarmed at the situation, the Surat Municipal Corporation undertook a stringent programme of cleaning the city. Within a year after the plague, the level of (daily) solid waste collection increased from 30 to 93 per cent, and 95 per cent of streets are cleaned daily. Market areas, major roads and litter-prone spots are cleaned twice a day.

Surat is now identified as one of the cleanest cities in the region.

A basic answer is public health. UN-Habitat data shows significant increases in the incidence of sickness among children living in households where garbage is dumped or burned in the yard. Typical examples include twice as high diarrhoea rates and six times higher prevalence of acute respiratory infections, compared to areas in the same cities where waste is collected regularly.[20] Therefore, providing comprehensive waste collection is an equity issue.

Uncollected solid waste clogs drains and causes flooding and subsequent spread of waterborne diseases. Infectious diseases such as cholera or the plague do not respect wealth – not collecting waste in the slums may also cause sickness in the richer parts of the city. In one small city in Egypt, 89 per cent of villagers living downwind of the burning dumpsite were suffering from respiratory disease.[21] Contaminated liquids, or leachate, leaking from dumpsites may also pollute a city's drinking water supplies.

The modernization challenge facing a low- and middle-income country city includes how to extend collection coverage to unserved parts of the city where there is less infrastructure and the ability to pay is lower. This is something that few cities in Europe or North America have to think about, and it is another major source of increasing costs. But without providing comprehensive collection, these cities are not fulfilling their responsibility to protect public health – not just for the poor, but for all their citizens.

Cleaning up campaigns in Quezon City, Philippines, contributing to improving health conditions for the children of the Barrangays (local communities)

© Quezon City

MOVING TOWARDS SUSTAINABLE SOLUTIONS

Solid waste and the Millennium Development Goals (MDGs)

The Millennium Development Goals (MDGs) were ratified by 189 heads of state at the United Nations Millennium Summit in September 2000, with the overall objective of halving world poverty by 2015. Improving solid waste management systems will contribute to achieving many of them, in spite of the fact that solid waste is never explicitly mentioned in the MDGs.

There are several places where the MDGs and the modernization of waste management come together, as is shown in more detail in Table 2.4:

- MDGs 1 and 7, on livelihoods and poverty, on the one hand, and on environment, on the other, point to the urgency of inclusive policies in waste management so that the role of the informal waste sector in cleaning up cities and recovering resources is recognized, while working conditions and livelihoods are improved. Recent work suggests that the informal sector both contributes to a city's recycling rates and substantially reduces its costs for managing solid waste.[22]
- Improving the coverage of waste collection services contributes to the health-related MDGs 4, 5 and 6, and will reduce both child diseases and mortality.
- MDG 8, on global partnerships, is a blueprint for cities to work with private formal and informal actors, on the one hand, and to join with communities in participatory planning and problem solving, on the other. Partnerships can improve governance, bring about financial sustainability and support proactive policy formulation.

Modernization of solid waste management in the West started when recycling rates had declined to a very low level, and has included a drive to rebuild recycling through the municipal waste system. Most developing and transitional country cities still retain their informal recycling systems, which provide a source of livelihood to vast numbers of the urban poor. Building on this existing system makes good sense.

The integrated sustainable waste management (ISWM) framework

When the current modernization process started in developed countries during the 1970s, solid waste management was seen largely as a technical problem with engineering solutions. That changed during the 1980s and 1990s when it became clear that municipalities could not successfully collect and remove waste without active cooperation from the service users. Cities also learned that technologies depend on institutional, governance and policy frameworks, which are highly varied and complex, and directly related to local conditions.

There is now broad international consensus for what has come to be known as ISWM: integrated sustainable (solid) waste management. As is shown in Figure 2.2, ISWM identifies three important dimensions that all need to be addressed when developing or changing a solid waste management system – namely, the stakeholders, the elements and the sustainability aspects.

ISWM is designed to improve the performance of solid waste system and to support sound decision-making. It does this by framing the solid waste process, and balancing short-term crisis management and long-term vision. It helps municipal officials and other stakeholders to understand how the different parts of the system relate to each other.

The examples from Denmark or Japan – which some would regard as world icons of good waste management practice – suggest that a sustainable, affordable waste management system consists of a stable mixture of technolo-

gies and institutions, which function flexibly under a clear policy umbrella.

Such systems mimic an ecosystem, which is robust and resilient when there is a mix of unique niches and competition for resources. If one species falls out, others move in to take its place. In low- and middle-income countries, there is often a variety of formal and informal, public and private systems already operating, so the basis for a stable mixed system is already in place. What most low- and middle-income cities miss is organization – specifically, a clear and functioning institutional framework, a sustainable financial system, and a clear process for pushing the modernization agenda and improving the system's performance. As long as there is no umbrella framework, the mixture remains a cluster of separate parts that do not function well together – or at all.

Sustainability in solid waste management is possible

The severity of the local solid waste management problem may lead a city mayor to grab at whatever is offered that sounds like a solution, particularly if it appears to solve an urgent problem in a politically comfortable way. But if a solution seems 'too good to be true', it's probably not true.

There are few global controls on the claims made by individuals or companies seeking to do business with cities. Marketing representatives travel the world over and offer mayors and city councillors the one 'right answer', the magic

Trucks going via the weighbridge to the landfill, Nicaragua

© UN-Habitat
Jeroen IJgosse

bullet to slay the solid waste dragon. But just as solid waste isn't really like a dragon, a magic solution isn't really possible. Solid waste is part of modern daily life, not something unexpected. If there was one thing to learn from the Naples, Italy, waste strike in 2007 to 2008, it is that no matter what the politicians do, the solid waste keeps coming. And the public who generate it and the politicians and officials responsible for managing it need to understand what they are doing and be able to make good decisions based on sound local knowledge.

This means that waste collection and disposal technologies need to be both appropriate and financially sustainable under local conditions. For example, large waste-compaction collection vehicles designed to collect low-density, high-volume wastes on broad suburban streets built to withstand high axle-loading rates in Europe or North America are unlikely to be suitable for use in a developing country city. There the vehicles have to be smaller, lighter and narrower to allow collecting much denser wastes from narrow streets and transporting it over rutted roads going up and down steep hills – even well-surfaced main roads tend to be designed for lower axle-loading rates. In many cases, a small truck, a tractor or even a donkey fits local collection needs, while a 20 tonne compactor truck does not.

While the latest European Union (EU) standards for sanitary landfill are required and

Quezon City Barangay primary collection system, Philippines

© Quezon City

affordable in France, they are unlikely to be either appropriate or financially affordable in Ouagadougou, simply because people have lower incomes and can't pay much for waste removal. A modern waste-to-energy incinerator designed for high-heating-value Japanese or European waste is likely to require supplementary fuel to burn a typical high-organic and relatively wet waste in a transitional country; while the costs and skills required to operate, maintain and regulate the state-of-the-art air pollution control equipment required to protect public health are likely to restrict such technologies to a few of the most advanced cities. And a novel waste treatment technology, which has not yet found a buyer in a European market, is a risky choice for the low- and middle-income country mayor who needs a guarantee that his wastes will be collected, treated and disposed of reliably, 365 days a year.

It is this need to keep going, day in and day out, that makes it so critical to shift from the term 'solid waste management' to 'sustainable waste management'. 'Sustainability' is a long word for 'common sense', and there are some relatively simple ways to improve the performance and sustainability of waste management systems. Magic technology doesn't work at least in part because it tries to reduce the problem to a purely technical one, whereas the key message of ISWM is that all stakeholders need to be engaged *and* all sustainability aspects need to be addressed. It is the transparent processes of users talking to providers, communities sharing responsibility for planning, and recycling businesses working with cities that make for sustainability.

And as long as that is true, solutions have to be the result of citizens, leaders, and the waste and recycling sector working together to come up with approaches to make decisions. This book provides some wonderful examples, from the experiences of the reference cities to other stories and anecdotes, from which to obtain inspiration.

Table 2.4

Relevance of improved SWM to the Millennium Development Goals

Sources: Gonzenbach et al (2007); Coad (2006); Hickman et al (2009)

MDGs	Achieving MDGs through improved sustainable waste management
1. Eradicate extreme poverty and hunger	Informal-sector self-employment in waste collection and recycling currently provides sustainable livelihoods to millions of people who would otherwise have no stable source of income and would be most susceptible to extreme poverty and hunger. City authorities can both promote recycling and create more opportunities for the informal sector to provide waste collection services in unserved areas and thereby help eradicate extreme poverty and hunger.
2. Achieve universal primary education	Waste management activities contribute indirectly to education through income generated by the parents. Many waste-pickers earn sufficient income to send their children to school and do so with pride. The poorest waste-pickers do engage their children for picking and sorting waste; but in instances where NGOs are involved, classes are organized for these children, after their working hours, and parents are informed about the need and the benefits of primary education.
3. Promote gender equality and empower women	A substantial percentage of informal-sector waste collectors and waste-pickers are women. Efforts to improve solid waste management services and enhanced recycling can include improvement and equal working conditions for men and women by creating financial and other arrangements that build capacity and empower women.
4. Reduce child mortality	Effective solid waste collection and environmentally sound disposal practices are basic public health protection strategies. Children living in households without an effective waste collection service suffer significantly higher rates of, for example, diarrhoea and acute respiratory infections, which are among the main causes of childhood deaths. Cooperation with informal-sector waste collectors and recyclers will improve their livelihoods and reduce child labour and, hence, direct contact of children with the wastes.
5. Improve maternal health	Almost all women waste-pickers have no maternal healthcare available to them. Enhanced recycling may directly/indirectly improve maternal health through achieving improved living standards among households engaged in the sector.
6. Combat HIV/AIDS, malaria and other diseases	Originally, municipal waste management activities started due to public health concerns. The reasons are almost self-evident: uncollected waste clogs drains, causes flooding and provides breeding and feeding grounds for mosquitoes, flies and rodents, which cause diarrhoea, malaria, and various infectious and parasitic diseases. Mixing healthcare wastes with municipal solid waste and its uncontrolled collection and disposal can result in various infections, including hepatitis and HIV. Reliable and regular waste collection will reduce access of animals to waste and potential for clogging of drains. Proper waste management measures can practically eliminate risks associated with healthcare waste.
7. Ensure environmental sustainability	Few activities confront people with their attitudes and practices regarding sustainability as waste management does. Reduce, reuse, recycle is yet to realize its full potential as a guiding principle for environmental sustainability through conservation of natural resources and energy savings, as well as through reduction of greenhouse gases (GHGs) and other emissions.
8. Develop a global partnership for development	Through cooperation and exchange, developed and developing countries can develop and implement strategies for municipal services and job creation where unemployed youth will find decent and productive work and lead a dignified and good life.

Dare to innovate

Most books on solid waste view developing and transitional country solid waste systems as imperfect or incomplete copies of an 'ideal' system that operates in developed countries such as Canada or Sweden. Many, if not most, waste interventions seek to perfect or improve the copying process and spread the ideal. Or, at most, low- and middle-income countries have, until now, sought to adapt the models from developed countries to their local circumstances.

This book takes a different view, responding to a growing global consensus that cities in low-income, middle-income and transitional countries need to take charge of the modernization process and develop their own models for modern waste management that are more and other than simply 'imperfect copies' – models with focus and approaches that fit their own local conditions.

Daring to innovate, or to 'think outside the box', helps us to understand, for example, how solid waste is different from many other public utility functions, as the following example shows. The closest public service to solid waste, in terms of its regularity and complexity, is perhaps

the postal service. In a sense, waste management could be viewed as a kind of 'postal system in reverse' – indeed, some researchers have classified waste management as 'reverse logistics'. The postal service runs quite well in most countries of the world, whereas the waste management system does not. Why is this?

Simply put, we value our post. We make sure we put the right number of stamps on our letters or packages, and we ensure that they are placed in the letterbox or deposited at the post office. And because we value the cargo, we have no problem paying for this service.

On-time collection in residential areas in Ghorahi, Nepal. Household containers of waste are loaded directly into the truck, thus improving both public health by avoiding intermediate storage in the open air and cost effectiveness by avoiding multiple manual handling of the wastes

© Bhushan Tuladhar

KEY SHEET 4
RECYCLERS AND CLIMATE CHANGE

The world's waste-pickers

We recyclers and other recycling workers in the informal economy are environmental entrepreneurs performing with high efficiency and have generated a climatic debt for our history and current contribution to the reduction of greenhouse gases and the reduction in costs for waste management.

Material recovery and recycling are for us the best options for managing urban waste. Therefore, we don't consider the extraction of landfill gases to produce energy, or incineration projects or the production of derivated fuels to be recycling or recuperation operations.

The industrialized countries must reduce their consumption of natural resources, limit the generation of waste, increase recycling and avoid all exports of waste and technologies contributing to climate change. We call upon the United Nations Framework Convention on Climate Change (UNFCCC) and our local governments to:

- Recognize the critical and productive role that the recyclers contribute to the mitigation of climate change, and invest resources in programmes for recovery at source that ensure a dignified way of life for all workers and traders from the recycling industry.
- Study and remove the support for all projects that divert recyclable waste to incineration or landfilling.
- Establish mitigation mechanisms that are directly accessible by recyclers and which are significant in terms of financial and technical support.
- Consult the recyclers first in relation to energy from waste generation.
- Support projects and technologies that divert organic waste from landfills by means of composting and methane production, and which should be adopted as options due to the reduction of methane.

Bonn, Germany
8 June 2009

Box 2.3 Integrated sustainable waste management (ISWM)

Integrated sustainable waste management (ISWM), as shown in Figure 2.2, is a framework that was first developed during the mid 1980s by WASTE, a Dutch non-governmental organization (NGO), and WASTE's South partner organizations, and further developed by the Collaborative Working Group on Solid Waste Management in Low- and Middle-Income Countries (CWG) in the mid 1990s. Since then it has become the 'norm'.[23]

ISWM is a systems approach that recognizes three important dimensions, which all need to be addressed when developing or changing a solid waste management system. The dimensions, shown in Figure 2.2, correspond to three key questions:

1   The *stakeholders* – the people or organizations with a 'stake' or interest in solid waste management: *who* needs to be involved?
2   The *elements* – the technical components of a waste management system: *what* needs to be done?
3   The *aspects* which need to be considered as part of a sustainable solution: *how* to achieve the desired results?

Stakeholders. The main 'recognized' stakeholders include the local authority (mayor, city council, solid waste department), the national environment and local government ministries, and one or two private companies working under contract to the municipality. Often unrecognized stakeholders include (female) street sweepers, (male) workers on collection trucks, dumpsite 'waste-pickers', some of whom may actually live on or at the edge of the dumpsite, and family-based businesses that live from recycling.

Other key stakeholders include the waste generators: the users of the waste management service provided by the city, including households, offices and businesses, hotels and restaurants, institutions such as hospitals and schools, and government facilities such as airports or the post office.

Elements. These are the technical components of a waste management system. Part of the purpose of using the ISWM framework is to show that these technical components are *part of* the overall picture, not *all of* it. In Figure 2.2, the boxes in the top row all relate to removal and safe disposal, and the bottom row of boxes relate to 'valorization' of commodities. Solid waste management consists of a variety of activities, including reduction, reuse, recycling and composting, operated by a variety of stakeholders at various scales.

Aspects. For a waste management system to be *sustainable*, it needs to consider *all* of the operational, financial, social, institutional, political, legal and environmental aspects. These form the third dimension in Figure 2.2, in the lower box. The aspects provide a series of analytical 'lenses', which can be used, for example, for assessing the situation, determining feasibility, identifying priorities or setting adequacy criteria.

'Integrated' in ISWM refers to the linkages and interdependency between the various activities (elements), stakeholders and 'points of view' (sustainability aspects). Moreover, it suggests that technical, but also legal, institutional and economic linkages are necessary to enable the overall system to function.

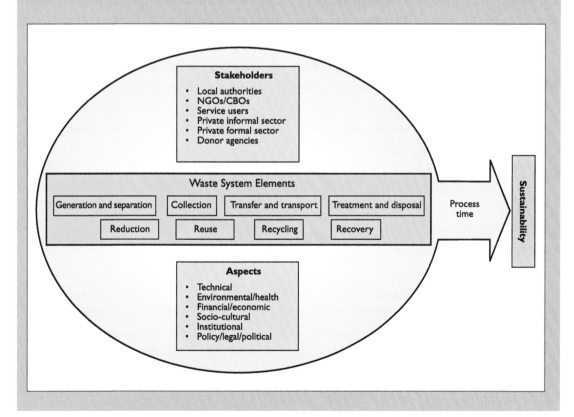

Figure 2.2

The integrated sustainable waste management (ISWM) framework

Source: WASTE (advisers on urban environment and development), Gouda, the Netherlands

This is the crucial difference between waste management and other utilities and public services. Most people don't care where their waste goes, as long as it is not next to their house. They may be willing to pay for removal of their waste from the immediate vicinity of their house, but often not for its subsequent treatment and disposal. And when they are not willing to pay, someone else generally suffers. Whereas an individual misses their post, the individual opting out of a waste management service doesn't notice much 'personal' impact. It is much easier, as well as much more harmful, to burn or dump your own waste than it is to generate your own electricity, or, indeed, to deliver your own letter to your family in a distant village.

Another important difference between waste and utility services such as water or electricity is that the impact of waste management is not as direct; you will not receive a collection service that collects your waste as soon as you generate it. You will have to manage it yourself for some time, before giving it to the collection service that passes by once or twice a week. Availability of communal containers and/or informal collectors that provide 'private' services can sometimes create the notion that a waste collection service is always available to take the waste out of sight of the place of generation, without necessarily attending to the issue of public health protection.

Waste management isn't as technically complex as energy or housing, but it does have its own set of issues and solutions, and these deserve attention. This book is about that attention, and it supports the real work of supporting decision-makers and practitioners in the daily work of figuring out what is right for each city's particular climate, economy and citizens.

The book's authors trust that by identifying good and innovative practices from cities at all stages of developing their waste management systems, this book will contribute to helping cities find innovative and workable solutions that are appropriate to their own particular circumstances.

It is also part of the ambition of this book to encourage decision-makers to think beyond the short term. An effective collection system serving the whole city and a safe and environmentally sound disposal site are essential components of an ISWM system. But so are effective systems to address the 3Rs: reduce, reuse, recycle (i.e. to reduce the quantities of waste generated, and to build on the existing, largely informal sector systems for reuse and recycling).

NOTES

1   OECD Environmental Data Compendium 2006–2008, www.oecd.org/dataoecd /22/58/41878186.pdf.
2   Hoornweg and Thomas, 1999.
3   Basel Convention, www.basel.int/.
4   Wilson et al, 2003.
5   Prüss et al, 1999.
6   Coffey, 2009.
7   Key Note Publications Ltd, 2007.
8   US Census Bureau, Population Division, International Data Base, www.census.gov/ipc/ww w/idb/worldpop.php.
9   Scheinberg and IJgosse, 2004.
10 For example, average MSW growth rates in England averaged –0.4 per cent per annum for the five years to March 2008 (Defra, 2008).
11 Wilson, 2007.
12 Strasser, 1999.
13 'Any rags, any bones', Strasser, 1999, Chapter 2, pp69–109.
14 More on climate change in Chapter 5, and in Key Sheet 13.
15 We use the term 'developed' interchangeably with 'higher income' because it is generally used to refer to higher-income countries. But we use it with some regret, as some so-called 'less developed countries' are more developed in some dimensions than the so-called 'developed countries'.
16 Scheinberg, 2003.
17 Hoornweg and Thomas, 1999.
18 World Bank (undated).
19 Sanjay Gupta, own experience.
20 UN-Habitat, 2009, p129
21 SWM case study for Sohag City, as part of the UK Department for International Development (DFID)-funded Support for Environmental Assessment and Management (SEAM) project in Egypt, 1999.
22 GTZ/CWG, 2007.
23 Anschütz et al (2004).

3

PROFILING THE REFERENCE CITIES

Because of the difficulties in obtaining comparable information from cities, this Global Report is based on profiling and presenting 20 reference cities. This chapter introduces both the cities and the methodology that has been created to stimulate their participation, and to increase the comparability and accuracy of the data that has been collected.

Presenting information in a consistent way helps to understand how things work within and across countries. Solid waste management is fragmented across cities and countries, as well as within them. It seldom has an academic disciplinary home and, as a result, French African countries, for example, measure their waste differently from Balkan former socialist states, while high-income European countries each use different categories. In some countries municipal waste includes waste from commercial enterprises and shops; in others, it includes institutions such as schools but excludes commercial waste. In some countries waste is classified by where it comes from, in others by where it is allowed to go. The point here is that establishing a consistent frame allows patterns to emerge.

For this reason, this Global Report profiles a group of reference cities in a consistent way, asking research questions about the nature and sustainability of waste management and recycling in a globalizing world.

SELECTING THE REFERENCE CITIES

The goal for working with 20 cities was a need for:

- a qualitative understanding of what drives the system, how it works and who is involved in it;
- hard data and facts from official and reported sources, framed and validated by the visual presentation of a process flow;
- information on what works and what doesn't, both in individual cities and across cities.

The cities were selected by combining criteria on representativity with indications of ease of access; the group of people working on the report arrived at a group of 20 cities that form the core of this book.

Two sets of criteria were used:

1   Criteria for the mix of cities:
    - a range of sizes, from mega-city to small regional city;
    - a range of geographic, climatic, economic and political conditions;
    - the distribution of cities to include most in low- and middle-income countries, with a significant number in Africa;

- at least one from each continent, including a few from high-income countries.

2   Criteria for each city:
- a good illustration of one or more of the main topics and main messages around which the Global Report is structured;
- a city that is willing to participate;
- a city willing to invest in preparing the materials and providing information;
- a city willing to share both good and not-so-good practices;
- someone from or working closely with the city who is willing to take responsibility for collecting data from that city and preparing it in the form desired;
- the more close the contacts with the city, the more favourable it is to include it.

UNDERSTANDING THE REFERENCE CITIES

One element that is interesting about the reference cities (see Table 3.1) is that they are so varied: the smallest is Cañete, Peru, with less than 50,000 people, and the largest Delhi, with a population of more than 13 million. The two sub-Saharan African cities have the highest growth rates at about 4.5 per cent.

The 20 reference cities used in the book provide a reasonable cross-section across the world, but meeting all possible selection criteria is challenging. It is hoped that similar city profiles will be prepared and published in the future; priorities for inclusion would include cities from the former Soviet Union/Newly Independent States; Middle East; English-speaking West Africa; Portuguese-speaking Africa and an island city state.

Table 3.1

The reference cities

This table shows the range of cities profiled for this Report. The smallest in area is Curepipe, more of a town than a city; the largest in area are cities or regions: Kunming, Tompkins, Adelaide.

Note: NR = not reported.

Source: original data for this report supplemented by information from UNDP

City	Size of city (km²)	Population	Growth rate	Country GDP (US$ millions) (UNDP, 2007; HDR, 2009)	Human Development Index (HDI) (UNDP, 2009)
Adelaide, Australia	842	1,089,728	3.3%	821,000	0.97
Bamako, Mali	267	1,809,106	4.5%	6900	0.37
Belo Horizonte, Brazil	331	2,452,617	1.2%	1,313,400	0.81
Bengaluru, India	800	7,800,000	2.8%	3,096,900	0.61
Canete, Peru	512	48,892	2.7%	107,300	0.81
Curepipe, Republic of Mauritius	24	83,750	0.8%	6800	0.80
Delhi, India	1,483	13,850,507	1.5%	3,096,900	0.61
Dhaka, Bangladesh	365	7,000,000	1.7%	68,400	0.54
Ghorahi, Nepal	74	59,156	4.0%	10,300	0.55
Kunming, China	2,200	3,500,000	NR	3,205,500	0.77
Lusaka, Zambia	375	1,500,000	3.7%	11,400	0.48
Managua, Nicaragua	289	1,002,882	1.7%	5700	0.70
Moshi, Tanzania	58	183,520	2.8%	16,200	0.53
Nairobi, Kenya	696	4,000,000	4.5%	24,200	0.54
Quezon City, Philippines	161	2,861,091	2.9%	144,100	0.75
Rotterdam, Netherlands	206	582,949	−0.2%	765,800	0.96
San Francisco, USA	122	835,364	1.0%	13,751,400	0.96
Sousse, Tunisia	45	173,047	3.3%	35,000	0.77
Tompkins County, USA	1,272	101,136	0.1%	13,751,400	0.96
Varna, Bulgaria	80	313,983	−0.1%	39,500	0.84
Average	510	2,462,386	2.2%	2,013,905	0.72
Median	310	1,046,305	2.7%	87,850	0.76

METHODOLOGY

How is it possible to research and understand 20 cities in a short period of time? Some basic instruments have been derived from the integrated sustainable waste management (ISWM) framework, with a focus on three system elements and three governance aspects, and include:

1 using a process flow approach to understanding the entire waste and recycling system through the construction of a process flow diagram (PFD);

2 developing and requesting unusual data points and indicators as a way of extending the boundaries of what can be understood and compared;

3 designating a person who has worked in the city and knows it well, named hereafter the 'city profiler'.

Profilers draw on their own practical knowledge of the city, in addition to consulting key stakeholders, newspaper reports, plans, photos, and/or using records of stakeholder meetings or other events. These will be discussed in more detailed, with special attention given to the process flow diagram.

The process flow approach and the use of a process flow diagram

This report has been prepared by collecting original data from 20 cities. The information includes text, tables and diagrams produced by the city profilers, who are identified in the Acknowledgements section. The 'profile' is a long data form used in compiling this book, with roughly 45 pages of instructions and forms to fill in; this is not a readable document but UN-Habitat or WASTE will make it available upon request. Most of the information comes from the 'city presentation' or simply 'presentation', a report of around 15 pages which has been used by the cities to present their city to the project

team. These 'city presentations' will be published in a follow-on volume. The presentations have also been used as the basis for the two-page city inserts in this volume.

First, the cities have been asked to diagram their solid waste and recycling system – including formal and informal elements and operations – in a process flow diagram (PFD). A PFD turns out to be a relatively powerful way of presenting the system as a whole in a comprehensive but concise way. A combination of process flow and materials balance was used in the GTZ/CWG (2007) study *Economic Aspects of the Informal Sector in Solid Waste* in order to understand the relationships between formal and informal

Involving households in data collection in Managua, Nicaragua, to obtain an insight into waste generation and characterization information of domestic waste

© UN -Habitat, Reymar Conde

Interviewing informal recyclers in Managua as part of the profiling process

© UN -Habitat, Reymar Conde

sectors. The instruments developed here are in part based on that experience. A PFD approach is useful because it:

- gives a fast picture of what is happening to which streams;
- is a good way of ensuring that the whole system is included in the analysis;
- makes it clear(er) where the system boundaries are and provides a structure for analysing the materials that 'escape' from the system;
- shows where the materials actually end up, including highlighting where leaks and losses are occurring;
- provides a check on data provided in other ways (e.g. a waste stream that has been left out of the composition analysis may well be shown on the PFD);
- allows for and, indeed, facilitates understanding linkages between formal and informal activities, actors and steps in the chain of removal, processing, valorization or disposal;
- shows in a concrete way the degree of private-sector participation in the system and in the management of different materials;
- is a reliable way of estimating recovery rates for specific materials and mixed streams;

- allows for comparison of costs and efficiencies between different operations and for the system as a whole;
- shows the degree of parallelism and mixing in the system.

Cañete is the smallest and simplest of the cities, and its PFD is also one of the clearest and easiest to understand. By looking at the PFD, it is possible to see where mixed waste, recyclables, organic waste and residuals are coming from in the city, who handles them, where they go, and what is lost or leaks from the system on the way. Moreover, a few things are immediately clear. The first is that while all the waste from formal collection goes to the La Arena dumpsite, not all of it stays there. The second aspect is that substantial waste is not collected. Belonging neither to formal nor informal sectors, this 4.23 tonnes per day nevertheless finds its way to small dumpsites. The reader is asked to look first at the process flow for Cañete shown in Figure 3.1. What can be seen at first glance about organic waste and recyclables? How many different kinds of operations compete with each other for different kinds of waste from households? These are the elements that help us to understand the city's waste system in a very basic way.

Although Cañete is a very small city, its PFD shows the continuum between large and small, motorized and manual, formal and informal, and consists of varied elements that are both intricately connected and in constant flux. By following the flow of materials, processes and people through the system, a framework for understanding the city's waste system is provided. Process flow diagramming – with the materials balanced for each step – ensures a good basis for decision-making. The PFD, for example, shows that the nearly 10 tonnes recovered goes to recycling and not to organics recovery, and all of it is collected with non-motorized transport. However, two-thirds of a tonne is 'lost' each day from households, which suggests minor amounts of animal feeding or home composting.

Collection of organic wastes from parks and gardens was identified as an important waste source using the process flow approach in Cañete, Peru

© IPES,
Humberto Villaverde

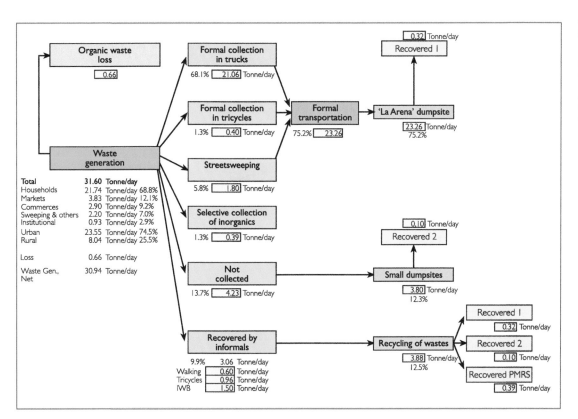

Figure 3.1

**Process flow
diagram for
Cañete, Peru**

Source: presentation for
Cañete, Peru. IPES, Lima,
Peru

A PFD is also the best way to understand the amount of parallelism and mixing in the system, which is an indicator of state and type of modernization.[1] 'Parallelism' here is used to mean that there are competing options or paths for materials to move along the same place in the chain, so that in Lusaka formal collectors compete with informal collection service providers to collect waste from households. In Varna, in contrast, there are franchises for different sub-municipalities for waste collection; but there is only one main route that mixed waste follows when it leaves the household.

For organic waste and paper, there are alternative paths: many people use kitchen waste to feed animals at their village houses, and they burn paper in small woodstoves in the winter. This would be called mixing rather than parallelism.

Process flow diagrams are methodologically useful for additional reasons. By looking at the paths of materials, it is possible to know a great deal about transactions and relations between different stakeholders, as well as understanding how formal and informal systems relate to each other. PFDs tell rather a lot about rela-

tionships between stakeholders, and they help with understanding financial and governance issues about the system as well. In addition to diagramming, some cities went so far as to research and calculate the materials balances for all of the process steps, showing how many tonnes go in, get transformed and go out. This kind of modelling clarifies the situation still further and improves the quality of the information significantly.

Process flow diagramming introduces facts that can be easily visually communicated to policy-makers. The work done on Lusaka for the GTZ/CWG (2007) informal study showed the advantages of using a PFD: through diagramming it became clear that more than 30 per cent of Lusaka's waste was collected by informal service providers, who were considered as unregistered or illegal. At first, this activity was 'written off' as a loss or leak; but the fact that the PFD showed it so clearly stimulated a discussion about the value of this activity, which could then be framed as informal but organized. This, in turn, influenced the city's attitude towards – and recognition of – this unusually large informal service sector.[2]

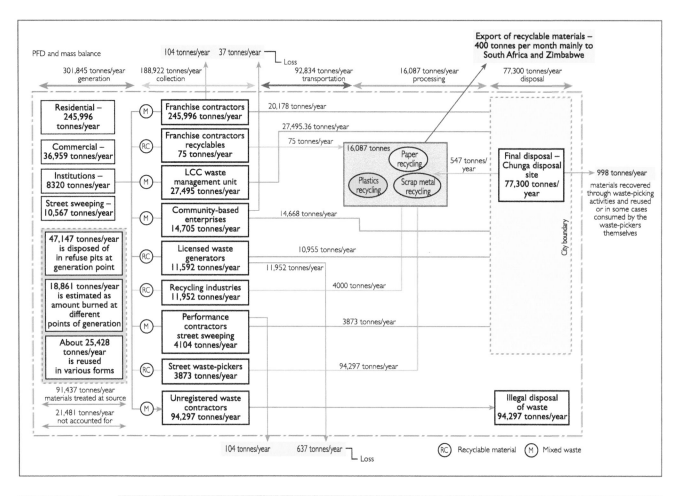

PFD and mass balance

301,845 tonnes/year generation | 188,922 tonnes/year collection | 104 tonnes/year | 37 tonnes/year | 92,834 tonnes/year transportation | 16,087 tonnes/year processing | 77,300 tonnes/year disposal

Export of recyclable materials – 400 tonnes per month mainly to South Africa and Zimbabwe

Residential – 245,996 tonnes/year

Commercial – 36,959 tonnes/year

Institutions – 8320 tonnes/year

Street sweeping – 10,567 tonnes/year

47,147 tonnes/year is disposed of in refuse pits at generation point

18,861 tonnes/year is estimated as amount burned at different points of generation

About 25,428 tonnes/year is reused in various forms

91,437 tonnes/year materials treated at source

21,481 tonnes/year not accounted for

Franchise contractors 245,996 tonnes/year — 20,178 tonnes/year

Franchise contractors recyclables 75 tonnes/year — 27,495.36 tonnes/year — 75 tonnes/year

LCC waste management unit 27,495 tonnes/year

Community-based enterprises 14,705 tonnes/year — 14,668 tonnes/year

Licensed waste generators 11,592 tonnes/year — 10,955 tonnes/year — 11,952 tonnes/year

Recycling industries 11,952 tonnes/year — 4000 tonnes/year

Performance contractors street sweeping 4104 tonnes/year — 3873 tonnes/year

Street waste-pickers 3873 tonnes/year — 94,297 tonnes/year

Unregistered waste contractors 94,297 tonnes/year

16,087 tonnes — Paper recycling — Plastics recycling — Scrap metal recycling — 547 tonnes/year

Final disposal – Chunga disposal site 77,300 tonnes/year — 998 tonnes/year

materials recovered through waste-picking activities and reused or in some cases consumed by the waste-pickers themselves

City boundary

Illegal disposal of waste 94,297 tonnes/year

104 tonnes/year | 637 tonnes/year — Loss

(RC) Recyclable material   (M) Mixed waste

Figure 3.2

Process flow diagram for Lusaka, Zambia, from the GTZ/CWG (2007) report

Source: Riverine Associates and the City of Lusaka Solid Waste Department; the Process Flow was originally prepared for GTZ/CWG, 2007 (draft)

In Lusaka, Zambia, the process flow diagram identified both formal and informal actors in the collection service provision. The photo shows the formal, municipality, collection service

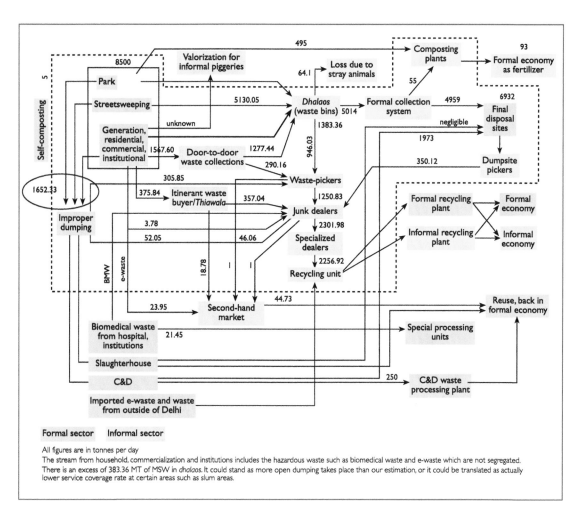

Figure 3.3

Process flow diagram for Delhi, India

Source: Chintan-Environmental, Delhi, India

Process flow diagramming, when combined with materials balances, also fundamentally changes the international discourse on informal recycling. First, it provides indications of the size, richness and impact of informal recycling activities. Second, it illustrates quite clearly the many types and intensities of interrelationship between households, providers, and formal and informal economic actors. Cañete is small, but its process flow diagram shows that a substantial amount of material leaves the formal disposal facility via informal channels, and then ends up in the formal recycling supply chain. In Delhi there is, in effect, no formal primary collection system serving households: the informal system functions as the connection between households and the *dhalaos*. In the large majority of the cities, there is, in effect, a formal–informal continuum, with different categories of actors who interact, overlap and may themselves change category in response to changing circum-

stances. This is supported by the fact that so many of the cities, alongside their trucks and tractors, still have a considerable portion of their waste, recyclables and/or organic materials moved with animal or human muscle power.

The process flow for Delhi, India, suggests quite clearly that the only waste that is being collected is moving through the informal sector. Everything else is either thrown onto streets, where it is captured by sweepers, is taken to containers or is discharged in the park. In this example, a world-class city depends upon its waste-pickers for keeping a basic level of cleanliness.

In the process flow diagram for Quezon City, the Philippines, the materials recovery facilities (MRFs) in the Baranguays, play a key role in processing materials for valorization. The process flow shows that key role quite clearly, as many streams of materials are shown to come together there.

Former rag-picker engaged in door to door primary waste collection in Delhi, India, en route to a secondary collection point. The importance of such informal sector collectors, now officially recognized by the City, was highlighted through the process flow approach

© Sanjay K. Gupta

Information and indicators

A short set of indicators was, secondly, prepared based on the six ''themes' of ' good practice in ISWM components that form the focus of this report, as follows.

Three drivers and physical elements:

1   Public health/collection.
2   Environment/disposal.
3   Resource management.

Three ISWM governance aspects, which include:

1   Inclusivity.
2   Financial sustainability.
3   Sound institutions and proactive policies.

These indicators are useful for analysing how processes work within a city and comparing across cities. The point is not so much to see how one city 'scores', but how things cluster and what this tells about the city. These indicators are presented in the insert, which includes two pages per city of key information and the short indicator set.

One of these new indicators was inspired by the experience of Delhi/New Delhi, where the profiler and Chintan-Environmental, the host NGO, were astounded to find out how challenging it was for the city officials to find or provide information. This led to the creation of a relatively new governance indicator: the age of the most recent reports that are available.

In the comparative tables distributed throughout this Global Report, as many cities as possible will be included in the comparison based on the availability of information per city. In cases where information is not reported, the abbreviation NR will be marked, and in cases where information is not available, the abbreviation NA is used.

The role of city profilers

Third, the individuals who described the cities for the book (the 'city profilers') are mentioned; these city profilers collected examples, stories, photos, newspaper articles and other qualitative information. Together with the profilers, the co-authors and editors of the book used their collective experience to really understand the 'story' of solid waste in each city, how the drivers have influenced solid waste, and how to understand both successes and problems. Some examples of 'stories' include the following:

Municipal staff identifying the different roles in solid waste management within the organizational chart in Managua, Nicaragua

© UN-Habitat Jeroen IJgosse

- Cities with good collection at the sub-municipal level, such as Bamako or Nairobi or Managua, may have distant, limited or no controlled disposal simply because there is no one at the city council level who 'owns' the problem or is committed to proactively seeking a solution.
- Or consider the paradox of Curepipe, Adelaide and Rotterdam: too much moderately priced disposal reduces incentives for both users and providers to work on source separation and recovery of recyclables and organic waste – even when there is a policy commitment. The result is missed opportunities and disappointing recycling performance.

Other ways of understanding the cities include comparative tables, photos, diagrams and stories; these have come from the city profilers; from their sources (both in terms of reports and in terms of talking to people) in the cities; and from the collective professional memory of all the writers and teams working on the Global Report.[3] The sum of all these parts is designed to give a three-dimensional insight into the cities that builds understanding about ISWM in specific places, and also in its totality.

City Profiler involved in engaging and acknowledging waste-pickers in the formal primary waste collection system in Buldana, India

© Sanjay K. Gupta

INFORMATION QUALITY

If knowledge is power, than a city without knowledge of its solid waste system may lack the power to make positive changes. Solid waste information is subject to a number of widely encountered structural weaknesses. In many cities, information on solid waste is:

- old – more than 10 years' old and, in some cases, more than 15, while changes in the composition of the waste stream, population and behaviour are continuously occurring;
- orphaned – neither owned nor recognized by the city itself, particularly in low- and middle-income countries, where a donor, or a state, provincial or national government paid for the study or financed the consultant, and did not ensure that the information was useful for the city; or where there is no central archiving system in the city;

- secret – considered to be secret or proprietary because of the involvement of private-sector actors or investors;
- estimated – estimated based on national or regional figures, without verification in field assessments;
- political – highly politicized and subject to distortions in support of the policy ambitions of particular stakeholders;
- not permanent – because it related only to a specific period of a government administration and experiences from previous administration are seen as 'useless';
- missing – missing or incomplete because there is no party willing to invest in gathering accurate information on such a dirty subject; and/or
- inaccessible – because it might not be written in the language of the municipality, but rather in the language of the consultant hired by the donor.

On the other hand, city governments or solid waste agencies that consider waste to be a prior-

Table 3.2

Values of a short set of indicators in the reference cities (percentage).

There is more discussion on the indicators shown in this table in the introduction to the City Inserts (pages 41 to 45).

Note: NA = not available. Italics = estimated. Curepipe, Delhi, Ghorahi and Quezon City do not have a municipal waste fee. Belo Horizonte: 70% of slum populated was covered in 2008.

| | Drivers for solid waste management | | | | Governance | | | |
| | Public health collection/ sweeping coverage (%) | Public health/ environment Controlled disposal/ incinerated of total disposed /incinerated (%) | Environmental control Waste captured by the waste system (%) | Resource management Materials prevented or recovered (%) | Inclusivity | | Financial sustainability Population using and paying for collection as percentage of total population | Institutional coherence Degree of institutional coherence |
					Degree of user-inclusivity	Degree of provider-inclusivity		
Adelaide	100%	100%	100%	54%	HIGH	HIGH	100%	HIGH
Bamako	57%	0%	57%	85%	MEDIUM	MEDIUM	95%	LOW
Belo Horizonte	95%	100%	100%	1%	HIGH	HIGH	85%	HIGH
Bengaluru	70%	78%	90%	25%	MEDIUM	MEDIUM	40%	MEDIUM
Canete	73%	81%	83%	12%	MEDIUM	HIGH	40%	HIGH
Curepipe	100%	100%	100%	NA	LOW	LOW	0%	HIGH
Delhi	90%	100%	76%	33%	HIGH	MEDIUM	0%	LOW
Dhaka	55%	90%	56%	18%	MEDIUM	MEDIUM	80%	HIGH
Ghorahi	46%	100%	88%	11%	MEDIUM	LOW	0%	MEDIUM
Kunming	100%	100%	100%	NA	MEDIUM	MEDIUM	50%	HIGH
Lusaka	45%	100%	63%	6%	MEDIUM	MEDIUM	100%	MEDIUM
Managua	82%	100%	97%	19%	MEDIUM	LOW	10%	MEDIUM
Moshi	61%	78%	90%	18%	MEDIUM	LOW	35%	MEDIUM
Nairobi	65%	65%	70%	24%	MEDIUM	HIGH	45%	LOW
Quezon City	99%	100%	99%	39%	MEDIUM	MEDIUM	20%	HIGH
Rotterdam	100%	100%	100%	30%	HIGH	LOW	100%	HIGH
San Francisco	100%	100%	100%	72%	HIGH	LOW	100%	HIGH
Sousse	99%	100%	100%	6%	LOW	LOW	50%	MEDIUM
Tompkins County	100%	100%	100%	61%	HIGH	MEDIUM	95%	HIGH
Varna	100%	100%	100%	27%	LOW	LOW	100%	HIGH
Average	82%	90%	88%	30%			57%	
Median	93%	100%	98%	25%			50%	

ity have the tendency to invest in monitoring and documentation of waste information and reap the benefit of good data. And cities that have a strong resource management driver and are seeking to achieve high recovery rates are often willing to invest more in detailed waste characterization studies, so that they really understand what can be recovered. As a result, quality of information may serve as an indicator of commitment.

Data from 2008 or later was available from Adelaide, San Francisco, Tompkins County, Belo Horizonte, Managua and Varna. Other cities reported a variety of sources, or their sources were undated.

Quick look at the main indicators in the reference cities

The short set of indicators used for a rapid assessment of the 20 cities can be seen in Table 3.2.

 ## CITY INDICATORS

Interpreting the data

Each city has a series of indicators that are representative of different aspects of a city's solid waste system. Behind these indicators are the overarching 'drivers' for the modernization of the solid waste management system, which include improving public health, reducing impacts to the environment, and increasing resource recovery through minimizing waste generation combined with increasing materials recycling. These three drivers should be considered linked; addressing impacts upon the environment necessarily includes addressing potential impacts upon human health. Similarly, reducing waste generation and subsequent disposal through waste prevention, reuse and recycling has quantifiable benefits to both human health and the environment.

An integrated and sustainable waste management approach to solid waste necessitates addressing these three elements; but this is done within the context of government institutions. The modernization of the solid waste management system often sees establishment of new policies, regulations and possible restructuring of management and administration to better address the minimization of public health and environmental impacts while maximizing the recovery of resources from the waste stream. Thus, good governance becomes a driver for a sustainable and adaptable solid waste system.

An 'indicator' suggests that a data set has been chosen to provide an indication of how a city has addressed one of the aforementioned drivers. The chosen indicators in Table 3.2 should not be considered as the only lens through which one would assess the movement towards a modernized solid waste management system. They were selected as representative of specific drivers in order to provide the reader with a quick summary of the state of an integrated sustainable solid waste management (ISWM) system for the representative cities. It should also be realized that ISWM is a process, so these indicators only reflect a snapshot in time. All of the cities reviewed for this Global Report are involved in a continued process of evaluating, planning and implementing new initiatives.

Table 3.2 summarizes the indicators for each of the cities and summarizes the information in the charts in the two-page city inserts.

Description of each indicator

Collection/sweeping coverage: percentage of the city that receives a regular service of waste collection and street sweeping. The driver is public health, involved with keeping garbage and the associated vectors from waste accumulating within the city.

Controlled disposal: percentage of the waste that ends up in a disposal facility with basic controls. The driver could be considered both public health, especially with labour associated with disposal sites, as well as environmental protection of soil, water and air resources.

Waste captured by the system: percentage of waste that enters the formal waste management system via any of the possible paths in the process flow diagram, including but not limited to collection. Although there are obvious public health benefits, the overall impact of non-managed waste is driven by activities such as open burning and disposal in watercourses, with direct environmental and ecological consequences.

User inclusivity: extent to which the users of the system have access, control and influence on how the system works. This aspect of governance can be considered from two perspectives. There is the inclusivity of the users of solid waste services – that is, to what degree are these stakeholders included in the planning, policy formation and implementation processes? The second perspective on user inclusivity refers to the performance of the system, and the extent to which it serves all users equitably and according to their needs and preferences.

Provider inclusivity: extent to which the economic niches in service delivery and valorization are open and accessible to non-state actors, especially the private formal and informal sectors, micro- and small enterprises (MSEs) and community-based organizations (CBOs). Inclusivity can also be considered from the perspective of the waste service provider, which includes both the informal and formal sector. This indicator signals the degree to which the formal authorities allow and enable non-state providers to be integrated within an overall solid waste collection, transfer, materials recovery and disposal strategy. For the four developed country cities, this is what the indicator shows. For the rest, the indicator points to this aspect of enabling and including informal service providers, and/or recognizing and protecting the value of the informal recycling sector with overt policies that institutionalize and integrate these service providers.

Financial sustainability: percentage of system costs recovered from user fees and payments. This is based upon the economic tenet that the user pays for the service. It also implies that systems subsidized are susceptible to changes in external dynamics, which in turn can affect the sustainability of a solid waste management system.

Institutional coherence: percentage of total solid waste budget that falls in the budget line of the main organization in charge of solid waste. Institutional coherence signals the ability of the designated local governing entity to control the overall budget for solid waste management. This ensures that other activities associated with institutional stability and the ability to implement proactive policies are not necessarily influenced by external factors not directly associated with the services provided.

NOTES

1   Spaargaren et al, 2005.
2   GTZ/CWG, 2007;
     Lusaka city report and
     workbooks.
3   In addition to the refer-
     ence cities, examples,

photos and stories from
other cities around the
world will be presented
based on this collective
memory.

CITY INSERTS

This City Inserts section represents the reader's introduction to the 20 reference cities. Each of the 20 cities was 'profiled' for this report. The instruments for profiling included a long data form for collecting information, entitled 'City profile', and a shorter report template for presenting a summary of the data for each city, which was entitled the 'City presentation document'. The city presentation documents, in particular, provide a wealth of information on the state of solid waste management in a wide range of cities and form a good comparative tool. The information they provide is presented in many tables in Chapters 2, 4, and 5 of this report.

Due to the length and number of the city profile and city presentation documents, they are not included in this report, but will be published as a separate book in 2010. For this reason, a rather small extract of the information is presented in this chapter, in the form of a two-page 'city insert' that includes an introduction to each city and an overview of the solid waste system, seen through the integrated sustainable waste management (ISWM) framework of three physical systems and three governance aspects. Additional source information on the 20 cities is available from their own websites, which are listed in the References, and on the website of WASTE at www.waste.nl.

Interpreting the charts

Each city insert has a bar chart that presents a short series of indicators representing different aspects of a city's solid waste system. Behind these indicators are the overarching 'drivers' for the modernization of the solid waste management system, which include improving public health, reducing impacts to the environment and increasing recovery of materials through prevention, recycling or separate organics management. While conceptualized separately, it is clear that the system aspects represented by these indicators are connected to each other, and no indicator should be seen as completely separate from the whole system. An ISWM approach to solid waste necessitates physical systems for management of materials, but other approaches for managing behaviour of users, performance of providers and financial resources that make the whole system work. Thus, 'good governance' becomes a driver for a sustainable and adaptable solid waste system.

The purpose of the chart is to provide a visual representation, or 'snapshot', of the overall current condition of modernization in each city. Since all the cities are actively planning, implementing and moving to improve the sustainability of their solid waste systems, the image provided by the chart will continually change over time. The charts provide an *indication* of how a city has addressed the solid waste management system as a whole: a glimpse of how a city may be faring with regard to modernizing and managing their solid waste system. The eight indicators were chosen because they were based upon data and written information that were provided for almost every city in the study.

Number of indicator	Indicator name	Indicator	
1	Collection and sweeping coverage	Percentage of population who has access to waste collection services.	
2	Controlled disposal	Percentage of total waste destined for disposal that is deposited in an environmental landfill or controlled disposal site, or any other formal treatment system, including incineration.	
3	Waste captured by the system	Percentage of waste collected by the formal and informal sector or deposited by households in containers or depots. The final destination is not relevant.	
4	Materials prevented or recovered	Percentage of total waste which is prevented and recovered – that is, which fails to reach disposal because of prevention, reuse or valorization.	
5	Provider inclusivity	Composite score on a set of quality indicators allowing a yes for present and a no for absent. Represents the degree to which service providers (and waste recyclers) are included in the planning and implementation process of waste management services and activities.	
6	User inclusivity	Composite score on a set of quality indicators allowing a yes for present and a no for absent. Represents the degree to which users of the solid waste services are included in policy formation, planning, implementation and evaluation of these services.	
7	Financial sustainability	The percentage of households who both use and pay for waste collection services.	
8	Institutional coherence	Composite score of low, medium or high. Combines a percentage indicator for the degree to which the solid waste management budget is directly controlled by the agency, or entity, formally designated to manage the solid system within the city, combined with a qualitative assessment and the organogram.	

In addition to the key indicators in the chart, discussed in the following table, the City Inserts include, for each city, a table of solid waste benchmarks. These are discussed further on in this introduction and represent somewhat more 'normal' solid waste indicators than the chart.

The table below explains the quantitative and qualitative indicators utilized in the chart. These indicators are cross-referenced to the 'drivers' of modernization and good practice solid waste management system: improving public health; reducing impacts upon the environment; and improving resource management and recovery of materials. Also in the table are indicators reflecting the three key aspects of governance in waste management: inclusivity of both users and providers of waste services, financial sustainability, and the degree of institutional coherence.

	Driver and relation to the indicator
	Public health: the driver is public health associated with removing waste and preventing it from accumulating within the city. Alternative indicators for this driver could be the number of households served by collection services, or, spatially, the percentage of total street kilometres or surface area where collection and sweeping services are present.
	Public health in combination with environmental protection: controlled disposal indicates the portion of waste generated that ends up in a disposal site with a minimum degree of management – that is, there is gate control, fencing or other forms of control. As a system modernizes, such management reduces the potential of water, soil and air pollution associated with disposal of wastes. Usually, controlled disposal also implies that waste is managed with some protection of worker health and safety; but this specific indicator does not necessarily guarantee that this is included in the definition of controlled disposal used by each city.
	Environmental protection: this indicator provides information on the proportion of waste that enters the waste system and is processed. The opposite of this indicator is percentage of total waste generated that is a 'loss' to the waste management system through burning, burying, evaporation, and dumping in watercourses, empty lots or other unofficial places. It excludes both waste prevention and reuse by the generator (e.g. neither home composting nor animal feeding of kitchen waste would be considered to be captured by the system).
	Environmental protection in combination with resource management: this indicator is one commonly used by cities as their 'recovery percentage' and sometimes it is even referred to as the 'recycling rate'. It is a combination indicator because preventing waste generation and subsequent management through animal feeding, composting, reuse, remanufacturing and recycling has quantifiable benefits to both human health and the environment, in addition to conserving resources and reducing impacts of natural resource extraction.
	Governance/inclusivity: this indicator is designed to communicate how open the system is for participation of private- and community-sector providers, or the informal sector. The composite is based on a 'yes' value for present or 'no' value for absent, relating to the following set of qualitative indicators: • laws at national or local level in place that encourage private-sector participation, public–private partnerships (PPPs), or community-based organization (CBO) participation; • organizations or platforms prevalent that represent the private waste sector (formal and informal); • evidence of formal occupational recognition of the informal sector active in waste management practices or recycling; • evidence of protection of informal-sector rights to operate in waste management; • little, or no, institutional or legal barriers for private-sector participation in waste management in place; • institutional or legal incentives for private-sector participation in waste management in place.
	Governance/inclusivity: this indicator is based on a set of qualitative indicators: • laws at national or local level that require consultation and participation with stakeholders outside the bureaucratic structures; • procedures in place/evidence of citizen participation in the siting of landfills; • procedures in place/evidence of customer satisfaction measurement of waste management services at municipal or sub-municipal level; • procedures in place/evidence of feedback mechanisms between service provider and service user; • citizens' committees in place that address waste management issues.
	Governance/financial sustainability: this is a composite indicator from percentage coverage/access to services plus percentage paying for services. It represents sustainability because such a system is self-supporting. It is based on the implicit hypothesis that systems subsidized from outside sources are vulnerable to failure due to changes in external economic and political circumstances that neither users nor providers nor authorities can control.
	Governance/institutional coherence: solid waste is often an 'orphan' or 'child of many parents' in the waste management system. By comparing the fragmentation or distribution of budget in relation to administrative responsibilities, the indicator seeks to reflect institutional coherence, defined as the ability of the designated local governing entity to control the overall budget for solid waste management. Alternative indicators that might have been considered in assessing institutional coherence within a city include: • establishment of mechanisms for user feedback; • implementation of formal performance evaluation procedures; • ability to maintain and operate all facilities and equipment; • development of a formal solid waste and materials recovery plan; • development of rate structures based on total system cost accounting.

Overview of indicators used in the bar charts

Each city insert has a section entitled 'Some basic facts'. These include basic geographic, social, demographic, political and economic facts that are provided to orient the reader. Because not every city has provided precisely the same data, there is some minor variation between cities on what appears in this section. For some cities this section may include a prose introduction, provided by the city itself, to give additional flavour for the reader.

The next section is entitled 'The solid waste story'. This section follows the thematic organization of this Third Global Report and discusses, in brief, three key ISWM physical systems and three key ISWM governance features in the city, as drawn from that city's presentation document. This consists of five main topics, which, again, may not be precisely the same for each city:

1   The main driver for solid waste modernization, which consists either of public health, environment or resource management, or some combination. In general, where public health is the driver, the main focus of physical systems will be on collection. Where environment is the driver, the main focus usually shifts to disposal, and when environment is fully institutionalized, resource management is included both as a second level of environmental improvement and the basis for economic activity in the formal and informal sectors.
2   A brief description waste collection and its relation to the public health driver.
3   A brief description of waste disposal and its relation to the environment driver.
4   A brief description of resource management, including recycling, composting, other forms of recovery, prevention and organized reuse.
5   One good practice or special feature that the authors believe characterizes the 'personality' of the solid waste system in that city.

Finally, throughout the text there will be information about the solid waste governance aspects in the city – that is, inclusivity, financial sustainability and institutional coherence.

What follows is a table entitled 'Key benchmark numbers'. Whereas the bar chart has derived indicators, the benchmarks are, above all, descriptors: quantitative information provided by the city itself. This group of numbers is presented in order to complement and anchor the solid waste story. Unlike the chart, there are slight variations in the benchmarks per city.

Interpreting the key benchmark descriptors

Benchmark definition	Definition
Total tonnes municipal solid waste (MSW) generated per year	Reported tonnes of municipal solid waste generated, which includes residential, commercial and institutional wastes, unless noted.
Generation per capita in kilograms per year	This was either directly reported by the city or calculated based upon tonnes generated and population reported.
Percentage coverage	Percentage of a city's population covered by waste collection services.
Percentage disposal in environmentally sound landfills or controlled disposal sites	Percentage of the total waste generated that has been diverted to an environmental landfill or controlled disposal site.
Percentage municipal waste incinerated	Percentage of the total waste generated that has been diverted to incineration, whether or not there is energy recovery associated with such an option.
Percentage of waste prevented and valorized	Percentage of the total waste generated that has been diverted by the formal and/or informal sector through waste prevention or reuse, or valorized through recycling, composting or other methods.
Percentage of waste valorized by informal sector	Percentage of the total waste generated that has been diverted through recycling, composting or other methods by the informal sector.
Percentage of waste valorized by formal sector	Percentage of the total waste generated that has been diverted through recycling, composting or other methods by the formal sector.
Goals for waste collection coverage as percentage of population	Goals stated by the city, either formalized by regulations or included in strategic plans. These may reflect percentages of coverage and/or dates for reaching some benchmark.
Goals for environmentally sound (safe) disposal	Goals stated by the city, either formalized by regulations or included in strategic plans. These may reflect percentage of the amount of waste, number of households or population served and/or dates for reaching some benchmark.
Goals for valorization of waste materials through recycling (or diversion from disposal)	Goals stated by the city, either formalized by regulations or included in strategic plans. These can be through initiatives to increase waste prevention or reuse and/or by developing the infrastructure to recover materials through recycling, composting or some other means.
Prevented	Percentage of the total waste generated and includes waste prevention strategies or activities undertaken by individuals, businesses or institutions to reduce the volume and toxicity of material discarded.
Reused	Percentage of the total waste generated and includes use of waste materials or discarded products in the same form without significant transformation, and may include a system developed to repair/refurbish items.
Recycled	Percentage of the total waste generated and indicates extraction, processing and transformation of waste materials and their transfer to the industrial value chain, where they are used for new manufacturing. For some cities, recycling is only considered to have occurred when materials have been sold since the actual use of materials into the manufacturing process may require exporting to other regions or even countries.
Composted/agricultural value chain	Percentage of the total waste generated and indicates extraction and processing of organic waste materials at a location that incorporates a process which manages and controls decomposition of material to provide a soil amendment utilized by agriculture. It could also include diverting organic wastes as feed for livestock, or technologies such as anaerobic digestion.

The city insert closes with a section on main references used to create the presentation document and profile, if provided by the city profilers, as well as websites or literature that will offer the reader additional information on that city. The full set of references is included in the References chapter, and the full list of individuals working on gathering city data is included in the Acknowledgements.

ADELAIDE

South Australia, Australia, Australasia (Oceania)

34.93°S 138.59°E
40m above sea level

Andrew Whiteman (Wasteaware) and Rebecca Cain (Hyder
Consulting, Australia)

Some basic facts

Located in south-central Australia, the Adelaide metropolitan
area extends approximately 20km from the eastern coast of
Gulf St Vincent to the foothill suburbs of the Adelaide hills
(Mount Lofty Ranges) in the east. The area also extends about
90km from north to south. The Adelaide Metropolitan Area is
comprised of 19 councils, or municipalities.

Topography: sea level in the west, increasing to an altitude of
approximately 400m, 40km to the east. Mean maximum
temperature = 21.8°C; mean minimum temperature =
12.0°C. Mean rainfall = 529.2mm. Size of city/urban area:
841.5km². Population: 1,089,728. Population density: 1295 persons/km². Population growth rate (2008): 3.3% (between 2001 and
2006). Average household size: 2.4.

The solid waste story

Main driver

South Australians are highly environmentally conscious. Since the adoption of container deposit legislation (CDL) over 30 years ago,
which imposed a deposit fee on packaging such as beers and soft drinks, and also due to the acute water shortages in the state,
South Australians are used to working for the environment and expect the same standards from their industry and government. The
sophisticated nature of the industrial sector in Australia, combined with the tendency towards large nationally operating companies
mean that all stages of the waste management process are well developed and regulated, and are capital/technology intensive rather
than labour intensive.

Resource management represents a major policy priority. In July 2003 a new government body, Zero Waste South Australia (ZWSA),
was established to drive forward waste reduction, recycling and reuse practices. Setting up Zero Waste SA was a key development
underpinning the South Australian government's commitment to establish a new legislative framework for state and local
government to work together under an integrated strategy. One of the most innovative aspects of Zero Waste SA is that their
revenue stream is linked ('hypothecated') to the landfill tax revenue receipts of state government. Out of every dollar of landfill tax
charged, 50 cents is made available to Zero Waste SA for initiatives which divert waste from landfill.

Public health/collection

The waste collection system in Adelaide is highly modernized, and 100 per cent of households in the Adelaide metropolitan area
receive a high-quality kerbside waste collection service, usually on a weekly basis. The high standards of collection and street and
public place cleansing services and customer care are consistent regardless of the socio-economic status of the area. Approximately
70 per cent of the population receive kerbside collection services that are operated by the private sector under contract to local
councils, and 30 per cent by a public company set up by a group of councils. The majority of collection services operate as three-bin
systems for separate collection of recyclables, green organics and residual waste.

Environment/disposal

Landfilling has been carried out to a high standard of environmental protection for decades. Yet, public opposition to a large landfill in the late 1990s, in combination with recognition that the national target of reducing landfilling to 50 per cent of 1990 levels by 2000 had not been achieved, stimulated the enactment of the Zero Waste Act in 2003. The goal is now to promote waste management practices that, as far as possible, eliminate waste or its consignment to landfill, advance the development of resource recovery and recycling, and are based on an integrated strategy for the state.

While there is considerable landfill space available to dispose of Adelaide's waste, diversion is encouraged by a combination of material-specific bans, increases in landfill tax, and innovation in recycling and recovery through grants and research programmes. Whereas many cities in the world are striving to bring all waste into controlled disposal, South Australia is striving to make disposal as irrelevant and unnecessary as possible.

Resource management

The recycling system is designed to collect source-separated comingled plastic, steel, cardboard, paper and aluminium packaging, and to transport these materials to a materials recovery facility (MRF). After manual pre-sorting, automated sorting, compaction and baling, the materials are transported to processing facilities for recycling or exported to overseas markets. Beverage containers collected through container deposit legislation are also mechanically sorted before being transported to processing facilities.

Garden organics and food waste as well as 75 per cent of recovered timber and wood products are processed into soil conditioner, compost, potting mixes and mulches, which are sold for residential and commercial use. The demand for composted organics has shifted from soil conditioner to mulches, possibly due to a greater awareness of the need to conserve water in domestic gardens under drought conditions. Approximately 25 per cent of recovered timber is used as fuel in cement kilns. The use of timber as a fuel in cement manufacture began in 2004/2005 and has utilized significant quantities of timber previously disposed of to landfill.

Special features

Adelaide and South Australia's waste and resources management system is in some respects global best practice. South Australia has demonstrated a high level of political commitment and willingness to 'stick its neck out' and implement some policies and legislation upon which other administrations take a more conservative position. The Zero Waste Act and Plastic Bag Ban are two excellent examples of South Australia's politicians showing leadership by putting in place the institutional structures, financing mechanisms, organizational capacity, and actions to support a major drive towards the 3Rs (reduce, reuse, recycle).

Key benchmark numbers

Total tonnes municipal solid waste (MSW) generated per year	742,807 tonnes*	Percentage valorized by informal sector of total MSW generated	None
Generation per capita in kilograms per year for South Australia	490kg	Percentage valorized by formal sector of total MSW generated in South Australia**	54%
Percentage coverage	100%	Goals for waste collection coverage as percentage of population	100%
Percentage disposal in environmentally sound landfills or controlled disposal sites of total waste generated	46%	Goals for environmentally sound (safe) disposal	Moratorium on additional landfills
Percentage municipal waste incinerated of total waste generated	0%	Goals for valorization of waste materials through recycling (or diversion from disposal)	25% by 2014
Percentage diverted and valorized of total waste generated (2007–2008) across South Australia	54% of MSW 76% of C&I waste 72% of C&D waste	Prevented Reused Recycled Composted/agricultural value chain	54% of MSW (recycled) 7.68%

Notes:* Commercial and institutional municipal waste is not included here because no data are available for this waste stream (commercial and institutional waste is reported together with industrial waste as 1,251,935 tonnes generated per year).

** Based on the total of 742,807 tonnes, divided by the South Australia population of 1,514,337 according to the 2006 Census.

Figures in italics are estimates.

BAMAKO
Mali, West Africa, Africa

12°39'N, 8°0'W,
approximately 350m above sea level

Modibo Keita (CEK), Erica Trauba (WASTE intern), Mandiou Gassama, Bakary Diallo and Mamadou Traoré (all of Cabinet d'Études Kala Saba, CEK, Bamako, Mali)

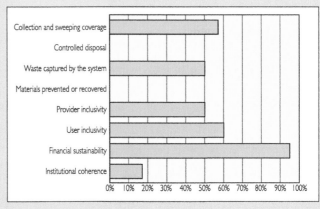

Some basic facts

The district of Bamako is situated in the Koulikoro region in southern Mali. It is the country's capital and largest city. It is made up of seven territorial collectivities that include six communes and the district mayor's office.

Topography: Niger River Valley surrounded by hills that extend from the Manding Mountains; there is a dry season from October to May, and a rainy season from June to September. Average rainfall: 919.3mm/year (the Malian weather service measures rainfall between 1 May and 31 October). Size of city/urban area: 267km². Population: 1,809,106 (2009 Census). Population density: 6331 persons/km² (city). Population growth rate (2008): 4.5%. Average household size: 6.7. The population is made up of 615,836 households that live in 85,728 concessions, or residential compounds. Human Development Index (2009): 0.371.

The solid waste story

Main driver

The main development driver in Bamako is the resource value of waste, especially kitchen and compound organic waste for vegetable farming. While public health and environmental aspects of waste management are important to Malians, they have not yet had the same power to mobilize waste management activities as the economic value of waste. The system is characterized by the public health and living environment (*le cadre de vie*) concerns, as the city is struggling to keep the city's commercial centre, selected big roads, the airport and central market areas clean from litter and waste piles. In 2002, when Mali hosted the Coupe d'Afrique de Nations (CAN) football championship, a mass clean-up effort was made, unfortunately with no lasting effects.

Public health/collection

Primary collection services – based on private-to-private arrangements and provided by micro- and small enterprises called *Groupements d'Intérêt Économique* (GIEs) – cover some 57 per cent of households in the district of Bamako. Most collection is done with 2-cubic-metre donkey-drawn carts that are staffed by one person. The average GIE has four carts and five donkeys. An increasing number of GIEs are currently investing in motorized vehicles to improve their collection capacities. After collection, the GIEs transport waste to one of the secondary collection points or sites (*depots de transit*), spread throughout the city. There are 36 such officially designated secondary collection sites, but only 14 are in use and only 1 is functioning as intended. The unused sites may either be open space or may be used for housing. This has resulted in over 75 large unauthorized secondary collection sites and hundreds of small waste piles in the city. Although there are institutions responsible for controlling waste management activities, enforcement is not realistic as there are hardly any alternatives. Residents who are not served by a GIE will either rely on informal collectors or take their own waste to collection sites. The Direction des Services Urbains de Voirie et d'Assainissement (DSUVA), the waste management department of the district of Bamako, is responsible for secondary waste collection and transport.

Environment/disposal

Approximately 40 per cent of household waste collected by GIEs within the district of Bamako remains within the city limits, at the depots or informal waste collection sites. The rest is transported by the DSUVA outside the city, where, depending on the season, it is either sold to farmers due to its high organic content or dumped in open spaces. The main challenge that the district of Bamako faces is the lack of an adequate final disposal site. There is insufficient space to build it within the limits of the district of Bamako; thus it must be built on the territory of neighbouring municipalities, which is the responsibility of the national authorities to regulate. There is a site designated for a controlled landfill about 30km outside the city limits in Noumoubougou, but has remained mostly unused because there is no clear body responsible for paying for its operation or the transportation costs involved, which are prohibitively high. There are ongoing discussions for developing landfills or waste-to-energy facilities with foreign firms, but none has materialized to date.

Resource management

The practice of *terreautage* at transit sites is common: the GIEs or the DSUVA take waste from the transit depots and sell it to crop farmers (*céréaliculteurs*) or drop it on open land or in rivers. The waste that has broken down (called *fumure*, or *terreau*) is sold to *maraîchers*, the vegetable farmers who grow their crops in the floodplain of the Niger River. The vegetable growers prefer to take organic waste from the collection sites that has already undergone some decomposition and is, thus, considered more valuable. The collection activities increase right before the rainy season because the demand from the crop farmers is very high. Recycling of non-organics is done entirely by the informal sector and at a relatively low level; however, direct reuse of products is widespread.

Special features

There is a lively market for both fresh and partially decomposed raw waste, which consists of 40 per cent sand and grit from floor sweepings and a large additional percentage of organic waste, and is a source of nutrients.[1] The practice of *terreautage* is well established, and is one of the barriers to making and selling compost formally; farmers do not see why they should pay a higher price for compost than they are used to paying for unprocessed waste. This waste valorization system partially explains why a city the size of Bamako avoids developing a landfill.

Communication about waste management in the district of Bamako occurs, among other means, through a local platform structure. Since 2000, platforms have been established in each of the six communes in the district of Bamako, in parallel with general efforts to decentralize government services. The platform model has allowed neighbourhoods to have more input in decisions about local waste management practices.

Key benchmark numbers

Total tonnes municipal solid waste (MSW) generated per year	462,227 tonnes	Percentage valorized by informal sector of total waste generated	54%
Generation per capita in kilograms per year	265kg	Percentage valorized by formal sector of total waste generated	NR
Percentage coverage	57%	Goals for waste collection coverage as percentage of population	None
Percentage disposal in environmentally sound landfills or controlled disposal sites of total waste generated	None	Goals for environmentally sound (safe) disposal	None
Percentage municipal waste incinerated of total waste generated	None	Goals for valorization of waste materials through recycling (or diversion from disposal)	None
Percentage diverted and valorized of total waste generated	*31% as terreautage* to agricultural supply chain and *21% recycling*	Prevented Reused Recycled Composted/agricultural value chain	NA NA 25% *31% as terreautage*

Notes: NR = not reported.
NA = not available.
Figures in italics are estimates.

Note

1   UWEP (Urban Waste Expertise Programme) *Urban Waste Expertise Programme 1996–2003 Report*

BELO HORIZONTE

Minas Gerais, Brazil, South America

19°55'S 43°56'W
600m–1600m above sea level

Sonia Maria Dias (independent consultant), Jeroen IJgosse (independent consultant) and Raphael T. V. Barros (UFMG)

The authors acknowledge the collaboration of the team of the Planning Department of the Superintendência de Limpeza Urbana (SLU).

Some basic facts

Belo Horizonte is the capital of the state of Minas Gerais, south-eastern Brazil, in a region with rugged and hilly topography. The metropolitan region of Belo Horizonte is Brazil's third largest, with an estimated population of 5.4 million (IBGE, 2007). The main municipality of Belo Horizonte has 2.4 million inhabitants (IBGE, 2007), including almost 0.5 million living in 140 *villas* and *favelas* (urban low-economy-class settlements).

Size of city/urban area: 331km². Population (2009): 2,452,617. Population density: 7291 persons/km². Population growth rate (2008): 1.18%. Average household size: 3.1. Human Development Index (2000): 0.839.

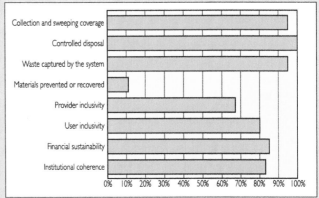

The solid waste story

Main driver

Public health was the main driver beginning in 1900, but by the late 20th century, socio-environmental concerns, such as upgrading existing systems and income generation for the poor, have catalysed improvements in solid waste management. Belo Horizonte has a strong and extensive tradition in municipal planning, including waste management services. In 1993 an integrated system was established, including upgrading of operations, implementation of recycling programmes for construction waste and organics, environmental education, upgrading working conditions of formal workers and integration of informal recyclers within the formal system.

Public health/collection

The Superintendency for Urban Cleansing (Superintendência de Limpeza Urbana, or SLU) is responsible for municipal waste management in Belo Horizonte. 95 per cent of the urban population and 70 per cent of the slum population received a collection service in 2008. Sweeping, weeding and other services cover 85 per cent of paved roads. Almost 95 per cent of municipal waste from households, institutions and commercial establishments is collected in urbanized areas by compactor trucks operating conventional house-to-house collection routes. For urban slums (*favelas*) and areas that are difficult to access, a variety of alternatives are in use. These range from so-called special kerbside collection using open trucks (less than 5 per cent), to secondary collection from 135 communal skip containers, or *caçambas* (10 per cent); to cleaning up of illegal dumps (over 15 per cent). 10 per cent of waste is self-hauled to disposal by larger commercial, industrial and institutional waste generators themselves.

Environment/disposal

Between 1975 and 2007 all the municipal solid waste from Belo Horizonte went to the designated disposal site Centro de Tratamento de Resíduos Sólidos (CTRS BR-040), a treatment centre for solid waste, in Bairro Califórnia, 12km from the city centre. This 115ha centre hosts a small composting plant, a recycling facility for construction waste, a storage for used tyres, a temporary storage facility for long-life packaging material, and a unit for environmental education. In the course of its use, the disposal site was upgraded into a sanitary landfill. A carbon credit certification project for landfill gas recovery has been recently approved for a total amount of 16 million Brazilian reais for exploration of biogas at the CTRS BR-040 site. The site was closed in 2007 and a new site is being sought. Since the closure of the landfill in December 2007, part of the old landfill has been transformed into a transfer station for transferring waste to more distant privately operated facilities. A new regional landfill is in the planning.

Resource management

The adoption of integrated solid waste management in 1993 made the recovery of recyclables a key feature of waste management in Belo Horizonte. Recovery of construction and demolition waste is the most significant recovery programme of the municipality and is operated through a mix of informal waste collectors, and municipal secondary collection service and recycling plants. Recyclable non-paper, plastic, glass and metals are collected as follows:

- since 1993, a drop-off system consisting of 150 delivery sites with 450 containers, which are emptied weekly by the SLU staff;
- since 2003, a kerbside collection system by the SLU, currently targeting almost 354,000 residents;
- door-to-door collection of recyclables by co-operatives of waste-pickers from commercial establishments and offices in the downtown area, using hand-drawn push carts.

In addition, recyclables are collected from big generators such as industries and in public offices using vehicles owned by the co-operatives. This channel resulted in over 50 per cent of all recyclables collected.

The collected material is taken to warehouses run by co-operatives of semi-formal waste-pickers where the materials are sorted, processed and stored for sale to industry in Belo Horizonte.

Special features

Waste management in Belo Horizonte has been a priority since 1900 when various advanced technologies were applied. Only later, during the 1960s, did the city resort to open dumping. Since 1990, the city has been at the centre of solid waste management development in Brazil again, leading the movement for inclusion of the informal recycling sector. In 1990, the city included a clause in its Organic Law stating that the collection of recyclables would preferably be the work of co-operatives – the organized informal sector – and that they should be the beneficiary of all collected recyclables. In 1993 the city partnered with this first waste-pickers' co-operative in the implementation of its municipal recycling scheme.

Since 2003 waste-pickers' co-operatives and informal collectors of debris have joined forces in the Belo Horizonte Waste and Citizenship Forum, which has been an important institutional medium in which to discuss guidelines for the integration of all these organizations within solid waste management (SWM).

Key benchmark numbers

Total tonnes municipal solid waste (MSW) generated per year	1,296,566 tonnes*	Percentage valorized by informal sector of total waste generated	1%
Generation per capita in kilograms per year	529kg	Percentage valorized by formal sector of total waste generated	10%
Percentage coverage	95%**	Goals for waste collection coverage as percentage of population	95% centre/ 80% elsewhere
Percentage disposal in environmentally sound landfills or controlled disposal sites of total waste generated	89%	Goals for environmentally sound (safe) disposal	100%
Percentage municipal waste incinerated of total waste generated	None	Goals for valorization of waste materials through recycling (or diversion from disposal)	1050 tonnes per day, an increase from 7% to 24% of collected MSW
Percentage diverted and valorized of total waste generated	11%	Prevented Reused Recycled Composted/agricultural value chain	0.07%

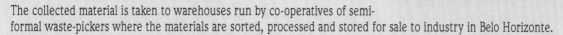

Notes:* Official definition of municipal waste in Belo Horizonte also includes construction and demolition waste, which constitutes an additional 777,634 tonnes per year.
** Coverage of population in slum areas was 70 per cent in 2008.
Figures in italics are estimates.

References

SLU (Superintendência de Limpeza Urbana) (2008) *Annual Report 2008 of the SLU*, Belo Horizonte, Brazil IBGE (2007)
IBGE (Instituto Brasileiro de Geografia e Etatistica) (2007)

BENGALURU
Karnataka State, India, South Asia

12°58'N, 77°38'E
920–962m above sea level

Sanjay K. Gupta (Water, Sanitation and Livelihood, Smt. Hemalatha, KBE) and Mahangara Palike

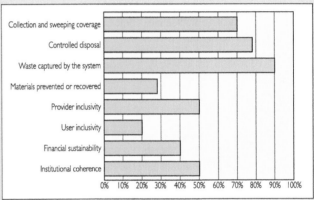

Some basic facts

India's pride, Bengaluru, or Bangalore, as it used to be known, is nearly 500 years old and has grown from a small-time settlement where Kempe Gowda, the architect of Bengaluru, built a mud fort in 1537 and his son marked the city boundaries by erecting four watch towers. Today Bengaluru has grown well beyond those four towers into a sprawling metropolis of nearly 8 million people that is referred to as the Silicon Valley of India – accounting for more than 35 per cent of India's software exports. Bengaluru's temperate climate, high-quality educational, scientific and technology institutions, coupled with a thriving information technology (IT), biotechnology and manufacturing industry, make it one of the most sought-after global destinations.

Topography: flat except for a ridge in the middle running north-north-east–south-south-west. The highest point in Bengaluru is Doddabettahalli, which is 962m and lies on this ridge. Climate/rainfall: temperatures ranging between 33°C and 16°C, with an average of 24°C. The summer heat is moderated by occasional thunderstorms and squalls. Bengaluru receives adequate rainfall of about 860mm from the north-east monsoon as well as the south-west monsoon. The wettest months are August, September and October. Size of city/urban area: approximately 800km². Population (2009): 7.8 million. Population density: 9750 persons/km². Population growth rate (2008): 2.8%. Average household size: 3.5. Human Development Index: 0.70 for the state of Karnataka.

The solid waste story

Main driver

The main driver in Bengaluru is a combination of public health and environment. The 2000 Municipal Solid Waste Rules are the policy guidelines for managing waste and enhancing recycling; but it was the pressure to maintain the city image as the Silicon Valley of India and to attract foreign investment that prompted a citizens' business and municipal partnership, which accelerated the pace of modernization of solid waste in the Bengaluru. Non-governmental organizations (NGOs) also played an important role in promoting better practices in solid waste management.

Public health/collection

There is strong political commitment to improving and modernizing collection, with high-level performance goals and a mixed system approach. A mix of the municipality and private operators provide a direct, daily door-to-door primary collection system to 70 per cent of Bengaluru citizens in high-income, middle-income and some low-income and slum areas. Private contractors provide services in the central business district and in the better-off residential areas. The handcarts used by the door-to-door collectors are directly unloaded into large vehicles, including auto-tippers and state-of-the-art compactors, for transportation to the processing or disposal sites, making most of the city effectively bin free. The exception is a few low-income areas on the city boundaries and in old settlements, where the municipality provides less frequent and less regular waste collection services from community waste collection.

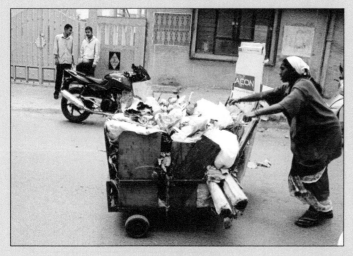

Environment/disposal

It is estimated that nearly 10 per cent of waste still goes for open dumping. Two designated controlled dumpsites closed in 2007, and the rest of Bengaluru waste now goes to two new modern landfills constructed near processing plants in Mavallipura Hesaragatta Hobli and Mandur BideraHalli Hobli. The expected useful life is at least until 2030, provided that the processing plants function at full capacity. The landfill sites are operated by the private sector based on public–private partnership (PPP) models.

Resource management

The 2000 Municipal Solid Waste Rules require cities to recover all recyclables and compostable materials and only allow rejects and inerts to be landfilled. Even though this has not been operationalized, in Bengaluru, 25 per cent of the total waste is being valorized by informal and formal activities, and there are plans for more. An active informal sector recovers around 15 per cent of the city waste and feeds the regional industrial recycling supply chain, while supporting livelihoods of more than 30,000 waste-pickers in the city and at the landfill site, who sell to junk dealers, sorters and recycling units, which comprise an additional estimated 10,000 workers. Now that a formal door-to-door collection system has been extended to the majority of the population, the formal-sector waste collectors and transfer station workers also retrieve recyclables and sell to small scrap dealers. Segregated organic waste from hotels and fruit and vegetable markets is directly taken to four composting plants, operated by the formal private sector on PPP models. The compost is sold to the fertilizer agencies who market it mostly to coffee planters and large farmers. There is still not enough demand for compost; hence, the processing plants produce compost when they have buyers, and the rest goes to the landfill sites. There are plants under construction to make refuse-derived fuel (RDF) from the rejects from compost processing, which contains some organic material and often a high percentage of paper and plastic.

Special features

As Bengaluru has an intensive IT industry, there are NGO initiatives to decrease relatively dangerous informal e-waste recycling and to collect e-waste separately and send it for formal recycling and recovery. The IT industry is also a key player in the Bangalore Agenda Task Force (BATF) initiative, organizing the modernization of waste management and upgrading of collection.

Key benchmark numbers

Total tonnes municipal solid waste (MSW) generated per year	2,098,750 tonnes	Percentage valorized by informal sector of total waste generated	1%
Generation per capita in kilograms per year	269kg	Percentage valorized by formal sector of total waste generated	13%
Percentage coverage	65%	Goals for waste collection coverage as percentage of population	100%
Percentage disposal in environmentally sound landfills or controlled disposal sites of total waste generated	70%/20% (90% in total)	Goals for environmentally sound (safe) disposal	20%–30% of waste to landfills
Percentage municipal waste incinerated of total waste generated	None	Goals for valorization of waste materials through recycling (or diversion from disposal)	50% of organic waste to be valorized by 2010 with 70% of recyclable recovery from waste
Percentage diverted and valorized of total waste generated	14%	Prevented Reused Recycled Composted/agricultural value chain	10%

Note: Figures in italics are estimates.

References

Bengaluru City Development Plan, 2007

CAÑETE

San Vicente de Cañete, Peru, South America

13°09'S 76°22'W

33m above sea level

Oscar Espinoza and Humberto Villaverde (city profilers)
Jorge Canales and Cecilia Guillen (municipality contact persons

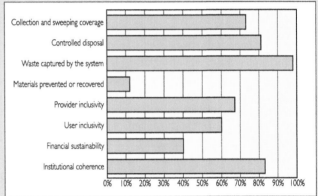

Some basic facts

San Vicente de Cañete district is the capital of Cañete Province and is located on the central coast of Peru, 140km south of Lima City (capital of Peru). The district is close to the Pacific Ocean and is also part of the watershed of the Cañete River.

Topography: the topography is flat. Climate: during winter, from May to November, the temperature varies between 14°C and 18°C. In summer, from December to April, the temperature varies between 20°C and 28°C. Size of city/urban area: 512.16km². Population (2009): 48,892. Population density: 90.72 persons/km². Population growth rate (2008): 2.7%. Average household size: 4.4. Human Development Index: 0.6783.

The solid waste story

Main driver

While collection and disposal are still being modernized in both urban and rural areas, there is a real commitment to resource management as part of the modernization process. Part of this process is the successful inclusion of the informal workers. Valorization is deployed as a way of diverting waste from landfill.

Public health/collection

Currently the municipal waste collection services are provided to 93 per cent of the urban population and only to 15 per cent of the rural population in Cañete. The service in the urban area is provided daily. Currently there are initiatives to expand services in rural areas.

Environment/disposal

All the wastes collected are transported to a controlled dumpsite, Pampa Arena, located 15km to the south of Cañete. This dumpsite has been in use by the municipality for 20 years. The municipality is currently organizing decentralized solid waste services using several other small dumpsites in rural settlements.

Resource management

There is a separate collection for inorganic recyclables with a focus on plastics, metals, paper, cardboard and glass in about 15 per cent of the city. The collectors are four women and three men – formalized waste-pickers who used to work at the dumpsite. About 10 per cent of the waste is recovered by informal itinerant waste buyers and waste-pickers, including those at the dumpsite. This percentage is increasing due to the municipal commitment to source separation, which is based on inclusion of the informal sector.

There are 176 people working in informal solid waste and recycling in the Cañete district: 121 men, 50 women and 5 children (between 12 and 15 years of age). Women and children are mainly engaged in street waste-picking. The conditions of work of these people are poor in terms of security, occupational safety and security of income.

Special features

There are new programmes for valorization that include recycling in schools and demonstration composting facilities. The municipal authorities have made a decision to develop as active as possible a separate collection system for recyclables in order to eliminate or substantially reduce the need for post-collection sorting. There are high prices and levels of demand for polyethylene terephthalate (PET) bottles in the market for recyclables in Peru, which command better prices than other materials. Consequently, they constitute 30 to 40 per cent of the income in Cañete's Selective Collection Programme. The second highest value waste material is paper (mainly white paper) and then the other plastics – polypropylene (PP), polystyrene (PS) and plastic bags. In the municipal environmental demonstration centre there are several activities to valorize organic waste: composting, vermi-composting, a nursery and a vegetable garden.

Key benchmark numbers

Total tonnes municipal solid waste (MSW) generated per year	12,030 tonnes	Percentage valorized by informal sector of total waste generated	11%
Generation per capita in kilograms per year	246kg	Percentage valorized by formal sector of total waste generated	1%
Percentage coverage	73% (94% in urban areas and 15% in rural areas)	Goals for waste collection coverage as percentage of population	90% in 2011
Percentage disposal in environmentally sound landfills or controlled disposal sites of total waste generated	72%	Goals for environmentally sound (safe) disposal	80% landfilled in 2012
Percentage municipal waste incinerated of total waste generated	None	Goals for valorization of waste materials through recycling (or diversion from disposal)	30% of households for 2010
Percentage diverted and valorized of total waste generated	12%	Prevented Reused Recycled Composted/agricultural value chain	NA NA NA None

Note: NA = not available.
Figures in italics are estimates.

CUREPIPE
Island of Mauritius off the coast of Africa

20°19'S 57°31'E
400m above sea level approximately

Professor Edward Stentiford (University of Leeds, UK) and
Professor Romeela Mohee (University of Mauritius, Mauritius)

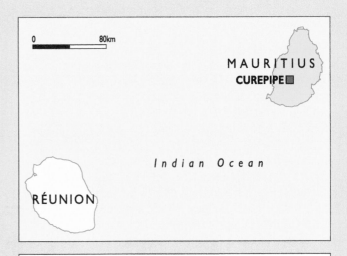

Some basic facts

One of the major industries on Mauritius is tourism and the image of Mauritius is that of a beautiful tourist destination. Although Curepipe is 17km inland from the coast, a large proportion of the population works in the tourist industry, which places a high value on cleanliness.

Topography: located in the south-central highlands of the island of Mauritius. Climate/rainfall: the average rainfall at Curepipe Point is about 4000mm/year. The average rainfall decreases rapidly to about 1600mm/year one third of the distance from Curepipe Point to the coast, and then decreases more gradually to 800mm to 900mm/year at the coast. Size of city: 23.7km^2. Population: 83,750. Population density: 595 persons/km^2. Population growth rate (2008; estimate for whole of Mauritius): 0.78%. Average household size: 4. Housing stock: unknown. Human Development Index (2005): 0.804.

The solid waste story

Main driver

There is an interesting lack of a single driver reported in Mauritius; the level of education in Mauritius has meant that waste management has always been seen as an important aspect of quality of life; therefore, both collection and disposal are well organized. It is difficult to pinpoint one event – a crisis, a policy decision, legislation, or anything else – that has led to major changes in the waste management system as these changes have been largely incremental. The government of Mauritius is the main funder of the waste management system via central taxation, with the money being reallocated for the principal cities, such as Curepipe, to manage.

Public health/collection

Curepipe has well-managed waste collection, waste disposal and street cleaning, which results in an overall 'tidy' city. Household wastes are collected door to door on a weekly basis by a mixture of private contractors (40 per cent) and the municipality (60 per cent). There are parallel waste collection systems for households and for the commercial institutions, and coverage is reported to be 100 per cent.

Environment/disposal

The waste from both residential and commercial collection systems goes to the transfer station at La Brasserie, where wastes are compacted before being transported to the landfill at Mare Chicose. The landfill became operational in 1997 and since then has been taking waste from all parts of the island. The expectation was that it would have no further capacity after 2010 and this has led to current plans for much of the organic fraction of the waste to be composted and used in agriculture. In the meantime, the landfill has been extended to keep receiving waste until 2015, unless more waste is recycled than is now the case.

The construction of the landfill was funded by the national government as a strategy for controlling disposal and closing the unspecified number of unauthorized dumps that threatened the image of Mauritius as a beautiful tourist destination. The operation of the site was let out by tender to a private operator. The modern landfill has an appropriate liner system, leachate treatment facilities, landfill gas collection and flaring system. The waste is covered on a daily basis, normally using soil.

Resource management

Mauritius has very little in the way of primary industries to take recycled materials. An informal sector collects bottles, tins, aluminium and metal for shipping to recyclers off the island. Even though there is a very good awareness of the importance of recycling, there are few local markets and therefore currently no formal recycling of materials from the domestic waste stream. Curepipe has set up a project with a non-governmental organization (NGO), Mission Verte, to put recycling bins around the city to collect general waste, paper, cans and plastic, even though there are no recycling facilities on the island. This material, after being separately collected, is also mixed in the transfer station.

Construction and demolition wastes are collected by the municipality and private contractors as requested by the waste generator; but the majority of the construction and demolition wastes are recycled back to operational construction sites. Separate collection of green wastes has taken place for many years in Curepipe; but all the collected materials are mixed together at the transfer station before being transported to the landfill site. Nevertheless, this provides a basis for the planned expansion of composting as a form of landfill avoidance.

Special features

Care for the environment has been a key element of Mauritian law for many years, with one of the main drivers being to maintain a thriving tourist industry, which is dependent on a high-quality natural environment. The need to protect the coral reefs and the surrounding ocean has been instrumental in the construction of high-quality wastewater treatment systems and the construction of a state-of-the-art containment landfill for the whole island at Mare Chicose. Being an island of 1.3 million inhabitants, strategic decisions on solid waste management are taken at island level, rather than at municipal level.

Key benchmark numbers

Total tonnes municipal solid waste (MSW) generated per year	*23,760 tonnes*	Percentage valorized by informal sector of total waste generated	NA, expected to be small
Generation per capita in kilograms per year	284kg	Percentage valorized by formal sector of total waste generated	NA
Percentage coverage	100%	Goals for waste collection coverage as percentage of population	100% (achieved)
Percentage disposal in environmentally sound landfills or controlled disposal sites of total waste generated	Almost 100%	Goals for environmentally sound (safe) disposal	NA
Percentage municipal waste incinerated of total waste generated	None	Goals for valorization of waste materials through recycling (or diversion from disposal)	NA
Percentage diverted and valorized of total waste generated	NA	Prevented Reused Recycled Composted/agricultural value chain	NA Construction and demolition (C&D) waste only NA none

Notes: NA = not available.
Figures in italics are estimates.

DELHI

National capital territory of Delhi, India, South Asia

28°61'N 77°23'E
239m above sea level

Irmanda Handayani (student intern), Malati Gadgil,
Bharati Chaturvedi and Prakash Shukla (Chintan Environmental
Research and Action Group) Jai Prakash Choudhury (Santu)
and Safai Sena

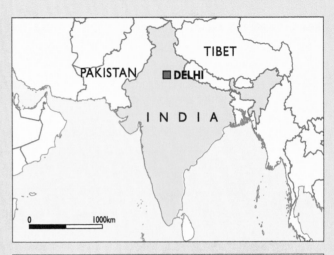

Some basic facts

Located in the northern part of India, the National Capital Territory (NCT) of Delhi is a metropolitan city and the seat of India's central government. It is located between 200m and 300m above the sea level. The Yamuna River, a tributary of the Ganges, crosses the city from the north to the south. Administratively, the city is divided into three statuary towns governed by three different municipalities: the Municipal Corporation of Delhi (MCD), the New Delhi Municipal Committee (NDMC) and the Delhi Cantonment Board. Almost 97 per cent of the current estimated 17.7 million inhabitants (Delhiites) live in MCD municipality. As a major political and commercial centre, Delhi attracts about 600,000 commuters daily.

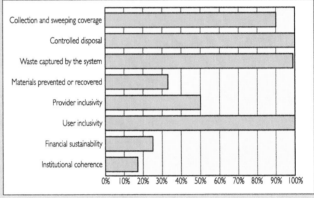

Climate/rainfall: humid subtropical climate, 827.2mm, mostly during monsoon period (June to September). Size of city/urban area: 1483km^2. Population (2001): 13,850,507; rural: 6.82%; urban: 93.18%. Population density (2001): 9340 persons/km^2. Population growth rate (2009 estimate): 1.548%. Average household size: 5.1. Human Development Index (2001): 0.737.

The solid waste story

Main driver

The main driver is the public image of the city, propelled by public interest litigation during the late 1990s. In the last two years, further steps have been taken at great speed in time for the 2010 Commonwealth Games that will be held in New Delhi. These have accelerated the development of infrastructure in the city and the modernization of solid waste management services as the city authorities are making efforts to present Delhi as a clean world-class city with advanced technology.

Public health/collection

Delhi has opted to maintain a community container infrastructure, and most residents and other waste generators bring their waste to temporary storage points, called *dhalaos*. In a service organized by non-governmental organizations (NGOs) and the New Delhi Municipal Committee, in coordination with resident welfare associations (RWAs), private door-to-door waste collection service is provided by waste-pickers at a fixed monthly fee to 80 per cent of high-, middle- and low-income households in New Delhi municipality. The service providers take the waste to the *dhalaos*, extracting the valuable materials before discharge. Since 2005, four private companies have been servicing these secondary collection points under a public–private partnership (PPP) agreement with the city authorities; currently, they serve approximately 50 per cent of the total area and the municipalities do the rest.

Environment/disposal

Delhi relies on well-organized waste disposal in three controlled disposal sites, at Okhla, Bhalaswa, and Gazipur, but without environmental protection measures. The three sites have all reached capacity, but are still in operation as the future landfills are not yet ready. A feasibility study has been prepared for gas recovery and utilization after closure of these sites to reduce greenhouse gas emissions.

Resource management

The resource management activities in Delhi are a rich mixture of government, private, informal and formal recovery and valorization. Formal service providers direct the organic waste to the two privately managed composting plants in Okhla and Bhalaswa. In addition, there are small-scale composting units initiated by NGOs and RWAs at community level, such as in Defence Colony, a neighbourhood in South Delhi in the MCD area. Catering establishments such as hotels give away or sell organic waste to the informal sector, which sells it on to piggeries as feed. At least 150,000 waste-pickers throughout the Delhi waste management system divert over 25 per cent of all waste generated in Delhi from disposal and into recycling of materials, thus saving very substantial funds for the municipal authorities. They upgrade the materials and sell them into the recycling supply chain, a pyramid going from pickers to the small junk dealers (*kabaris*), on the boundaries of formal and informal recycling. The *kabaris* sell to specialized dealers and end-users, and some material is also exported.

Special features

The contribution of the informal sector to solid waste management and resource recovery has been explicitly and officially recognized in several documents, including the National Action Plan on Climate Change of June 2008. Numerous organizations, such as Chintan Environmental Research and Action Group, advocate rights and organize the activities of these informal waste professionals, who are increasingly part of the global movement for integration of the informal sector.

Key benchmark numbers

Total tonnes municipal solid waste (MSW) generated per year	2,547,153 tonnes	Percentage valorized by informal sector of total waste generated	27%
Generation per capita in kilograms per year	184kg	Percentage valorized by formal sector of total waste generated	7%
Percentage coverage	90%*	Goals for waste collection coverage, as percentage of population	100%
Percentage disposal in environmentally sound landfills or controlled disposal sites of total waste generated	50%	Goals for environmentally sound (safe) disposal	New sanitary landfills meeting criteria which were set by CPCB will be opened in Jaitpur in 2010
Percentage municipal waste incinerated of total waste generated	None	Goals for valorization of waste materials through recycling (or diversion from disposal)	33.33% of waste stream collected by 2014
Percentage diverted and valorized of total waste generated	33%	Prevented Reused Recycled Composted/agricultural value chain	 0.8% 26.5% 6.5%

Notes: * Coverage reflects non-slum area of Delhi: 50 per cent of the population live in slums in Delhi.
CPCB = Central Pollution Control Board
Figures in italics are estimates.

DHAKA

Bangladesh, South Asia

24°40'–24°54'N 90°20'–90°30'E

0.5m–12m above sea level; 70% of the total area within 0.5m to 5m

Andrew Whiteman (Wasteaware), Monir Chowdhury
(Commitment Consultants), Shafiul Azam Ahmed (consultant),
Dr Tariq bin Yusuf (DCC), Professor Ghulam Murtaza
(Khulna University), Dr Ljiljana Rodic (Wageningen University)

Some basic facts

Dhaka is the capital of Bangladesh, the most densely populated city in one of the most densely populated countries on Earth. With an official population of over 7 million, and a daily population reported to be more than 12 million, Dhaka's residents live in an immensely challenged urban environment. As the country's economic centre, Dhaka continues to attract migrant workers and families from other parts of the country. The population is continuing to grow at a rate of 3 per cent per annum. Keeping pace with infrastructure and housing needs is a massive challenge: over 3 million people live in slums, and more than 55 per cent of people live below the poverty level.

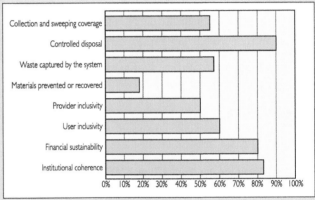

Climate/rainfall: Dhaka enjoys a subtropical monsoon climate. The annual rainfall varies from 1429mm to 4338mm. Size of city/urban area: 365km^2 (metropolitan area). Population: 7 million. Population density: 19,178 persons/km^2. Population growth rate (2008): 1.7%. Average household size: 5. Human Development Index (2009): to 0.543.

The solid waste story

Main driver

Environment is the main driver, but public health is still important. In 2000, a new stage in development of solid waste management in Dhaka began through a technical cooperation initiative of the government of Japan. The development intervention consisted of assessing the existing solid waste management situation, formulation of a master plan, and follow-up implementation of some of the priority issues in waste management.

Public health/collection

The main model of waste collection in Dhaka is based on door-to-door primary collection of waste by micro-enterprises, who take the waste to designated points on the roadside or collection/transfer points. The municipal authority, Dhaka City Corporation (DCC), services these secondary collection points. Ward-based waste management schemes are in place in several wards of Dhaka City. Citizens form an active part of the waste management efforts, receive training, promote public awareness, monitor system performance, and help in troubleshooting. Primary collectors in the ward are also given training, which facilitates integrating primary collection with the secondary collection by DCC.

Environment/disposal

Collected waste is disposed of at one of the two disposal sites: Matuail and Aminbazar. Matuail is the first ever sanitary landfill in Bangladesh, upgraded from a dumpsite previously in operation for many years. The resulting engineered disposal facility opened in October 2007 as a part of a long-standing partnership between DCC and Japan International Cooperation Agency (JICA). Matuail now has year-round access, gate controls, a computerized weighbridge, perimeter drainage, cellular emplacement, and leachate and landfill gas control. DCC now faces the more challenging task of operating and maintaining the facility as a sanitary landfill, and to date the facility is being managed well. Matuail is operated 24 hours a day in three shifts. It receives around 1200 tonnes of waste per day, arriving at both day- and night-time. DCC is currently in the process of upgrading the Aminbazar disposal site.

Resource management

The concept of 'waste' is only a relatively recent phenomenon in Bangladesh, and many of the materials which are generally considered to be waste in other countries are stripped out of the waste stream for extraction of resource value. Newspapers, glass bottles, metal cans, plastic items – practically anything of value – are reused or sold by their owners or informal waste-pickers. Resource recovery in Dhaka is carried out and managed by multiple and complex chains of informal and formal, public, private and multinational actors. About 120,000 people are involved in the informal recycling trade chain in Dhaka City. None of these activities receive any public funding support. The organic fraction, more than 65 per cent of municipal waste, was traditionally not viewed as valuable, and composting in the 1990s failed to spread due to a lack of markets. A new compost plant located just outside Dhaka started operation in 2008 by WWR Bio Fertilizer Bangladesh Ltd (a joint venture company of Waste Concern, Bangladesh, World Wide Recycling BV, FMO and Triodos Bank, The Netherlands). This activity is accompanied by collection services of market (organic) waste for this composting plant. The project is registered and approved by the executive board of the Clean Development Mechanism under the Kyoto Protocol to the United Nations Framework Convention on Climate Change (UNFCCC) and is gradually scaling up.

Special features

Dhaka offers an excellent example of a waste management plan being prepared to a high standard, and then being implemented in management cycles with the support of development partners. Some of the achievements are the establishment of a waste management department; implementation of a ward-based approach to waste management; and construction of the first ever sanitary landfill in Bangladesh. In 2008 a new Waste Management Department was inaugurated at DCC, combining the conservancy and mechanical engineering departments into a single department, putting Dhaka City Corporation at the forefront of modernized institutional arrangements for solid waste management in South Asia.

Key benchmark numbers

Total tonnes municipal solid waste (MSW) generated per year	1,168,000 tonnes	Percentage valorized by informal sector of total waste generated	18%
Generation per capita in kilograms per year	167kg	Percentage valorized by formal sector of total waste generated	None
Percentage coverage	43%	Goals for waste collection coverage as percentage of population	None
Percentage disposal in environmentally sound landfills or controlled disposal sites of total waste generated	44%	Goals for environmentally sound (safe) disposal	Construction (upgrade) of second major sanitary landfill in process
Percentage municipal waste incinerated of total waste generated	None	Goals for valorization of waste materials through recycling (or diversion from disposal)	None
Percentage diverted and valorized of total waste generated	18%	Prevented Reused Recycled Composted/agricultural value chain	NA NA 18% 0.19%

Notes: NA = not available.
Figures in italics are estimates.

GHORAHI

Dang District, Mid-Western Region, Nepal

27°59'24"N 82°32'42"E

668m above sea level

Bhushan Tuladhar (Environment and Public Health Organization, ENPHO, Nepal)

Some basic facts

Topography: valley. Climate/rainfall: average temperature 23.1°C. Annual rainfall: 1271mm. Size of city/urban area: 74.45km²/8.56km². Population (2009, estimate based on 2001 Census): 59,156. Population density: 795 persons/km². Population growth rate (2001): 4%. Average household size: 4.82. Housing stock: 12,273 (number of households based on average household size of 4.82). Human Development Index (2001): 0.409.

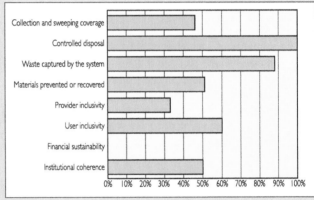

The solid waste story

Main driver

The main driver in Ghorahi is a combination of public health and environment, translated into civic pride in a clean city and a safe environment. Nepal's national legislation and regulations are less of a driver than local initiative, and Ghorahi has formulated its own local guidelines on health, sanitation and environment in order to manage solid waste and solve other environmental problems.

Public health/collection

Ghorahi relies on both motorized and non-motorized collection to keep the city clean. On-time collection using a tipper and compactor in some parts of the city collects about 50 per cent of the waste, while the other half is collected through roadside collection using a sequence of open dumping, sweeping, manual loading of rickshaws, and manual transfer to a larger mechanized vehicle. This involves handling the waste several times, which is inefficient and increases public health risks.

Environment/disposal

While waste collection still has a long way to go in the modernization process, the city has done an impressive job in addressing the environmental issues related to waste disposal. The municipality's own Karauti Danda Sanitary Landfill is one of only three landfill sites in the country. Environmental protection is achieved through selection of a site with thick deposits of natural clay, a leachate collection and treatment system, as well as a buffer zone that includes a small forest, a flower garden and bee farming.

Resource management

The resource management system is a mixture of formal, informal, public and private. The municipality has a plastic sorting facility at the landfill site, where about 20 tonnes of plastics are recovered annually and marketed to a recycling industry about 350km away in Chitwan district. Compost pits are also present on the site of the landfill; but efforts to sell the compost have proved disappointing and demand for the compost remains weak, probably due to glass contamination. Collection of recyclables is done by about 35 informal sector itinerant waste buyers (IWBs) who sell plastics, metal, bottles and paper to four (formal) private recycling businesses. The total amount of recyclable materials recovered is estimated to be between 100 and 500 tonnes per year. Most of the waste from rural areas is reported to be recycled within the households; but there is little information about either recycling or composting outside of the city.

Special features

Governance is the main special feature in Ghorahi: first, because of the priority given to safe disposal and recycling, and, second, because the city has set up appropriate institutional mechanism for managing waste. This mechanism also involves other stakeholders: the landfill management system includes a committee involving local community leaders and members of the business community. Moreover, the city has supported the formation of 218 community-based organizations. There is room to improve cost recovery to ensure that the waste management system as a whole is sustainable. The city has just started to collect service fees for waste management; but the rates are fairly low and the municipality's target is to raise only about 10 per cent of the total waste management costs from users.

Key benchmark numbers

Total tonnes municipal solid waste (MSW) generated per year	7285 tonnes total; 3285 tonnes* in urban areas	Percentage valorized by informal sector of total waste generated	9%
Generation per capita in kilograms per year	167kg	Percentage valorized by formal sector of total waste generated	2%
Percentage coverage	46%	Goals for waste collection coverage as percentage of population	None
Percentage disposal in environmentally sound landfills or controlled disposal sites of total waste generated	67%	Goals for environmentally sound (safe) disposal	None
Percentage municipal waste incinerated of total waste generated	None	Goals for valorization of waste materials through recycling (or diversion from disposal)	None
Percentage diverted and valorized of total waste generated	11%	Prevented Reused Recycled Composted/agricultural value chain	NA NA NA NR

Notes: NA = not available.
NR = not reported.
* Urban waste generation used as reference for calculating other figures.
Figures in italics are estimates.

References

SWMRMC (2007) *Baseline Information on Solid Waste Management of Tribhuwannagar Municipality of Nepal*, Solid Waste Management and Resource Mobilization Centre, Ministry of Local Development, Kathmandu

SchEMS, SWMRMC and Ghorahi Municipality (2005) *Initial Environmental Examination (IEE) Proposed Karauti Danda Sanitary Landfill Site, 2005*, Solid Waste Management and Resource Mobilization Centre, Ministry of Local Development, Kathmandu

CEN, ENPHO and SWMRMC (2003) *Solid Waste Management in Tribhuwan Nagar Municipality, 2003*, Solid Waste Management and Resource Mobilization Centre, Ministry of Local Development, Kathmandu

PAN and EU (2008) *Best Practices on Solid Waste Management in Nepalese Cities*, Practical Action Nepal, Kathmandu

KUNMING
Yunnan Province, People's Republic of China, Asia

25°04'N 102°41'E
1890m above sea level

Ljiljana Rodic and Yang Yuelong (Wageningen University,
The Netherlands)

Some basic facts

The rich history of Kunming dates more than 2000 years, back
to the year 279 BC when a general of the Chu Kingdom formed
a settlement near Dian Lake. Ever since, this area has been
inhabited and has attracted settlers from other Chinese
provinces. It is now the capital city of Yunnan Province in
south-west China and is known as a spring city due to a mild
and pleasant climate of 15°C to 25°C throughout the year, and
sufficient rainfall for abundant vegetation. In terms of its
current size and level of development and income, Kunming is
a typical medium-sized Chinese city which is undergoing rapid
development, following the central government's Western
Region Development Strategy, 2000–2010.

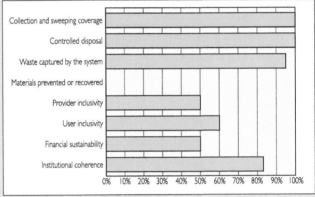

Topography: flat, on a high plateau (1890m above sea level). Climate/rainfall: 1031mm; rainy season is May to October. Size of
city/urban area: 21,473km^2; 2200km^2 in the four districts under study and 220km^2 in the main urban area. Population: 6.8 million;
3.5 million live in the four districts. Population density: 1590 persons/km^2 in the four districts. Average household size (2007): 3.16.
Human Development Index: 0.772.

The solid waste story

Main driver

Solid waste management in Kunming City is well on the way to modernization. Environment is the main driver, symbolized and
driven by concerns about the city's international image in the context of preparing the World Horticulture Exhibition in 1999, with a
focus on improving the situation with regards to pollution in Dian Lake. These preparations prompted development of the first sanitary landfill in 1997, funded by the World Bank. The Western Region Development Strategy of 2000, which promotes the
development of Yunnan and other less developed western provinces of China, contributes to the 248 ongoing projects for construction of waste landfills and sewage treatment plants in Yunnan Province. These installations are funded by state and provincial
governments, a large public utility company, Water Investment, and some private investors. Of those 248 projects, 31 are in the
larger Kunming area, 10 of which are waste landfills.

Public health/collection

Consistent with the priorities of socialist states elsewhere, Kunming has long had virtually 100 per cent collection coverage and city
streets are clean. Collection occurs daily in the central business district and less frequently elsewhere. Collection and sweeping are
operated by two levels of government under the city administrative level – the district level and the 'street neighbourhood' or ward
level. Kunming has an excellent waste collection system based on over 120 small transfer stations throughout the city and a combination of low-tech tricycles wand high-tech compaction vehicles; all are operated by public employees. Since 2006, street sweeping
has been progressively privatized in the city's districts and the trend towards privatization is likely to continue in solid waste services
in general.

Environment/disposal

With economic development and changes in lifestyles, waste composition has changed markedly, and the percentage of ash in municipal solid waste has decreased from over 50 per cent in 1997 to less than 25 per cent in 2004. Final waste disposal is controlled – municipal waste ends either in the incinerator or at one of the two landfills. The technology of choice for the near future is incineration once the existing landfills are closed, something which is expected soon. One of the landfills, Guandu, has an ongoing carbon credit project in cooperation with a European partner, by chance the same partner as in Belo Horizonte, Brazil.

Resource management

The existing – and thriving – recycling business, with the focus on metals, functions as commodity trading and is separate from the solid waste management system, as was the case in the US and Europe prior to the 1970s, the Philippines during the 1990s, and Lusaka, Moshi, Bamako and Nairobi up to the present. In contrast, the opportunities to develop valorization of organic waste have not yet been seized.

Special features

The more than 120 small transfer stations represent a global good practice and have provided a model for other cities worldwide for connecting non-motorized community collection with high-technology compacted motorized transport, for an effective waste collection service. Another special feature in Kunming is the city's successful involvement of the private sector in financing the development of new disposal capacity, including incineration.

Key benchmark numbers

Total tonnes municipal solid waste (MSW) generated per year	*1 million tonnes*	Percentage valorized by informal sector of total waste generated	NA
Generation per capita in kilograms per year	*258kg*	Percentage valorized by formal sector of total waste generated	NA
Percentage coverage	100% (in the urban area)	Goals for waste collection coverage as percentage of population	100%
Percentage disposal in environmentally sound landfills or controlled disposal sites of total waste generated	*Almost 100%*	Goals for environmentally sound (safe) disposal	100%
Percentage municipal waste incinerated of total waste generated	37%	Goals for valorization of waste materials through recycling (or diversion from disposal)	None
Percentage diverted and valorized of total waste generated	NA	Composted/agricultural value chain	500 tonnes per year/0.06%

Notes: NA = not available.
Figures in italics are estimates.

LUSAKA

Lusaka Province, Zambia, Southern Africa

15°30'S 28°10'E;
1200m–1300m above sea level

Michael Kaleke Kabungo (Lusaka City Council, Waste Management Unit) and Rueben Lupupa Lifuka (Riverine Development Associates)

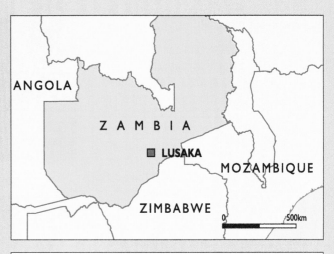

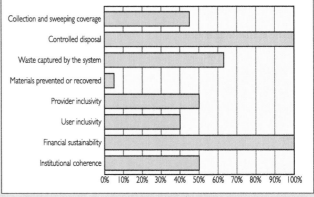

Some basic facts

Topography: mostly on a flat topography. Although small hills rise above the plateau surface, and rivers, especially towards the north, have cut valleys below its surface, there is no significant landscape diversity. Although Lusaka is the smallest town in Lusaka Province and one of the smallest in the nation, it has the biggest population and largest growth rates, and there is pressure from in-migration and growth in peri-urban settlements. Climate/rainfall: there are three main seasons in the city – cool and dry from May to August; hot and dry from September to October; and warm and wet from November to April. The average annual rainfall is 802mm. Size of city/urban area: 375km². Population: 1.5 million. Population density: 4166 persons/km². Population growth rate (2008): 3.7%. Average household size: 5.5. Housing stock: 300,000. Human Development Index (2005): 0.434.

The solid waste story

Main driver

The national environmental policies formulated by the Environmental Council of Zambia that are in place consider effective waste management as part of the strategy of environmental protection and pollution control; but the approach to municipal solid waste management in Lusaka is still focused on protecting public health. For most waste generators, the public health dimension is more persuasive than the environmental one and this becomes apparent during the rainy season, when most areas, particularly the overcrowded and densely populated peri-urban areas, experience outbreaks of epidemics such as cholera. Funding for solid waste management is generally limited although central government and the cooperating partners tend to realize additional funds to mitigate the spread of cholera through regular clean-up of uncollected waste in the communities.

Public health/collection

Lusaka is a city which has decided to actively design and manage a mixed collection system. The dual waste collection system is based on the city managing and monitoring a zonal monopoly system tailored to the demographics of different communities. All operators are responsible for marketing services, collecting fees, implementing collection, and meeting targets. Formal private-sector operators collect waste door to door, or provide skip buckets for larger generators or housing estates in the conversional (planned) areas. Micro-franchising of primary waste collection via contracts to community-based organizations (CBOs), based on International Labour Organization (ILO) area studies, is the main coverage strategy in the peri-urban (informal) areas. This produces a total official coverage rate of 45 per cent, which does not, however, include the more than 30 per cent of households who are served by unregistered informal collection service providers. The city council organizes street sweeping differently, with performance contracts by zone.

Environment/disposal

The enactment of the Environmental Protection and Pollution Control Act in 1990 has brought waste management to prominence in terms of environmental protection. Partly through support from the Danish International Development Agency (DANIDA) and other donors, Lusaka is a city where environmentally sound disposal is being developed concurrently with collection. The city boasts Zambia's only engineered sanitary landfill, with a weighbridge and gate controls and a tipping fee; but there is also widely reported illegal dumping in drains, quarries and open places. Also somewhat unusually for a low-income country, the previously controlled dumpsite has been officially decommissioned.

Resource management

Lusaka has a very small informal recycling sector and a relatively active formal one, which together achieve a recycling rate of about 6 per cent of total generated. Five formal recycling companies organize their own collection depots and pick-up service for paper, plastic and metals, which are transported to Zimbabwe or South Africa.

Special features

The high participation of informal service providers in Lusaka provides both a challenge and an opportunity for the city council, the waste management unit and policy-makers. In terms of good practice, Lusaka is an unusually clear example of an African capital city which has made a clear decision – backed up by practical steps – to take the controlling role in managing a mixed waste system. This attention to governance sets Lusaka apart from many other African cities as an example of global good practice.

Key benchmark numbers

Total tonnes municipal solid waste (MSW) generated per year	301,840 tonnes	Percentage valorized by informal sector of total waste generated	2%
Generation per capita in kilograms per year	201kg	Percentage valorized by formal sector of total waste generated	4%
Percentage coverage	45%	Goals for waste collection coverage as percentage of population	NR
Percentage disposal in environmentally sound landfills or controlled disposal sites of total waste generated	26%	Goals for environmentally sound (safe) disposal	NR
Percentage municipal waste incinerated of total waste generated	None	Goals for valorization of waste materials through recycling (or diversion from disposal)	None
Percentage diverted and valorized of total waste generated	6%	Prevented Reused Recycled Composted/agricultural value chain	NA NA NA NA

Notes: NA = not available.
NR = not reported.
Figures in italics are estimates.

Reference

GTZ/CWG (2007) *Lusaka City Report*, Annex 6

MANAGUA
Nicaragua, Central America, the Americas

12°01'–12°13'N 86°07'–86°23'W;
on average 83m above sea level

Jane Olley (technical adviser for UN-Habitat Improving Capacity for Solid Waste Management in Managua programme), Jeroen IJgosse (facilitator and international consultant for UN-Habitat Improving Capacity for Solid Waste Management in Managua programme) and Victoria Rudin (director of the Central American NGO ACEPESA, partner organization for UN-Habitat Improving Capacity for Solid Waste Management in Managua programme)

Supporters from the municipality of Managua include the Technical Committee for the UN-Habitat Improving Capacity for Solid Waste Management in Managua programme, Mabel Espinoza, Wilmer Aranda, Juana Toruño and Tamara Yuchenko

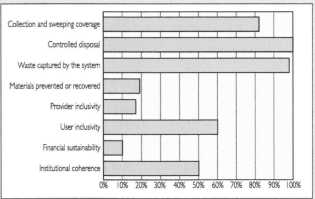

Some basic facts

Managua is the industrial, commercial and administrative centre of Nicaragua and home to nearly 20 per cent of the country's population. The city has grown horizontally along the main transport arteries to the east, south-east and south-west. The population has increased from 430,700 in 1971 to over 1 million in 2009.

Topography: Managua is located on the banks of Lake Xolotlán (1100km^2) and rises from between 43m above sea level at the lake side to 700m above sea level in the surrounding hills. Climate/rainfall: tropical climate. Temperatures range from 27°C to 38°C. Rainy season occurs during May to November, with a rainfall of 1100mm to 1600mm/year. Size of city/urban area: 289km^2/150.5km^2. Population: 1,002,882. Population density (2009): 3668 persons/km^2. Population growth rate (2008): 1.7%. Average household size (2005): 4.6. Housing stock (2005): 197,739. Human Development Index (2006): 0.699 for Nicaragua.

The solid waste story

Main driver

Public health is the main driver for improving the solid waste management (SWM) services for Managua City authorities, who currently struggle to provide adequate collection and disposal services. Environment is the main driver for the current central government, in power since January 2007 and in office until the end of 2011. It has prioritized the subject of SWM in its environmental agenda and is seeking to update and facilitate the implementation of this policy.

Public health/collection

The city reports 82 per cent area coverage for household waste collection service. However, only 65 per cent of the waste collected comes directly from the households. 23 per cent of all collected municipal wastes (including household, commercial and industrial) comes from the cleaning up of illegal dumpsites. Managua operates its own waste collection fleet made up largely of donor-supplied vehicles and equipment that is reported to be 'extremely costly to run'. The city has a number of unsuccessful experiences with micro- and small enterprises (MSEs) and community-based organizations (CBOs) for household waste collection, which contrasts with some other Central American countries such as Honduras or Costa Rica, where MSE private contractors for household waste collection are well established. A limited number of private contractors do provide waste collection services to some commercial and industrial generators in the city. The municipal government recognizes that the current arrangements are unsustainable, and with the support of UN-Habitat recently began a strategic planning process which seeks to provide the technical, financial and institutional framework for improving waste collection, treatment and disposal services in the city.

Environment/disposal

An estimated 90 per cent of municipal waste goes to the La Chureca controlled disposal site, which has been operating for nearly 40 years. Managua has not, to date, suffered any specific public health or environmental crisis that would lead to sustained citizen pressure for the upgrading of municipal solid waste (MSW) management services. In fact, the public's evaluation of this municipal service is consistently positive. This may account for the fact that for a long time there was a general lack of ownership of the disposal problem and that over the last 15 years, significant study and donor support for waste management failed to translate into sustained improvements. However, this situation is changing. The La Chureca site is being upgraded as part of the Spanish-funded Integrated

Development of the Acahualinca Neighbourhood Project. The plan is for it to be operated for a period of five to seven years while a new sanitary landfill site can be identified and developed.

Improving the environment is also a key driver for the current central government. The central government has prioritized the subject of solid waste management (SWM) in its environmental agenda and is in the process of updating and facilitating the implementation of the National Policy for Integrated Solid Waste Management passed in 2005.

Resource management

The increase in the number of private individuals and businesses earning a livelihood in waste recovery and recycling in Managua during the last ten years appears to be related to the high levels of under- and unemployment in the city, as well as the increases in global recycling market demand. Despite the recent global economic crisis, the number of intermediaries, exporters and recyclers dedicated to this activity remains fairly stable and the recent formation of a national association of recyclers, Associacion Nacional de Recicladores (ASORENIC), suggests that the upward trend in waste recovery and recycling activities will continue in the foreseeable future.

Special features

The First National Recycling Forum was held in Managua on 17 to 18 August 2009 with the aim of establishing the basis for developing waste recovery and recycling at the national level, and represents an important first step in the recognition and integration of private-sector recycling activities within the MSW management system. There are a number of current initiatives, including the United Nations Development Programme (UNDP) public–private partnerships (PPPs) for the urban environment and Basmanagua projects supporting the integration of the informal waste recovery and recycling sector within the municipal SWM system.

Key benchmark numbers

Total tonnes municipal solid waste (MSW) generated per year	420,845 tonnes	Percentage valorized by informal sector of total waste generated	2%
Generation per capita in kilograms per year	420kg	Percentage valorized by formal sector of total waste generated	17%
Percentage coverage	82%	Goals for waste collection coverage as percentage of population	None
Percentage disposal in environmentally sound landfills or controlled disposal sites of total waste generated	90%	Goals for environmentally sound (safe) disposal	Upgrading of dumpsite; siting, design and development of new sanitary landfill
Percentage municipal waste incinerated of total waste generated	None	Goals for valorization of waste materials through recycling (or diversion from disposal)	Being developed
Percentage diverted and valorized of total waste generated	19%	Prevented Reused Recycled Composted/agricultural value chain	NA NA NA 0.01%/1.99%

Notes: NA = not available.

Figures in italics are estimates.

MOSHI

Tanzania, Kilimanjaro Region, East Africa

3°18'S 37°20'E;
700m–950m above sea level from south to north

Alodia Ishengoma (independent consultant, formerly ILO),
Bernadette Kinabo (municipal director),
Dr Christopher Mtamakaya (head of Health and Cleansing
Department) Viane Kombe (head of Cleansing Section),
Fidelista Irongo (health secretary), Lawrence Mlay (environmental
health officer) (all of Moshi Municipal Council)

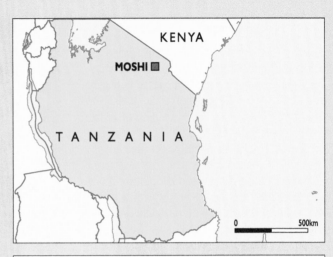

Some basic facts

Moshi is the tourist, commercial and administrative centre of
Kilimanjaro Region in northern Tanzania. It is located on the
southern slopes of Mount Kilimanjaro, between 950m and
700m above sea level, with an area of 58km². Moshi
Municipality has grown from a small urban area of 8048 residents in 1948 to a population of 144,000 according to the
2002 Census, plus an estimated 70,000 day residents who commute daily from the surrounding settlements of Mwika, Marangu,
Machame, Kibosho, Uru and Old Moshi (estimates for day residents vary with the day of the week, with less people coming on holi-
days).

Topography: on the slope of Kilimanjaro. Climate/rainfall: mean temperature of 25°C, 550mm per annum. Size of city/urban area:
58km². Population: 183,520. Population density: 3164 persons/km². Population growth rate (2008): 2.8%. Average household size:
4.1. Human Development Index (2007): 0.503.

The solid waste story

Main driver

The main driver is public health and its relation to waste governance. As early as the 1950s, solid waste management in Moshi was
introduced as a mandatory service for residents, with the goal to protect public health. In 1956, every householder was required to
provide a standard sanitary dustbin for their household. All of the activities – street sweeping, waste collection, transportation and
dumping – were financed by the central government. Although the Public Health Act was enacted in 1966, the infrastructure did
not keep pace with post-colonial growth, and collection coverage at the present time is 60 per cent. The United Republic of Tanzania
Ministry of Health Waste Management Guidelines (2003), which were followed by the enactment of the 2004 Environmental
Management Act to further modernize solid waste management. A national solid waste management policy is in the pipeline for
further modernization of solid waste management.

Public health/collection

Waste collection services are provided by the Moshi Municipal Council (MMC), a private contractor on a pilot basis, and community-
based organizations (CBOs). The private contractor provides services in one of three wards in the central business district (of 15 in
total in Moshi). The arrangement is that private contractors collect both waste and fee and pay 3 per cent of the total fee collected
to the municipal council. ThemmC serves the rest of the urban area and provides secondary collection in peri-urban areas where
CBOs and individuals are doing primary collection.

Environment/disposal

Moshi has a controlled dumpsite, which is one of the best managed in the country. The site is fenced and has a gate attendant.
About 15 waste-pickers work on the dumpsite.

Resource management

There are no records of how much waste is valorized, but the estimates are that 20 per cent – or one third of the organic waste generated – goes into the agricultural supply chain, either as animal feed or compost.

Special features

Moshi is a good example of stakeholder participation and both provider and user inclusivity. A stakeholder platform has been active since 1999, comprised of the municipality and citizens, as well as CBOs. Service coverage is designed to include all areas equitably and the implementation is ongoing. In order to provide basic collection and sweeping services in the low-income or peri-urban areas, which constitute one third of all households (36.5 per cent), primary waste collection is conducted by CBOs and the municipal council carries out secondary collection.

Key benchmark numbers

Total tonnes municipal solid waste (MSW) generated per year	62,050 tonnes	Percentage valorized by informal sector of total waste generated	NA
Generation per capita in kilograms per year	338kg	Percentage valorized by formal sector of total waste generated	NA
Percentage coverage	61%	Goals for waste collection coverage as percentage of population	100%
Percentage disposal in environmentally sound landfills or controlled disposal sites of total waste generated	49%	Goals for environmentally sound (safe) disposal	100%
Percentage municipal waste incinerated of total waste generated	None	Goals for valorization of waste materials through recycling (or diversion from disposal)	None
Percentage diverted and valorized of total waste generated	15–20%	Prevented	NA
		Reused	NA
		Recycled	NA
		Composted/agricultural value chain	20%

Notes: NA = not available.
Figures in italics are estimates.

References

Integrated Waste Management in Tanzania (1998) www.GDRC.ORG/uem/waste/matrix/project.htm

JKUAT (2001) 'Sustainable management of domestic solid wastes in developing countries', *Journal of Civil Engineering*, vol 6, pp13–26

Kironde, J. M and Yhdego M. (1997) 'The governance of waste management in urban Tanzania: Towards a community based approach', *Resources, Conservation and Recycling*, vol 21, no 4, pp213–226

Mato, R. (1997) *Environmental Profile for Tanzania*, Report prepared for JICA

Mato, R. and Kassenga G. (1996) *Development of Environmental Guidelines for Solid Waste Management in Tanzania*, Consultancy report to NEMC

MMC (1999) *Environmental Profile of Moshi Municipality*, Moshi

MMC (2006) *The Moshi Municipal Council Waste Management Bylaws*, Moshi

MMC (2008a) *Environmental Profile of Moshi Municipality*, Moshi

MMC (2008b) *Health and Cleansing Department Annual Report*, Moshi

MMC (2008c) *Comprehensive Council Health Plan*, Moshi

URT (2004) *The Environmental Management Act (EMA)*, Moshi

URT/Ministry of Health (2003) *Waste Management Guidelines*, Moshi

Yhdego, M. (1995) 'Urban solid waste management in Tanzania: Issues, concepts and challenges', *Resources, Conservation and Recycling*, vol 14, no 1, pp1–10

NAIROBI

Kenya, East Africa

1°25'S 36°55'E
1660m above sea level

Misheck Kirimi (NETWAS), Leah Oyake-Ombis (City Council of Nairobi/Wageningen University) and Ljiljana Rodic (Wageningen University)

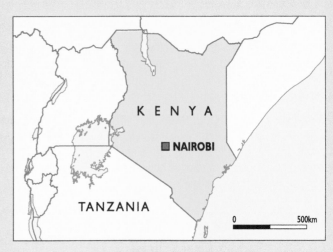

Some basic facts

'The Green City in the Sun', Nairobi, is the political and administrative capital of the Republic of Kenya and the largest metropolitan area in Eastern Africa. Established in 1895, its mild climate, abundant fresh water and absence of malaria promoted rapid settlement by Europeans during the early 20th century, and it was granted 'city' status in 1950. The city's strategic location has made it an important regional hub for commerce, industry, tourism, education and communication. Nairobi hosts the headquarters of two United Nations global programmes – namely, the United Nations Environment Programme (UNEP) and the United Nations Human Settlements Programme (UN-Habitat), besides many international organizations and missions.[1]

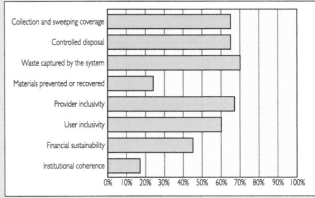

Size of area: 696km². Population: 4 million (including additional daytime population). Population density: 5746 persons/km². Population growth rate (annual): 4–5%. Average household size: 6. Human Development Index (2009): 0.541.

The solid waste story

Main driver

The main driver for solid waste management in Nairobi is public health: policy-makers, citizens and other actors in the city all share awareness that inadequate waste management is directly linked to poor human health. Persistent accumulated heaps of garbage have prompted a public outcry resulting in diverse actions from public-, civil- and private-sector actors to provide adequate waste collection services. Modernization has not yet begun in terms of moving up the disposal ladder.

Public health/collection

Nairobi is a city where, for the last two decades, the private sector has been leading the way in waste collection and materials recovery initiatives. The flourishing private waste collection sector consists of more than 100 companies, micro- and small enterprises (MSEs) and community-based organizations (CBOs) registered to collect waste, recyclables and compostables. The city authority's focus has been in policy development and other donor-driven initiatives. In the year 2001, the City Council of Nairobi (CCN) published a policy document on private-sector involvement in solid waste management (SWM) to define a systematic approach and provide a framework of operation. It further formulated a policy framework in 2002 to promote the private activities of non-state actors in composting and recycling. Acknowledging private collection efforts, the CCN instituted a formal registration process for collectors in 2006.

Environment/disposal

The flip side of successful private initiative is the *laissez-faire* attitude shown by public bodies in relation to disposal. In the nearly 15 years of solid waste modernization since the Japan International Cooperation Agency (JICA) began its waste management plan in 1996, the CCN, which is responsible for disposal, has not yet made it onto the disposal upgrading ladder. As a result, the CCN relies entirely on the Dandora dumping site, situated in a former quarry some 25km to the east of the city centre, for uncontrolled disposal of municipal waste.

Resource management

Recovery of materials occur in all stages of waste material flow through the city, but most extensively by about 1000 waste-pickers living on the Dandora dumpsite, with some coming from the neighbouring suburbs. The main items of importance are paper, textile, glass, metals and bones. Recycling provides informal employment and a means of livelihood to many informal recyclers, and reduces CCN's waste management costs. With a Chandaria paper mill and a large industrial base, combined with many commercial relationships with regional powerhouse South Africa, markets for recyclables in Nairobi are better than in most other East African countries. One research study showed that 20 per cent of the household wastes generated were recovered by either informal pickers or CCN employees. The CCN tends to overlook the informal recovery and recycling, making formalization and integration of the informal waste recycling activities within the formal SWM system difficult.

Special features

Nairobi reports 60 to 70 per cent collection coverage, with nearly 100 per cent in the central business district, and overall 54 per cent of generated waste being collected – quite an achievement for the private waste sector.

Key benchmark numbers

Total tonnes municipal solid waste (MSW) generated per year	876,000 tonnes	Percentage valorized by informal sector of total waste generated	NA
Generation per capita in kilograms per year	219kg*	Percentage valorized by formal sector of total waste generated	NA
Percentage coverage	60–70%	Goals for waste collection coverage as percentage of population	NA
Percentage disposal in environmentally sound landfills or controlled disposal sites of total waste generated	59%	Goals for environmentally sound (safe) disposal	NA
Percentage municipal waste incinerated of total waste generated	None	Goals for valorization of waste materials through recycling (or diversion from disposal)	NA
Percentage diverted and valorized of total waste generated	24%	Prevented Reused Recycled Composted/agricultural value chain	NA NA NA 43% of organic waste

Notes: NA = not available.
* Calculated by team, different from reported.
Figures in italics are estimates.

Note

1   City Council of Nairobi (2006) *2006–2010 Strategic Plan*, City Council of Nairobi, September

QUEZON CITY
National Capital Region, the Philippines,
South-East Asia

14°38'N 121°2'E; 8m–20m above sea level

Lizette Cardenas, Lilia Casanova, Honourable Mayor Feliciano
Belmonte, Honourable Vice Mayor Herbert Bautista, Quezon city
councillors, Quezon City Environmental Protection and Waste
Management Department headed by Frederika Rentoy, Payatas
Operations Group, Quezon City Planning and Development Office,
Andrea Andres-Po and Paul Andrew M. Tatlonghari of QC-EPWMD

Some basic facts

Quezon City, a former capital (1948 to 1976) of the Philippines
is located on the island of Luzon. The city has a total land area
of 161.12km^2 and is the largest of the 15 cities and 2 municipalities in the National Capital Region (also known as Metro
Manila). The city's average temperature is low at 20.4°C,
usually in January, and high at 34.9°C in May. The city has a
largely rolling landscape with alternating ridges and lowlands.
The tropical monsoon climatic zone has two pronounced
seasons: a relatively dry season from December to April and a
wet season from May to December.

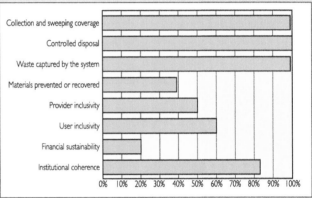

Rainfall: the city receives an annual rainfall of 2532mm with a maximum of 527mm in August and a minimum of 9mm in February.
Population (2009): 2,861,091. Population density: 166 persons/km^2. Population growth rate (2008): 2.92%. Average household size:
5. Housing stock: 495,823. Human Development Index (2009): 0.771.

The solid waste story

Main driver

Although there were already existing laws during the 1990s, it was the collapse of the Payatas waste disposal site in 2000 resulting
in the death of about 300 waste-pickers that accelerated the pace of modernization of solid waste in the Philippines, and in Quezon
City.

Public health/collection

Each of Quezon City's 2.86 million people (as of 2009) generates over 0.66kg of garbage every day (WACS, 2003). Organic or
biodegradable wastes represent the highest percentage of waste generated (48 per cent) by households and commercial
establishments. This is followed by recyclable wastes (39 per cent) and residual wastes (13 per cent), which are disposed of at the
Quezon City Controlled Disposal Facility in Payatas. Among the recyclables, paper and plastics are the most highly generated, with
17 and 16 per cent, respectively (SWAPP, 2006). The city's waste collection is done through both formal and informal systems.
Formal waste collection is undertaken by 5 city contracted hauliers, 13 *barangays* with their own collection trucks, and private
hauliers for commercial establishments. The 'package clean-up' system ensures that each contractor takes full responsibility for all of
the zones in which they work, including the provision of street sweepers; information, education and communication (IEC)
campaigners; riverways cleaning; bulky waste collection and a subsystem of garbage collection for inaccessible areas. The city collection gets about 365,390 tonnes; *barangay* collection is about 29,628 tonnes and private commercial collection is 75,880 tonnes.
Informal collection, by itinerant waste buyers (IWBs) focuses only on recyclables.

Environment/disposal

The Quezon City Controlled Disposal Facility is the oldest dumpsite in Metro Manila, operating for more than three decades. Once a
symbol of everything that was wrong with waste disposal systems, the dumpsite has been transformed into a model and pioneering
disposal facility. Through the Biogas Emission Reduction Project, initiated in 2007, accumulating biogas from the soil-covered
garbage mound is now being extracted, flared and converted into usable electricity. The project is registered under the Clean
Development Mechanism (CDM) and is projected to cut down greenhouse gas emissions by 1,162,000 tonnes of CO_2 throughout
its operating life.

Resource management

Collection of recyclables is done by both the formal and informal waste sector. For the formal sector, it is undertaken by the city collectors and *barangay* collectors. For the informal sector, it is done by itinerant waste buyers at the household level, garbage crews on the trucks, and waste-pickers at the dumpsite and the junk shops; in combination they get 241,195 tonnes. While itinerant waste buyers recover 73 per cent of the stated amount, the junk shops are the main system players that pool recyclables from the different sources for final recovery by the recycling industries and/or exporters. While some *barangays* have materials recovery facilities (MRFs), others have junk shops which have been registered and designated as fulfilling the MRF function. Biodegradable wastes that consist mostly of food wastes are collected from households and establishments by city collectors, *barangay* collectors, accredited kitchen waste collectors or private commercial hauliers. These are processed into composts, soil conditioners or feeds for animals. The total organics being generated at source is 383,499 tonnes per year. The buyers of the organics are livestock growers who get 12,723 tonnes and the compost producers and traders who get 1592 tonnes.

Special features

Quezon City has the highest recovery rate of any of the low- and middle-income reference cities, and the Philippines is active in the 3R Asia Regional Forum. The Philippines National Strategy for the integration of the informal sector, written in 2008 and supported by the United Nations Development Programme (UNDP), represents global best practice.

Key benchmark numbers

Total tonnes municipal solid waste (MSW) generated per year	736,083 tonnes	Percentage valorized by informal sector of total waste generated	31%
Generation per capita in kilograms per year	257kg	Percentage valorized by formal sector of total waste generated	8%
Percentage coverage	99%	Goals for waste collection coverage as percentage of population	100%
Percentage disposal in environmentally sound landfills or controlled disposal sites of total waste generated	61%	Goals for environmentally sound (safe) disposal	None
Percentage municipal waste incinerated of total waste generated	None, not allowed	Goals for valorization of waste materials through recycling (or diversion from disposal)	25%
Percentage diverted and valorized of total waste generated	39%	Prevented Reused Recycled Animal feeding/agricultural value chain	NA NA NA 2%

Notes: NA = not available.
Figures in italics are estimates.

References

HMR Envirocycle Inc (2009) *E-waste Price List*, HMR Envirocycle Inc, 5 June

Quezon City Environment Multi-Purpose Cooperative (2009) *Average Price List – Year 2008*, Quezon City Environment Multi-Purpose Cooperative, Quezon City

Quezon City Planning and Development Office (2009) *Quezon City Socio-Ecological Profile*, Quezon City Planning and Development Office, Quezon City

SWAPP (2006) *Economic Aspects of Informal Sector Activities in Solid Waste Management in Quezon City*, Solid Waste Management Association of the Philippines (SWAPP), Quezon City, 31 October

WACS (2000) *Terms of Reference on Solid Waste Collection, Cleaning and Disposal Services*, Quezon City Environmental Protection and Waste Management Department, Quezon City

WACS (2003) *Waste Characterization Study for Quezon City (ADB Study)*, Quezon City Environmental Protection and Waste Management Department, Quezon City, June

WACS (2008) *Quezon City 10 Year Solid Waste Management Plan*, Quezon City Environmental Protection and Waste Management Department, Quezon City

WACS (2009a) *Assessment of Quezon City's Compliance with Requirements of R.A. 9003 (With Door-to-Door Collection)*, Quezon City Environmental Protection and Waste Management Department, Quezon City

WACS (2009b) *Quezon City 2008 Accomplishment Report*, Quezon City Environmental Protection and Waste Management Department, Quezon City

ROTTERDAM
South Holland, The Netherlands, Europe

51°57'51"N 4°28'45"E;
altitude: below North Sea level

Frits Fransen, Joost van Maaren and
Anne Scheinberg

Some basic facts

Rotterdam is the second city in The Netherlands after
Amsterdam. Its motto, *Rotterdam durft*, means 'Rotterdam
dares' and, indeed, the city is known for taking risks and inter-
preting policy and law according to its own circumstances; the
text in the shield translates 'stronger through struggle'.
Internationally, Rotterdam is known for its international port
and industrial area, which is the largest in Europe (420 million
tonnes of cargo transferred in 2007). The city is also known for
its modern architecture. 75 per cent of the housing stock
consists of medium- to high-rise buildings, while 25 per cent
comprises low-rise or single-family houses. The centre of the

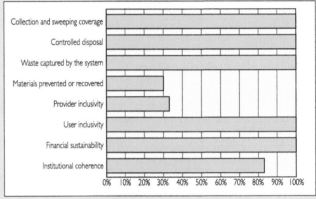

city was destroyed during World War II; the complete reconstruction has left the city with virtually no old neighbourhoods.

Topography: river delta area (Rhine and Maas rivers). Climate/rainfall: 760mm/year. Size of city/urban area: 121.7km². Population
(2008): 582,949; the urban region comprises approximately 1 million inhabitants. Population density: 4811 persons/km². Population
growth rate (2007–2008): 0.2%. Average household size: 1.95. Human Development Index (2009): 0.964.

The solid waste story

Main driver

The main driver has been the growing environmental awareness among the population and the increasing tendency to preserve the
resource values of waste. This awareness has resulted in an aggressive Dutch national policy framework that works to eliminate landfilling and maximize materials and energy recovery. Rotterdam's compliance is selective: the city chooses to maximize energy
recovery, while strict adherence to the spirit and letter of national policies would suggest a stronger emphasis on recycling, composting and prevention.

Public health/collection

ROTEB, the municipal waste management department, is run as a public company although its budget comes from the municipality.
It currently has a legal monopoly in the collection of household waste and waste from mixed residential/commercial zones, but is in
competition with the private sector with respect to commercial and institutional waste. The ROTEB was founded during the late
1800s, established as a municipal department, under political control and in response to public health concerns. Waste collection
operates according to a weekly routine, applying one (plastic bag, 240 litre container), two (plastic bag, 1100 litre container) or
three (3m³, 4m³ and 5m³ underground containers) collection services per week.

Environment/disposal

During the 1970s in The Netherlands, the awareness about the public health and environmental impacts of hazardous waste became
a political crisis when contamination was discovered at Lekkerkerk, a settlement built on a former dumpsite. This crisis and those
that followed prompted the modernization of disposal and the closing of dumps and older incinerators, also in Rotterdam. The high
water table in The Netherlands and its high degree of urbanization have pushed the country to opt for minimizing landfilling and
optimizing recycling, composting and incineration. The result is a dense network of processing and disposal facilities owned by both
private and public companies, and Rotterdam has more than its share of high-performance disposal facilities.

Resource management

During the 1980s The Netherlands was one of the most progressive and recycling-oriented countries in Europe, together with Germany and Denmark. The recovery strategy is based on research and analysis of the environmental footprint of 29 classes of products and materials. Producers and importers, working through their trade associations, have agreements ('covenants') with the environment ministry that establishes their responsibility for recovery and safe end-of-life management of their products. National policy goals regarding recycling and waste minimization are established in the recently updated National Waste Management Plan 2009–2021. The producers finance their obligations via advanced disposal fees (batteries, white and brown goods, automobiles) or direct producer payments (waste electrical and electronic equipment (WEEE), paper and packaging). In Rotterdam, the following waste streams are separately collected though depots, drop-off containers or house-to-house collection, and transported directly, without transfer, to upgrading/recycling enterprises:

- ferrous materials;
- non-ferrous metals;
- household goods (for reuse);
- 'white goods': kitchen appliances (washers and dryers, refrigerators, freezers, etc.);
- 'brown goods': audio and other appliances (TVs, audio equipment, toasters, coffee-makers, etc.);
- 'grey goods': information communication technology (ICT) equipment (mobile telephones, printers, electronic components of vehicles, etc.);
- plastic garden furniture;
- used textiles;
- paper/cardboard;
- window glass;
- glass packaging;
- homogeneous debris;
- mixed debris;
- wood (three qualities);
- car tyres;
- coarse garden waste;
- soil/sand;
- selected bulky waste (goes through post-collection mechanical separation);
- domestic hazardous waste;
- vegetable oils, fat;
- pressure vehicles.

Key benchmark numbers

Total tonnes municipal solid waste (MSW) generated per year	307,962 tonnes	Percentage valorized by informal sector of total waste generated	None
Generation per capita in kilograms per year	528kg	Percentage valorized by formal sector of total waste generated	28%
Percentage coverage	100%	Goals for waste collection coverage as percentage of population	100%
Percentage disposal in environmentally sound landfills or controlled disposal sites of total waste generated	Non-combustible, 0.01%	Goals for environmentally sound (safe) disposal	100%
Percentage municipal waste incinerated of total waste generated	70%	Goals for valorization of waste materials through recycling (or diversion from disposal)	43%
Percentage diverted and valorized of total waste generated	30%	Prevented	NA
		Reused	Via second-hand shops
		Recycled	28%
		Composted/agricultural value chain	1%

Notes: NA = not available.
Figures in italics are estimates.

SAN FRANCISCO

California, West Coast, US, North America

37°47'36"N 122°33'17"W;
at sea level with hilly areas

Portia M. Sinnott (Reuse, Recycling and Zero Waste consultant)
and Kevin Drew (special projects and residential zero waste
coordinator of the City and County of San Francisco)

Some basic facts

The city and county of San Francisco, California, is the financial
and administrative capital of the western US and a popular
international centre for tourism, shipping, commerce and
manufacturing. Quite small for a large city, 122km^2 with a
population of 835,364, it is located on a hilly peninsula separating San Francisco Bay from the Pacific Ocean. Blessed with a
mild Mediterranean-like climate, winters are wet and cool and
summers are dry, often foggy. The mean rainfall, most of which
occurs between November and April, is about 533mm. The city
is known for its diverse cosmopolitan population, including
large and long-established Asian, Italian, Irish and gay/lesbian
and transgender communities. Most residents live in small multi-family buildings and one third own their homes.

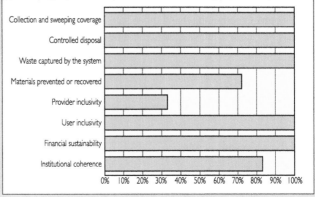

Topography: peninsula with hilly centre and flat areas adjacent to waterline. Climate/rainfall: 533mm. Size of city/urban area:
122km^2. Population: 835,364. Population density: 6847 persons/km^2. Population growth rate: 1%. Average household size: 2.3.
Human Development Index (2006): 0.950.

The solid waste story

Main driver

The main driver is the environment, in combination with resource management. The initiator of the United Nations Environment
Programme (UNEP) Urban Environmental Accords, San Francisco, is a national and international environmental leader. This willingness to take the lead is what drives San Francisco's environmental programme. The city's strong commitment to the precautionary
principle allows managers to develop progressive programmes and hire innovative staff. These practices, coupled with a motivated
collection contractor and an effective multi-departmental rate setting and service delivery system, compounded by escalating transport, processing and landfilling costs, are what motivates the solid waste programme to continue to push the frontier.

Public health/collection

Collection of waste in San Francisco has a long history, connected to rag-picking and informal recycling that dates back to the clean-
up following the 1906 earthquake and fire. The current private contractor, Recology, has a legal monopoly on collection in San
Francisco. The company is the result of a fusion of two competing firms, both of which dated back to federations of scavengers
(informal recyclers) during the 1930s. Working in partnership with Recology, San Francisco recently passed a mandatory recycling
and composting ordinance requiring all residents and businesses to separate their waste. The Fantastic 3 Program, initiated in 1999
and completed citywide in 2003, uses black, blue and green 240-litre wheeled carts. Generators segregate materials and split-
chamber trucks simultaneously pick up trash and recyclables. Single-chamber side-loading vehicles pick up compostables. Most
streets are swept mechanically at least once per week; several high-traffic areas are swept daily. Department of Environment and
Department of Public Works staff, working with business owners and residents, have developed innovative programmes to encourage
best practices and to implement clean-up projects at events such as street fairs, and to solve seasonal problems.

Environment/disposal

Since it does not have a landfill, San Francisco's discards are hauled 85km to Waste Management's Altamont Landfill and Recology's Jepson Prairie Compost Facility 96.5km away. Avoiding the expense of hauling and tipping fees at the landfill is a strong driver for diversion. Garbage rates have been set to strongly encourage recycling or composting. In the commercial sector they are discounted by up to 75 per cent off the cost of trash. In the residential sector, recycling and composting collection are provided at no additional cost. This 'pay-as-you-throw' system underpins San Francisco's diversion strategy and drives environmental programmes.

Resource management

The State of California's 50 per cent diversion goal, the city's ambitious 2020 zero waste goal and the stalwart commitment of elected officials and senior managers drive the focus on diversion, supported by a very strong willingness to participate by residents, businesses and government agencies. This focus is also supported by a very proactive non-profit sector, tight-knit regional and statewide professional associations, and robust recycling markets. The average San Franciscan generates 1.7kg of waste per day, of which 72 per cent is recycled. Three-quarters, 75 per cent, of the remainder could be diverted by existing programmes, and once this is realized, the city will achieve more than 90 per cent diversion.

Special features

'Zero waste or darn close'. The zero waste challenge is reflected in solid waste system support for reducing consumption, maximizing diversion and encouraging reuse, repair and green purchasing. It also involves banning troublesome goods such as plastic bags and superfluous packaging, and promoting alternatives such as recyclable or compostable take-out food packaging and reusable transport packaging. Most of these actions require ongoing outreach at homes, schools, businesses and events. In some cases, mandates and ordinances are required, such as mandatory segregation of recyclables and organics, and construction and demolition debris. One next major step includes supporting the passage of statewide legislation that holds manufacturers, businesses and individuals accountable for the environmental impact of the products that they produce and use.

Key benchmark numbers

Total tonnes municipal solid waste (MSW) generated per year	508,323 tonnes*	Percentage valorized by informal sector of total waste generated	NA
Generation per capita in kilograms per year	609kg	Percentage valorized by formal sector of total waste generated	72%
Percentage coverage	100%	Goals for waste collection coverage as percentage of population	100%
Percentage disposal in environmentally sound landfills or controlled disposal sites of total waste generated	28%	Goals for environmentally sound (safe) disposal	100%
Percentage municipal waste incinerated of total waste generated	Negligible >0.01%	Goals for valorization of waste materials through recycling (or diversion from disposal)	75% landfill diversion by 2010 and zero waste by 2020
Percentage diverted and valorized of total waste generated	72% (2008)	Prevented Reused Recycled Composted/agricultural value chain	20% composted

Notes: NA = not available.

* These are 2008 figures; the figures have not been validated by the State of California as of yet.

Figures in italics are estimates.

SOUSSE

Tunisia, North Africa

35°50'N 10°38'W;
at sea level

Verele de Vreede (WASTE) Tarek Mehri and Khaled Ben
Adesslem (Municipality of Sousse)

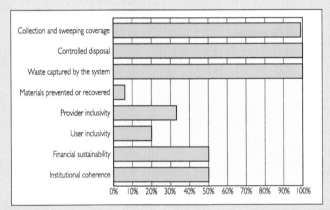

Some basic facts

Sousse is located on the coast of the Mediterranean Sea, and
the old part of the city is built on a small hill giving the medina
the form of an amphitheatre. The hinterland of Sousse is hilly
and largely covered by olive trees. The climate is mild, with hot
dry summers and mild wet winters. The hot days in the
summer are cooled by sea breezes. The Sahara Desert starts
approximately 260km south from the city, which makes the city
prone to dust from desert winds.

Size of city/urban area: 45km². Population: 173,047. Population
density: 3845.49 persons/km². Population growth rate (2008):
3.3%. Average household size: 4.03. Human Development
Index: 0.769.

The solid waste story

Main driver

The main driver operating in Sousse is the environment, and the main focus is on upgrading disposal. In 2005 the Ministry of
Environment and Sustainable Development established the Agence Nationale de Gestion des Déchets (ANGed) to be responsible for
waste management in Tunisia. Waste disposal became formalized, and cities began closing waste dumps and setting up controlled
landfills.

Public health/collection

In 1997 the city of Sousse decided to privatize part of the waste collection and contracted Seltene to collect some of the waste in
the city. Seltene did such a good job, and was so committed to its work, that it actually inspired the city government to be more
interested in solid waste. Because of this, over the years the percentage collected by Seltene increased, and today Seltene is responsible for the collection of 67 per cent of the waste. The municipality collects 33 per cent and plans to keep at least 25 per cent in
public hands so that it understands the costs, maintains competition, and maintains the human resource and institutional capacity to
operate collection in the future. This decision prevents the municipality from being fully dependent on one company for the collection of a whole city.

Environment/disposal

In 2005 the national government decided to open a controlled landfill serving each governorate. The aim was to close all waste
dumps within five years and to create landfills according to global standards for sanitary landfilling. The new state-of-the-art landfill
in the Sousse governorate opened at the end of 2008 and is run by a private company, ECOTI, which is a joint venture of DECO
from Italy and SECOPAD from Tunisia. ANGed monitors the developments at the site and remains responsible.

Resource management

Based on the extended producer responsibility (EPR) principle, the import of plastics for the packaging industry has been taxed and the income now supports several projects to keep plastics out of the waste stream. There is a national project called ECOLEF that buys collected plastics while stimulating entrepreneurship. In Sousse the municipality also took up the challenge to prevent plastics from entering the waste stream and set up the project SHAMS. The campaign is very intensive and includes door-to-door collection of the waste separated at the houses of about 5000 inhabitants. There is an informal recycling sector and some members have been hired to work in a project for the collection and sorting of recyclable wastes, mainly plastics, but also paper, glass, aluminium and Tetra Packs, within SHAMS. The work of the informal collectors is not hindered in any way and they can sell their collected material via intermediaries to the sorting site.

Special features

Producer responsibility, rare in Africa, as well as successful experience with privatization and the active development of plastics recycling are interesting features of Sousse. Attention is given towards increased recycling of plastics, via an EPR system called EcoLef. Activities are still in the pilot project phase. There is also some work on developing composting of green waste from parks, as well as further source separation activities.

Key benchmark numbers

Total tonnes municipal solid waste (MSW) generated per year	68,168 tonnes	Percentage valorized by informal sector of total waste generated	6%
Generation per capita in kilograms per year	394kg	Percentage valorized by formal sector of total waste generated	0.25%
Percentage coverage	99%	Goals for waste collection coverage as percentage of population	100%
Percentage disposal in environmentally sound landfills or controlled disposal sites of total waste generated	63%	Goals for environmentally sound (safe) disposal	100%
Percentage municipal waste incinerated of total waste generated	None	Goals for valorization of waste materials through recycling (or diversion from disposal)	Still being developed
Percentage diverted and valorized of total waste generated	6%	Prevented Reused Recycled Composted/agricultural value chain	NA NA 2% Seasonal green from markets; waste bread to farmers and urban poultry owners, 4%

Notes: NA = not available.
Figures in italics are estimates.

References

ANGed (2009) *Rapport d'activité 2008*, Tunisia

RDC Brussels (2008) *La gestion des déchets ménagers et assimilés au niveau de la municipalité de Sousse*, Brussels, Unpublished

TOMPKINS COUNTY
New York State, US, North America

42°45'N 76°47'W;
1100 feet above sea level

Barbara Eckstrom (Tompkins County solid waste manager), Kat
McCarthy (Tompkins County waste reduction and recycling
specialist) and Portia M. Sinnott (Reuse, Recycling and Zero
Waste consultant)

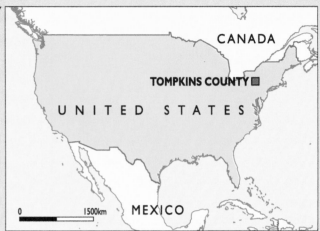

Some basic facts

Tompkins County and its largest city, the City of Ithaca, are
located in the centre of New York State's Finger Lakes Region, a
hilly rural area dominated by local agriculture, including vine-
yards and orchards, a domestic wine industry, outdoor
recreation, and institutions of higher education. The county is
comprised of the City of Ithaca, the villages of Dryden,
Freeville, Groton, Cayuga Heights, Lansing and Trumansburg,
and the towns of Caroline, Danby, Dryden, Enfield, Groton,
Ithaca, Lansing, Newfield and Ulysses. Ithaca, at the southern
end of Cayuga Lake, is host to Ithaca College and Cornell
University, perhaps the most famous US agricultural university,
which is also a centre for environmental and agricultural research.

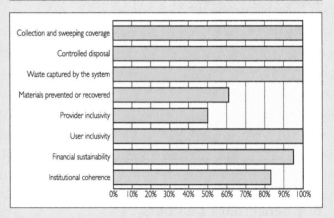

Climate/rainfall: the county has summer temperatures as high as 35°C and winter temperatures around −15°C. Rainfall is 899mm
per year and snowfall is 1709.42mm per year. Topography: rolling hills and farm land. Size of city/urban area: 1272km². Population
(2008): 100,628. Population density: 79 persons/km². Population growth rate (2008): 0.08%. Average household size: 2.32. Human
Development Index (2008): 0.956.

The solid waste story

Main driver

Resource management has been the main driver since the county closed the last unregulated landfill and began exporting its waste
to a privately owned landfill outside of its borders. This, combined with a strong environmental ethic, has accelerated the county
towards waste reduction-oriented planning, technical solutions, and steadily increasing recycling opportunities and performance
through the years.

Public health/collection

There are various approaches to managing garbage collection in the county. The City of Ithaca and the Village of Cayuga Heights
operate garbage collection, while the villages of Trumansburg and Dryden contract directly with a private-sector service provider. In
the remaining municipalities, individual households arrange for garbage collection with a private-sector service provider and an
unrecorded number choose to bring their own waste and recyclables to the Tompkins County Recycling and Solid Waste Center
(RSWC) in Ithaca. Collected or self-hauled waste can only be disposed of by paying a fee. For collected waste, payment is indicated
by a tag that is attached to the collection container or bag when it is placed at the kerb. All households receive two-weekly kerbside
recycling collection through a county contract with a private-sector service provider.

Environment/disposal

During the late 1980s, the county selected a landfill site that complied with New York State regulations and began the process of
developing a new landfill. It eventually became clear that the availability of much less expensive private disposal capacity outside of
the county represented a better option. The county had already made a commitment to develop the RSWC that focused on diverting
as much as possible from disposal, which worked well in combination with export.

Resource management

When the county's 20-year Solid Waste Management Plan was prepared in 1990, the State of New York required counties to include a goal of at least 50 per cent diversion from disposal. Tompkins County went further with a plan that called for gradually increasing goals and a shifting emphasis over the years from disposal to recycling, composting, organized and supported reuse, and prevention. The county, working in partnership with the private and non-profit sectors, is now diverting 60 per cent from disposal and has identified a 75 per cent diversion goal for 2015.

Special features

In its commitment to increase diversion, the county introduced a system of pay-as-you-throw in 1989. By law, all waste coming to the RSWC has to be separated from recyclables and yard waste. Disposing of non-recyclable waste requires payment of a fee; the recyclables – soon shifting to collection in a single stream – do not have to be paid for. Another special feature is the county's human resource commitment to promoting the 4Rs: reduce, reuse, recycle, re-buy. Whereas most New York and East Coast counties have one staff member in the position of recycling coordinator, Tompkins has, in addition to management and operational staff, four full-time equivalents working on increasing material diversion and recovery.

Key benchmark numbers

Total tonnes municipal solid waste (MSW) generated per year	58,401 tonnes*	Percentage valorized by informal sector of total waste generated	None
Generation per capita in kilograms per year	577kg*	Percentage valorized by formal sector of total waste generated	61%
Percentage coverage	100%	Goals for waste collection coverage as percentage of population	100%
Percentage disposal in environmentally sound landfills or controlled disposal sites of total waste generated	39%	Goals for environmentally sound (safe) disposal	100%
Percentage municipal waste incinerated of total waste generated	None	Goals for valorization of waste materials through recycling (or diversion from disposal)	75% total diversion by 2015
Percentage diverted and valorized of total waste generated	62%	Prevented Reused Recycled Composted/agricultural value chain	Less then 1% 61% NR

Notes: NR = not reported.
*These figures exclude commercial waste.
Figures in italics are estimates.

References

1990 Tompkins County Solid Waste Management Plan

New York State Department of Environmental Conservation (2008a) *Annual Recyclables and Recovery Facility Report*, New York State Department of Environmental Conservation, NY

New York State Department of Environmental Conservation (2008b) *Annual Transfer Station Report*, New York State Department of Environmental Conservation, NY

New York State Department of Environmental Conservation (2008c) *Household Hazardous Waste Collection and Storage Facility Year-End Report*, New York State Department of Environmental Conservation, NY

New York State Department of Environmental Conservation (2008d) *Planning Unit Recycling Report*, New York State Department of Environmental Conservation, NY

New York State Solid Waste Management Act of 1988

Tompkins County Charter and Code, Chapter 140

Tompkins County Solid Waste Management Division Annual Report (2008)

VARNA

Oblast (County) Varna, north-east Bulgaria, Europe

43°13'N 27°55'E,
39m above sea level

Kossara Bozhilova-Kisheva and Lyudmil Ikonomov (CCSD Geopont-Intercom)

Some basic facts

Varna is Bulgaria's third largest city after the capital Sofia and Plovdiv. It is the country's most important Black Sea resort city and attracts thousands of summertime vacationers each year, many from Russia and the Balkans, but also increasingly from Scandinavia and Western Europe. Within the city limits, and immediately outside, are several large resort complexes which have their own service arrangements, partially separated from those of the city itself.

The city occupies 205km² on verdant terraces, descending from the calcareous Franga Plateau (height 356m) in the north, and Avren Plateau in the south, along the horseshoe-shaped Varna Bay of the Black Sea, the elongated Lake Varna, and two artificial waterways connecting the bay and the lake and crossed by the Asparuhovo Bridge.

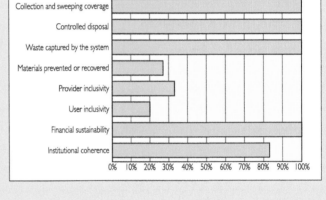

Rainfall: average rainfall is 500mm–550mm. Size of city/urban area: 80km², including five non-urbanized villages. Population (2007): 313,983. Population density: 1356.3 persons/km². Population growth rate (2002): –0.6%. Average household size (2002): 2.6. Human Development Index (2006): 0.834.

The solid waste story

Main driver

The main driver has been environmental improvement required for Bulgaria to become a member of the European Union in 2007.

Public health/collection

Like many post-socialist countries, public health measures resulting in 100 per cent collection coverage in cities were achieved during the socialist period. The tourist industry also puts pressure on the city for effective waste collection and street sweeping. Collection was concessionized during the early 1990s and is still operated by private companies, both domestic and joint ventures with German and other Western European firms. The only problem with collection is that villages are underserved (as they are virtually everywhere in the Balkans) and have a high degree of informal dumping. In this sense, it is not strictly true that there is 100 per cent coverage outside of the city limits.

Environment/disposal

The City of Varna has an active environmental department and many specific goals for sustainable development. The disposal site at Vaglen in an adjacent municipality was formerly a ravine that was first used in 1973 as an open dump. It has been continuously upgraded and expanded as the need grew and new technical information became available. It has recently achieved the status of a sanitary landfill, planned to serve the small region of Varna and two neighbouring communities. Since the harmonization of Bulgarian legislation with the EU, municipal administrations and mayors of the municipalities have a key role in planning and organizing the several municipal and construction waste management activities.

Resource management

The socialist era national recycling institution, Phoenix Resource, was privatized and split up during the late 1990s, and socialist era recycling initiatives now exist only in memory. There is an active informal sector in Varna, consisting mostly of ethnic Roma street-pickers who extract wastes from community containers. Recyclers use push carts, autos or, occasionally, horse-drawn vehicles for collecting cardboard and metal from businesses and a range of other materials 'donated' from households. A national initiative to add polyethylene terephthalate (PET) bottle recycling to this activity had modest success and resulted in one PET processor in Varna. But this activity remains largely separate from the formal solid waste infrastructure in spite of EU directives and the preparation of a recycling plan in a project during 2002.

Special features

Varna has an open and transparent policy setting for waste management. All types of non-state actors, including non-governmental organizations (NGOs), private consultancies, schools, resorts and city twinning projects with cities and regions in The Netherlands and other Western European countries, participate actively in planning and policy-setting. This, in combination with the concessionization of waste collection and disposal services, represents a quite exceptional degree of partnership.

Key benchmark numbers

Total tonnes municipal solid waste (MSW) generated per year	136,532 tonnes	Percentage valorized by informal sector of total waste generated	2%
Generation per capita in kilograms per year	435kg	Percentage valorized by formal sector of total waste generated	25%
Percentage coverage	100%	Goals for waste collection coverage as percentage of population	100%
Percentage disposal in environmentally sound landfills or controlled disposal sites of total waste generated	55%	Goals for environmentally sound (safe) disposal	NA
Percentage municipal waste incinerated of total waste generated	None	Goals for valorization of waste materials through recycling (or diversion from disposal)	10% recycling of collected x 10% recycling in 2005–2008 48% of packaging utilized/40% recycled by end 2010
Percentage diverted and valorized of total waste generated	27%	Prevented Reused Recycled Composted/agricultural value chain	NA NA 27% NA

Notes: NA = not available.

Figures in italics are estimates.

CHAPTER 4

THE THREE KEY INTEGRATED SUSTAINABLE WASTE MANAGEMENT SYSTEM ELEMENTS IN THE REFERENCE CITIES

This chapter discusses the three key physical elements necessary for an integrated sustainable waste management (ISWM) system. These three elements can be related to the three primary drivers (driving forces) for their development, as outlined in Chapter 2:

1 waste collection, usually driven by a commitment of authorities to protect and improve public health and reduce deaths and illnesses related to the presence of waste;

2 waste disposal, driven by the need to decrease the adverse environmental impacts of solid waste management; and

3 waste prevention, reuse, recycling and recovery of valuable resources from organic wastes, driven by both the resource value of waste and by wider considerations of sustainable resource management.

These considerations are supplemented by the current interest in improved energy efficiency, which is related to the emerging driver of climate change. All three of these key elements need to be addressed for an ISWM system to work well and to work sustainably over time. We can think of them as 'triangle 1', the first three main foundation stones of an ISWM system, which provide the physical basis for 'triangle 2', the governance aspects in the following chapter.

WASTE COLLECTION: PROTECTING PUBLIC HEALTH

Basic issues

Together with sanitation as the safe management of human excreta, effective removal and treatment of solid waste is one of the most vital urban environmental services. Waste collection represents both an essential utility function, together with electricity, gas and clean water, and a necessary part of urban infrastructure and services, alongside housing and transport, education and healthcare. In cities, poor solid waste management has a direct impact upon health, length of life and the urban environment. This matters and it is the basis for the idea that removing solid waste from urban centres is an essential function of the city authorities. Ever since the middle of the 19th century, when infec-

Stagnant water in open drain clogged by waste provides ideal breeding conditions for mosquitoes in Nigeria

© Kaine Chinwah

tious diseases were linked for the first time to poor sanitation and uncollected solid waste, municipalities have therefore been responsible for providing solid waste collection services to their citizens.

When solid waste is not removed, it ends up somewhere. That 'somewhere' is open spaces, backyards, public parks, alongside roads or pathways, and in nearby rivers or lakes. Waste is burned in a barrel or in a heap. Children, especially those living in slums, play in it and with it.

Poor waste management usually affects poor people more than their richer neighbours. Often the city centre receives a door-to-door collection several times per week and the peri-urban or slum areas rely on containers that are emptied so seldom that the area around them becomes an informal dumpsite, attracting insects, rats, dogs and grazing animals, and,

always, more waste. But providing a good collection service to the poor as well as the rich is more than just an equity issue – infectious diseases will affect the whole city. Figure 4.1 shows the global version of this, comparing data for non-slum and slum households.

Demographic and Health Surveys (DHS) data shows significant increases in the incidence of sickness among children living in households where garbage is dumped or burned in the yard. Typical examples include twice as high diarrhoea rates and six times higher prevalence of acute respiratory infections, compared to areas where waste is collected regularly.[1] Searching for food, cows, pigs, goats and horses eat the organic waste in many countries. This is not without a price: in Mali, cows regularly die if they are not operated on to remove tens of kilos of plastic bags accumulating in their stomachs.[2]

Uncollected solid waste clogs drains and causes flooding and subsequent spread of waterborne diseases. Blocked storm drains and pools of stagnant water provide breeding and feeding grounds for mosquitoes, flies and rodents. Collectively, these can cause diarrhoea, malaria, parasitic infections and injuries.

The annual floods in Kampala and other East African cities are blamed, at least in part, on plastic bags, known as '*buveera*' in Uganda, which block sewers and drains. In response to annual flooding in Mumbai, the State of

Figure 4.1

Collection coverage for non-slum and slum households

Source: Global Urban Observatory (2009); Demographic and Health Surveys

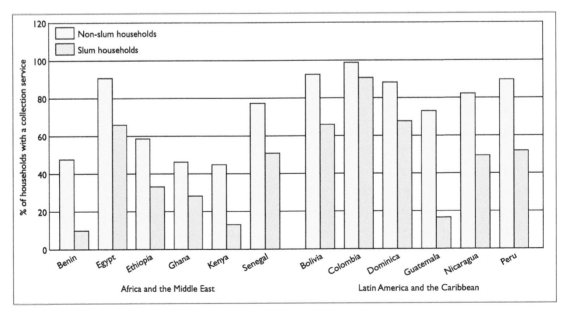

Maharashtra in India banned the manufacture, sale and use of plastic bags in 2005; unfortunately, poor enforcement means that the ban has so far been ineffective. In West Africa, floods are being blamed on the small plastic pouches for drinking water.

Uncollected waste has economic, social and technical costs for a city. A dirty and unhealthy city will make it difficult to attract businesses. In Tangier, Morocco, pollution of beaches by solid wastes was cited during the late 1990s as the leading cause of tourism decline that cost hotels in the area US$23 million per year in lost revenues.[4] In Costa Rica, the electric utility company has had so many problems with plastic litter clogging the turbines of their hydroelectric plants that they are financing plastics recycling in the catchment area behind their dams.

Even in Europe, uncollected waste can still hit the headlines, as in the 2008 example of Naples, Italy, where the collection service broke down due to a failure of governance and disagreements on the siting of a new waste disposal facility and financing of the system between elected officials, private companies and citizens.[5]

One thing is certain: even though the poor suffer more from inadequate waste collection services, the rich cannot afford to ignore the poor in their city – infectious diseases have no respect for wealth.

Insights from the reference cities and global good practice in waste collection

Effective waste collection is all about the city authorities understanding their citizens and their city, and making a focused and sustained effort to mobilize the human and financial resources. Many parts of the system need to work together to remove waste, serve households and keep the city clean. The authors of this Third Global Report generally agree with their colleagues worldwide that getting collection under control is the first step in climbing onto the modernization ladder. The reference cities show a wide variety of experience and give some new insights into how to do this efficiently, fairly and effectively. This section explores this under three headings, which echo the experiences of the cities and what they are proud of doing well:

- keeping the cities clean;
- improving cost effectiveness of the services;
- creating effective channels of communication between users and providers.

Cows feed on waste dumped in open spaces in Indian cities

©WASTE, Jeroen IJgosse

Maintaining streets and public areas clean through street sweeping in Quezon City

© Quezon City

KEY SHEET 5
EXAMPLES OF MUNICIPAL WASTE COLLECTION AND TRANSFER SYSTEMS

Manus Coffey Associates MCA

This Key Sheet illustrates three alternative municipal waste collection systems for a typical city with a waste generation of 1000 tonne per day and a distance of 40km from the city boundary to a landfill site east of the city. At peak traffic times there is severe traffic congestion in the city that restricts the hours of collection. The city covers an area of around 70 square kilometres (this is a composite of actual cases simplified to allow comparison of the three alternative collection/transfer systems).

EXAMPLE 1: EXISTING SYSTEM USING COMPACTOR TRUCKS DISCHARGING AT THE LANDFILL SITE

The wastes are collected by single-axle (4 × 2) compactor trucks that average 8000kg per load, with a gross vehicle weight of 17,000kg (although this exceeds the legal gross load limit of 15,000kg for 4 × 2 trucks in the country concerned) and are transported directly to the landfill. This is a very slow system as the trucks must travel long distances (average 6km × 2 = 12km return journey) within the city as well as long distances (40km × 2 = 80km return journey) to the landfill. Collection must take place at times when the city is free of traffic, taking 2.5 hours (including start-up times) with a crew of

four loaders. The traffic speed on the road to the landfill averages 35km per hour so that the return haul from the city boundary to the disposal site, including discharge time, also takes 2.5 hours. Only one 8000kg load is collected per day, and although it would be possible to collect a second load each day by working extended hours, the unions will not permit the driver to work a ten-hour day and it is not considered practical by the official concerned to retain the collection crews during the 2.5 hour waiting period that the truck is travelling to the disposal site. Double shift work is also not acceptable to the municipality.

It can be seen from this that the trucks have considerable distances to travel through the city, adding to the traffic congestion.

This system requires 125 trucks (plus standby vehicles), 125 drivers and 500 loaders.

EXAMPLE 2: PROPOSED LARGE TRANSFER STATION

It has been proposed by an international consultant that a conventional large transfer station (LTS), or possibly two LTSs to reduce haul distances within the city, should be introduced and two possible sites have been chosen. The sites are both on waste ground outside the urban boundary to the south and to the north of the city as no acceptable sites are available on the direct route to the landfill.

A ring road around the city enables the transfer trucks to reach the landfill without travelling through the city with a travel distance of 45km each way. With an average travel speed of 40km per hour, the return haul trip, including loading and discharge times takes, 2.45 hours. Each double-axle (6 × 4) transfer truck has a capacity of 15,000kg and can make three loads per nine-hour shift. By introducing some storage capacity at the LTS it would be possible to work two transfer shifts. However, storing wastes at conventional transfer stations can cause odour, insect and rodent problems.

The collection vehicles can now collect two loads working a seven-hour shift, avoiding city centre collections during peak traffic times (it would be possible to work two shifts, but the municipal council are reluctant to do this).

This collection system requires 63 collection vehicles (plus standby vehicles), 63 drivers and 250 loaders. The transfer system requires 22 trucks, 22 drivers and 22 drivers' assistants (plus standby vehicles), with a total of 85 drivers and 272 loaders and drivers' assistants.

EXAMPLE 3: TWELVE SMALL TRANSFER STATIONS ARE LOCATED IN THE RESIDENTIAL AND DOWNTOWN AREAS

Recent developments in small transfer stations (STSs) enable small transfer stations with a capacity of 120 tonnes per day to be located on sites of 26m × 10m, and such small sites are readily available. Since each transfer station is washed down at the end of each working day, there are no problems with smells, insects or rodents. Electronic weigh cells at the bottom of the transfer pits enable incoming loads to be recorded and transfer loads to be optimized with-

out overloading. It takes no more than five minutes for the transfer vehicle to pick up a container of wastes. These STSs can store up to seven 15,000kg loads (105 tonnes) in sealed containers between collection and transfer, enabling the collection to take place to suit the traffic conditions and transfer to be carried out during two shifts. Thus, each transfer vehicle can transport six loads or up to 90 tonnes per day (in Egypt in 2004, a 100-tonne-per-day STS costs US$122,000 to construct and a recent estimate for Nicaragua estimated a cost of around US$250,000; this is only a small fraction of the cost of a large transfer station).

Double-pit small transfer station (STS), with right-hand containers omitted for clarity. This STS can store seven containers (63 tonnes of waste) between collection and transfer, and can handle up to 100 tonnes of waste per day within a site of less than 20 × 10 metres. A STS using a 4 × 2 (double-axle) transfer vehicle with containers 9 metres long can store 105 tonnes of waste and transfer up to 150 tonnes per day.

© Manus Coffey

A small transfer station downtown in Ho Chi Minh City, Vietnam

© Manus Coffey

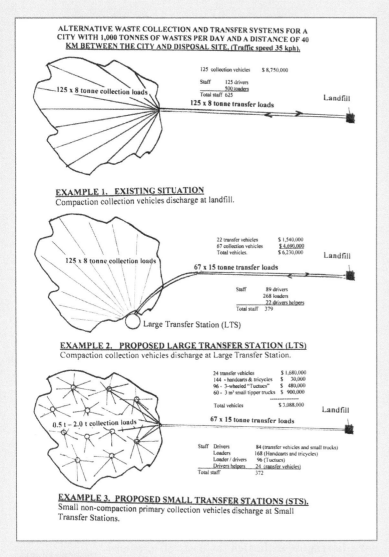

ALTERNATIVE WASTE COLLECTION AND TRANSFER SYSTEMS FOR A CITY WITH 1,000 TONNES OF WASTES PER DAY AND A DISTANCE OF 40 KM BETWEEN THE CITY AND DISPOSAL SITE. (Traffic speed 35 kph).

125 × 8 tonne collection loads

125 collection vehicles $ 8,750,000
Staff   125 drivers
         500 loaders
Total staff 625

125 × 8 tonne transfer loads   Landfill

EXAMPLE 1.   EXISTING SITUATION
Compaction collection vehicles discharge at landfill.

125 × 8 tonne collection loads

22 transfer vehicles $ 1,540,000
67 collection vehicles $ 4,690,000
Total vehicles. $ 6,230,000   Landfill

67 × 15 tonne transfer loads

Staff   89 drivers
         268 loaders
         22 drivers helpers
Total staff   379

Large Transfer Station (LTS)

EXAMPLE 2.   PROPOSED LARGE TRANSFER STATION (LTS)
Compaction collection vehicles discharge at Large Transfer Station.

24 transfer vehicles $ 1,680,000
144 - handcarts & tricycles $ 30,000
96 - 3-wheeled "Tuctucs" $ 480,000
60 - 3 m³ small tipper trucks $ 900,000

Total vehicles $ 3,088,000   Landfill

0.5 t – 2.0 t collection loads

67 × 15 tonne transfer loads

Staff   Drivers   84 (transfer vehicles and small trucks)
         Loaders   168 (Handcarts and tricycles)
         Loader / drivers   96 (Tuctucs)
         Drivers helpers   24 (transfer vehicles)
Total staff   372

EXAMPLE 3.   PROPOSED SMALL TRANSFER STATIONS (STS).
Small non-compaction primary collection vehicles discharge at Small Transfer Stations.

Routes travelled by the collection and transfer vehicles in the examples

© Manus Coffey

Twelve STSs, with an average capacity of 85 tonnes per day, will service the city. This gives a maximum collection distance of 2.8km from the STS and an average collection distance of around 1.5km. Collection is carried out by small vehicles, including handcarts, tricycles, three-wheelers and small non-compaction tipping trucks, each operating within its economic travel distance from the STS (note: one man with a tricycle was collecting 2200kg per day during a study in Kunming City in China). A typical collection fleet for each STS will be as follows:

- Twelve handcarts and tricycles collect 8 tonnes per day within a 0.5km radius of the STS (one crew).
- Eight three-wheelers (1.5-cubic-metre Tuctucs) collect 40 tonnes per day within a 1.5km radius (one crew).
- Five 3-cubic-metre tipping trucks collect 40 tonnes per day within a 3km radius (two crews) (each of the above vehicles will require backup vehicles to allow for breakdowns).

Transfer vehicles and primary collection vehicles for 12 STSs are as follows:

- Transfer vehicles: 12 work two shifts with 24 drivers and 24 assistants.
- Handcarts and tricycles: 12 × 12 = 144 with 144 loaders.
- Three-wheelers: 12 × 8 = 96 with 96 unskilled driver/loaders.
- Small tippers: 60 with 60 drivers and 60 loaders.
- Total labour requirements: 84 drivers and 324 unskilled workers.

It is proposed, however, that the primary collection should be franchised out to individuals who will operate the handcarts, tricycles and three-wheelers, collecting revenues directly from the householders with contracts for the collections from different streets. This will bring the contracting within the means of small family-based collection teams. The municipality will operate the transfer vehicles and can either operate or franchise out the small tippers.

KEY SHEET 6

EMERGING GLOBAL GOOD PRACTICE IN THE DESIGN OF SMALL-SCALE SOLID WASTE EQUIPMENT: THE SITUATION IN SOLID WASTE DIVISIONS IN CITIES IN LOW- AND MIDDLE-INCOME COUNTRIES

Manus Coffey Associates MCA

Solid waste collection will often account for by far the largest proportion of a city's municipal budget; however, it is perhaps the least researched in low- and middle-income countries. There has been an almost universal tendency in the past for international consultants working in these countries to assume that the technology from their own particular industrialized country will be appropriate wherever they go. Nothing could be further from the truth. The consultants are often followed by the salesmen for the latest solid waste collection vehicles who see this as an opportunity to promote their own particular vehicles, often with the support of their country's development aid programme, claiming they have 'state-of-the-art technology' or the 'internationally recognized way of collecting waste as used in New York, Hamburg or Tokyo'.

The manager responsible for the cleaning of the cities, in particular the smaller ones, may come from a very different discipline, perhaps he is the medical officer of health or the roads engineer who is also in charge of the landfill. It is often difficult for him to resist the advice of the consultant and high-pressure salesmen. He may look to the capital city of his country for advice without understanding the very different conditions pertaining in his own particular city. In the capital city, the system will often be the same throughout the city, catering for the conditions found in the commercial areas of the city centre as well as for low-, medium- and high-income urban and suburban areas, and also, perhaps, peri-urban areas. However, the collection requirements in any large city will vary greatly from district to district. Thus, systems developed for the city centre areas of the industrialized countries are being imposed on the very different situations in low- and middle-income countries in both city centre and residential areas in smaller cities with very different conditions.

KEY DIFFERENCES IN COLLECTION PARAMETERS

The first difference relates to *density and volume of the wastes at the point of collection*. These are not the same in industrialized and non-industrialized countries. This is especially true for waste collected after any recyclables or organics have been removed from the waste stream before collection.

Transfer station at Faraskour, Egypt. The design is suitable for up to 150 tonnes/day capacity

© Manus Coffey

The industrialized country has a waste density of 150kg per cubic metre and a generation per capita of 3kg per day, which is a daily volume of 20 litres per capita per day. It will use a compaction truck to collect the wastes in order to get an economic load into the truck.

The low-income country has a waste density of 500kg per cubic metre and a generation rate of only 0.2kg per capita per day, which is a daily volume of only 0.4 litres per capita per day. There is, therefore, a difference of 40 times in the waste volume produced by the inhabitants in the two different situations and there is no logical reason to use the costly compactor trucks in cities where the waste is already as dense as it will be after compaction in the earlier situation. Thus, the capital-intensive system that may be cost effective in an industrialized country will be unaffordable in low- or middle-income countries where labour is available and costs are low, but bank loans are not available to support the purchase of expensive trucks.

The second difference relates to the *distances which the wastes have to be transported* during and after collection. In an industrialized country, the landfill is often quite far away as disposal has been regionalized, so a truck is necessary. Many low- and middle-income cities still have the landfill in the middle of the city. Handcarts, tricycles and micro-trucks will be cost effective where primary haul distances are short. Tractors may be much more cost effective than trucks for short-haul secondary transport (typically up to 15km) in a low-income country. Haul distances will generally be shorter in small cities than in large ones.

The third important factor is *local availability of spare parts and servicing facilities*. This will greatly favour local manufacture or adaptation of appropriate vehicles in low- and middle-income countries where systems depending on imported vehicles and parts will be unsustainable. Actually, this is a similarity rather than a difference: no city, whether it is Kunming or New York or Ouagadougou or Singapore, wants to have to wait to import parts or get service from a distance. Low- and middle-income cities need local parts and service expertise too! The difference is that New York or Singapore make this part of their procurement specifications, whereas cities which are getting 'donations' don't feel that they can be so demanding. But without a guarantee that local parts and service are available, a donation may prove more expensive than buying it locally.

Something else to consider is *road and street conditions and axle load limits*. In some countries, vehicles axle loads of up to 12,000kg are allowed; but in other countries only 8000kg are permitted on urban roads. These regulations are seldom adhered to. A significant, although hidden, cost factor of heavy high-tech vehicles will be the damage to water and sewer pipes under the roads and road maintenance costs due to excessive axle loads.

SMALL TRANSFER STATIONS: A RECENT TREND

A recent trend, which started in China and spread to Vietnam, Egypt and now Nicaragua is to use small transfer stations located close to where the wastes are generated that facilitate very low-cost primary collection systems. In a typical system, handcarts will be used within a short radius of the transfer station, tricycles will be effective within a wider radius, and three-wheeled or four-wheeled micro-trucks within an even wider radius.

The small transfer station can enable the micro-privatization of the labour-intensive primary collection. Primary collection can be contracted or franchised to micro- and small enterprises (MSEs), community-based organizations (CBOs), non-governmental organizations (NGOs), and even down to individual collectors or family groups with a handcart, tricycle or micro-truck who have a contract for a specific collection route. This may be a franchise operation where the collector is paid by fees which he collects directly from each household that he services, thus avoiding the costs to the municipality of collecting the refuse charges. The collector may supplement his income by selling recyclable materials, which he collects from his area, to a recycler located alongside the transfer station.

The municipality then has only a monitoring role over the collection and can concentrate its resources on the efficient operation of the

Small Suzuki trucks for primary collection in narrow streets, designed for high tilting direct into larger vehicles for secondary transport

© Manus Coffey

transfer and disposal services, which require access to capital and which are beyond the resources of the small primary contractor. In this way, the capital requirements will be greatly reduced and municipal management and financial resources used effectively.

A weighing system at the transfer station can be used to monitor the performance of the primary collectors, as well as to control the weight in the transfer containers in order to ensure that the transport vehicles carry full loads without overloading, thus maximizing the transfer vehicle efficiency.

With this system, all wastes coming into the transfer station are fresh and no wastes remain in the transfer station after transfer has been completed. Thus, the whole transfer station can be washed down at the end of each day with a high-pressure hose so that there are no odour, rodent or insect problems.

Country	Year	Average (%)	Minimum (%)	Maximum (%)
Colombia	2005	97.2	89.0	100.0
Dominica	2002	83.6	78.2	85.8
Bolivia	2004	79.9	67.9	84.8
Peru	1991	70.8	59.1	85.6
Nicaragua	2001	64.7	56.1	80.8
Guatemala	1998	56.2	42.9	89.5
Egypt	2005	86.6	40.8	96.4
Senegal	1997	62.6	34.3	85.9
Ghana	2003	39.6	30.1	64.4
Ethiopia	2005	39.0	19.6	69.6
Kenya	2003	28.5	5.6	57.7
Benin	2001	27.3	12.4	47.4

Table 4.1

Waste collection coverage in urban areas (percentage) in selected Latin American and African countries[6]

Source: Global Urban Observatory (GUO) 2009; data compiled from national Demographic and Health Surveys

Table 4.2

Average collection coverage in the reference cities, in ranked order from 100 per cent

Note: Figures in *italic* are estimates. Belo Horizonte: 70% of slum populated was covered in 2008. Delhi: 75% including slums, based on weighted average.

■ Keeping the cities clean

Collection coverage (i.e. the extent to which collection services reach households) is considered to be the most basic indicator of a solid waste system's performance. There are major cities in all continents that have had formal collection services in place for a century or more. For example, The Netherlands Association of Municipal Waste Service Providers recently celebrated its 100th anniversary. Yet, it is also common that half the urban solid waste remains uncollected and half the city population unserved.

Table 4.1 provides data from UN-Habitat's Global Urban Observatory, based on Demographic and Health Surveys in 12 selected countries on waste collection rates in urban areas. Collection coverage – percentage of households receiving services (from either formal and informal, public and private providers) – varies widely, from 100 to less than 10 per cent.

In the reference cities, reported collection coverage rates range from 100 to 45 per cent of the city population. Waste collection services are not always evenly distributed throughout a city. While the city business district of Nairobi, Kenya, enjoys reasonably good services, and private collectors serve many housing estates, other areas are underserved. The difference is even more pronounced in cities such as Delhi and Bengaluru, India, where around 90 per cent of citizens receive good services, but some slum areas receive no services at all.

The situation is similar in rural areas within official administrative city borders. These rural areas receive fewer services than the urban area: Cañete, Peru, reports that most villages are without services; in Varna, Bulgaria, the 100 per cent coverage ends at the city borders and the city's five villages are each served with a single container that is collected once per month, if at all.

There are a number of things that collection coverage can tell us. Collection coverage tends to follow gross domestic product (GDP) – it is clearly higher in economically developed countries. Adelaide, Australia, has a 100 per cent waste collection coverage rate, with a consistent high standard of services regardless of the socioeconomic status of the area. Similarly, San Francisco, US, and Rotterdam, The Netherlands, provide services to all their citizens; but not all 13 sub-districts in Rotterdam score equally on cleanliness.

The coverage rate is also high in cities where authorities are concerned about the city image – for example, in preparation for a large international event happening in the city. The International Horticultural Exhibition held in 1999 in Kunming, Yunnan Province, the People's Republic of China, contributed to the fact that waste collection services are now provided to

City	Average collection coverage	Rank
Adelaide, Australia	100%	1
Curepipe, Republic of Mauritius	100%	2
Kumming, China	100%	3
Rotterdam, Netherlands	100%	4
San Francisco, USA	100%	5
Tompkins County, USA	100%	6
Varna, Bulgaria	100%	7
Quezon City, Philippines	99%	8
Sousse, Tunesia	99%	9
Belo Horizonte, Brazil	95%	10
Delhi, India	90%	11
Managua, Nicaragua	82%	12
Canete, Peru	73%	13
Bengaluru, India	70%	14
Nairobi, Kenya	65%	15
Moshi, Tanzania	*61%*	16
Bamako, Mali	57%	17
Ghorahi, Nepal	46%	18
Dhaka, Bangladesh	55%	19
Lusaka, Zambia	45%	20
Average	82%	
Median	93%	

100 per cent of its urban population. As Delhi is preparing to host the 2010 Commonwealth Games, activities are being intensified to improve the image of the city. In some cases, the measures taken did not have long-lasting effect, as in 2002, when Mali hosted the Coupe d'Afrique de Nations (CAN) football championship. Waste collection services are often improved upon in response to citizens' and political concerns about the investment climate or image, as was the case in Bengaluru.

High waste collection coverage is closely related to good governance since it demonstrates the commitment of city authorities to keep the city clean and healthy. Belo Horizonte, Brazil, and Quezon City, the Philippines, serve more than 90 per cent of all their citizens. The current solid waste management system in Belo Horizonte is the product of a gradual learning process in urban and environment management initiated a century ago. In Quezon City, solid waste management came together as a comprehensive programme within the mayor's vision to create a 'quality community' for city residents.

Waste collection coverage does not exceed 60 per cent in Dhaka, Bangladesh, where the city corporation cannot cope with an ever-increasing urban population of this already large city and the accompanying growing amounts of waste.

Bamako, Mali, as the fastest-growing African city, has a low rate of less than 60 per cent due to a generally low income level and consequent financial problems of a decentralized system based on community-based organizations (CBOs) and small- and medium-sized enterprises (SMEs).

Other cities in the poorest, or least developed, countries with low waste collection rates include small cities such as Ghorahi, Nepal, and Moshi, Tanzania, where local authorities struggle to finance the system, as citizens are either not charged at all (Ghorahi) or do not see the need to pay for the inadequate services (Moshi). Interestingly, in cases such as Lusaka, Zambia, services are offered but not accepted by all citi-

---

Box 4.2 Importance of waste for the Silicon Valley of India

During the 1990s, Bengaluru, India, was growing rapidly with its numerous information technology (IT) businesses and the city corporation was not able to deal adequately with the growing quantities of the city's solid waste. As the situation became acute during the early 2000s, a group of prominent citizens and IT industry leaders took an initiative to improve cleanliness. Working with the Bangalore Municipal Corporation (BMP) – the municipality – they formed the Bengaluru Agenda Task Force (BATF). As one of its main activities, the task force engaged in a strategic planning process for solid waste. It initiated collection and transportation through public private partnership and later followed with processing and disposal. Once its work was done, the BATF was disbanded and the products of its activities were transferred to operational divisions in the BMP. The plan that it produced still informs the decision-making process and the solid waste management system in Bengaluru.

---

zens, in part because they do not wish to pay. In Lusaka, more than 30 per cent have collection of waste through informal service providers who are not registered (and therefore do not use the controlled disposal facility).

Frequency of collection is often seen as a measure of good practice; but this is a complex issue relating to climate, socio-economic preferences, operational efficiency and the degree of source separation, as the example from Byala, Bulgaria, shows (see Box 4.3).

Many Dutch cities collect mixed waste only once every two weeks, alternating with source-separated organics. While cities in high-income countries often find that once per week is enough, low- and middle-income countries – particularly in the tropics – are convinced that once per day is necessary. Daily collection may be necessary and justified in your local circum-

---

Box 4.3 Collecting air in Byala, Bulgaria[7]

In 2002, Byala, Bulgaria, was a sleepy Black Sea fishing town with a few modest summer resorts and small hotels. Together with five extremely rural villages, it had a winter population of less than 2500. The economic transition was accelerating and fuel prices were rising rapidly. The cleansing department was using up its yearly fuel budget in the first four months of the year.

In the process of updating its solid waste plan, Byala invited an international consultant to help with cost reduction. During a visit in the off-season month of November, the consultant and staff conducted a field audit of the relationship between waste generated and frequency of collection. It turned out that 90 per cent of the 40 litre containers were less than 20 per cent full when they were collected three times a week.

Based on a simple calculation, the cleansing company was able to reduce off-season collection from three times per week to once per month for nine months of the year. This allowed the department to cover its fuel needs with the existing budget during the entire year, including the tourist season. The consultant is still welcomed in Byala as 'that girl who came from abroad to ask us why we were collecting empty containers'.

stances; but the question should be asked whether this is really the case. Closely monitoring the performance of the system will answer this question better than some global benchmark.

Municipal cleansing services – in other words, street sweeping – are intimately linked to waste collection. Many cities keep streets clean in the central business district but leave other areas unattended, which discourages visitors and investors. It has even been suggested that the visual cleanliness of the *whole* city can be used as a surrogate performance measure for city governance.

In the Philippines and Indonesia, cities annually organize a street cleaning competition among their communities to encourage active participation of their local residents, while national environmental agencies grant awards to

outstanding 'clean and green' cities. In Japan, street cleaning is a regular activity, and city authorities invite their residents to participate at least once a month. In the highly urbanized The Netherlands, many householders still regularly sweep and even wash the sidewalks in front of their houses.

The importance of recognition, image and municipal pride in keeping streets clear cannot be underestimated. Moshi is very proud of its title of the cleanest city in Tanzania; Adelaide is proud of its high recycling rate; Tompkins County has received awards for its high diversion. The opposite is also true: when a dirty city becomes a political issue, or when an international event is scheduled, the impossible becomes possible. Delhi now cleans up for the Commonwealth Games; Bamako worked hard on cleanliness when the World Championship Football was held there several years ago; Beijing cleaned up streets (and reduced air pollution) for the 2008 Olympics.

So what constitutes a good collection service? The answer is different in different places, but results talk. The collection service that serves all areas of the city on a regular basis, keeps streets clean and drains clear, hires collectors in a safe working environment for a living wage, and meets the needs of the users comes pretty close to the ideal.

■ Improving the cost-effectiveness of primary and secondary services

In most cities, regardless of size, waste collection services developed as a means of protecting public health in response to increasing urbanization. Where municipal authorities could not cope, it got worse before it got better, creating crises that were drivers for political action. In some cases it was local authorities who took up the challenge, such as occurred in Belo Horizonte, Kunming or Moshi; in some instances local authorities were strongly supported or even pushed by donors, such as in Dhaka, Managua and Lusaka. In some places, non-governmental organizations (NGOs) and individuals, so-called 'champions', took the initiative. Such an initia-

As the capital city of Dhaka, Bangladesh, was growing rapidly throughout the 1970s and 1980s, it was becoming increasingly difficult for Dhaka City Corporation to cope with the growing piles of waste in the city.

In 1987, an enlightened individual, Mahbub Ahsan Khurram, decided to take waste matters into his own hands. He organized the residents of Kalabagan neighbourhood and established a waste collection service by tricycle vans. For a small monthly fee, the tricycle driver collected garbage from the households and deposited it into the nearest community container. The result was immediately visible. The neighbourhood was quickly rid of garbage piles and became clean. It was such a remarkable success that the initiative was featured on national television. Learning from this experience, non-governmental organizations (NGOs) and community-based organizations (CBOs) started similar operations in other parts of the city.

Secondary collection point set up by the municipality in Caita la Mar, Venezuela

© Jeroen Ijgosse

tive has usually addressed an immediate problem of primary waste collection – removal of waste from houses to some kind of waste collection point in the neighbourhood.

Secondary collection of waste from those collection points and transport to the disposal site or a processing facility is usually done by a municipal department or a large private company that the city has contracted. But if a city is struggling to find the money to pay for its existing collection service, how can it hope to extend the service to unserved communities? When funds are limited, what can a city do to stretch resources that are urgently needed, for example, for hospitals and schools?

Part of the answer is to improve the cost effectiveness of current services in order to free up resources to expand the service. UN-Habitat has recently updated its seminal publication on waste collection in developing countries,[10] the key message of which is to design your system to be sustainable under local conditions.

In contrast, receiving 'free' collection vehicles from foreign donors is generally neither cost effective nor sustainable because a truck from elsewhere can be a 'Trojan horse'. Neither parts nor service are available for many types of donated vehicles. Moreover, vehicles from industrialized countries may not necessarily be

appropriate for collecting waste that tends to be wetter and denser in low-income countries than in Japan or Denmark. Donated vehicles may have electronic controls in a language which local drivers cannot read; and heavy vehicles may be a mismatch with local roads, which are seldom paved and may have lower legal weight limits for trucks, as they are built to a lower specification. For a donated vehicle to make sense, spare parts, specialized equipment and skilled labour have to be locally available for maintenance.

Where neighbourhoods cannot be served by large vehicles, a common approach is to provide primary collection using handcarts, tricycles, animal carts or small vehicles, which bring the waste to secondary collection points or small transfer stations for transfer to bigger vehicles. This makes sense, and in contrast to what many officials in low- and middle-income countries may think, 'modernization' does not necessarily mean 'motorization'.

Some of the most effective and reliable primary collection experiences are organized by

A resident handing waste over to a primary collector in Bengaluru, India

© Sanjay K. Gupta

Box 4.7 The motorization debate in Bamako, Mali

Since the district of Bamako, Mali, outlawed the use of donkeys on paved roads,[11] there has been a debate among the local stakeholder platforms and waste service providers about whether it is a good idea to replace donkey carts with tractors and trailers. While this appears to be an improvement, it does not occur widely – and for a very sensible reason: there is not enough money in the system to pay the fuel costs. In a situation where even the larger, more successful, service providers cannot manage to buy enough fodder for their donkeys, the chance that they will consistently be able to pay for fuel is very small.

At the same time, donkey collection is becoming less and less profitable, so the discussion continues.

community groups, micro- and small enterprises (MSEs), and the informal sector, using pushcarts, handcarts, wheelbarrows or wagons drawn by donkeys or horses.

In unplanned settlements in peri-urban areas, which are inaccessible by public providers, CBOs and MSEs are increasingly seen as an efficient and sustainable strategy to provide primary collection and bring waste from houses to collection points. Their effectiveness in removing waste depends on municipal vehicles or municipal contractors who collect waste from the collection points and take it to the final disposal.

In the reference cities, MSEs and CBOs are important links in the primary collection system in all sub-Saharan African cities; but their degree of integration with municipal operations varies. Lusaka stands out as a city in which 30 per cent or more of primary collection happens with unregistered informal service providers, and this appears to contribute to the 60 per cent of collected waste, which never reaches the controlled disposal site.

Box 4.8 A tale of two cities (in one)

In the New Delhi Municipal Council (NDMC), waste-pickers have been officially subsidized to pick up waste from the doorstep. The NDMC says that this is a win–win situation because a poor person has been made part of a system that benefits the residents, saves waste from reaching the landfill and improves segregation. For these reasons, it has also issued an order that facilitates the work of itinerant waste buyers. The Municipal Corporation of Delhi (MCD), on the other hand, has contracted out doorstep collection in two large zones to a large private company, displacing existing waste-pickers. In other areas, it has contracted similar companies to pick up the waste from transfer stations and to transport it to the landfill. All recyclables belong to the company. 'This privatization is a death sentence for us', declares an affected waste-picker of the MCD system.

In response to an acute situation with solid waste management in Bamako, Mali, in 1989, a group of young educated women founded the first waste collection co-operative in Bamako as a means of self-employment. Their organization, the Coopérative des Femmes pour l'Éducation, la Santé Familiale et l'Assainissement (COFESFA), was the precursor to the current-day *Groupements d'Intérêt Économique* (GIEs), which are private micro-enterprises that perform primary waste collection in the city. Today, there are more than 120 GIEs in Bamako, collecting an estimated 300,000 tonnes of waste per year. Alongside the formally registered GIEs, there are also informal service providers who work illegally and undercut the fees that the GIEs charge.

In Nairobi, CBOs and MSEs filled the gap in operations outside of the central business district (CBD) and wealthy Westlands area. A city council formalization and registration initiative in 2006 had some perverse impacts when some of the CBOs decided not to register and, as a result, were sidelined in their own areas.

Both Indian reference cities, Bengaluru and Delhi, have significantly improved their waste collection services in recent years, starting from apparently similar positions. Both cities engaged large private companies for secondary collection and transport in most of their areas, but with rather different approaches.

While Bengaluru involved 70 small- and medium-sized enterprises through annual contracts, Delhi has opted for a system in which the informal sector is engaged through NGOs and MSEs in providing door-to-door waste collection in an estimated 25 per cent of the city in all income classes. The shift from the use of community containers to a well-organized and well-coordinated door-to-door collection in most parts of Bengaluru is an example of especially good practice, as it has resulted in an immensely cleaner city. But the main point is that both cities have put in a lot of effort to come up with an affordable door-to-door primary collection service, resulting in cleaner streets and increased possibilities of diverting recyclables and organic waste from disposal. Both cities, therefore, have demonstrated inclusivity and cooperation with other stakeholders, and their selection of different strategies demonstrates that understanding local circumstances is essential to good practice.

Belo Horizonte was among the first cities anywhere to recognize the informal recycling sector and build a policy of inclusion of informal recyclers in their recycling strategy. A more recent example is Cañete, Peru, which is modernizing its waste collection service, beginning with seven waste-pickers who now have secure incomes.

Inclusion in East Africa began in the late 1990s, where the focus was and remains on primary collections services. The International Labour Organization (ILO) began to experiment with micro-franchising in 1998 in Dar es Salaam, and since then has disseminated this inclusive service model to many East African cities. Moshi, one of the reference cities, is starting to expand collection coverage with micro-franchising. In Nairobi, another of the reference cities, private collection companies and CBOs compete for collection zones in private–private arrangements, which the Nairobi City Council began to regulate in 2006 (city presentations and profiles of Moshi, Nairobi, Belo Horizonte).

Secondary collection, the removal of waste from communal collection or transfer points, is more often organized by the municipality or its contractors. There are at least three key principles of success – one technical and two organizational:

1 Use collection vehicles and transfer systems appropriate to the local waste characteristics, street and traffic conditions, and distances between collection and disposal points.

2 Keep costs down by limiting multiple manual handling of the waste. The ideal is for waste to be collected from household

City	Working fleet/ facilities (%)	Fleet size, motorized	Collection fleet, non-motorized	Fleet investment source
AUSTRALASIA				
Adelaide	*100%*	NR	N	NR
AFRICA				
Bamako	*25–30%*	176	Y	Donor
Curepipe	90%	20	Y	State, Private
Lusaka	100%/70–80%	NR	Y	Municipal
Moshi	50%	5	Y	Donor
Nairobi	15–30%	13	NR	NR
Sousse	40–80%	53	N	Municipal Budget
ASIA				
Bengaluru	85 / 100%	4682	Y	NR
Delhi	95%	1059	Y	Private
Dhaka	NR	NR	Y	NR
Ghorahi	100%	5	Y	Municipal Budget
Kunming	99%	NA	Y	Municipal Budget, State
Quezon City	100%	265	Y	Municipal Budget
EUROPE				
Rotterdam	100%	116	N	Municipal Budget
Varna	NR	82	N	Private
LATIN AMERICA				
Belo Horizonte	100%	363	Y	Municipal Budget, State
Canete	100%	11	Y	Donor, Other
Managua	75–80%	143	NR	Municipal Budget, Donor
NORTH AMERICA				
San Francisco	100%	428	N	Rates
Tompkins County	100%	NR	N	Rates

containers and transferred to a cart, from which it is tipped directly into a larger transfer container (or tipping vehicle) for direct transport to disposal without ever being tipped onto the ground and loaded by hand.

3 Ensure coordination of the primary and secondary collection services in order to ensure that the overall system works effectively and reduces the risk of illegal disposal by the primary collectors.

In relation to number 1, above, it is useful to look at the status of the fleet of collection vehicles in the reference cities. Most of the cities have both motorized and non-motorized transport. Of the fleet and solid waste facilities, more than 85 per cent are reported to be available for work, but not all of them are operational. The percentage of capital equipment which is actually operational is in some cities as much as 100 per cent and in others as little as 22.5 per cent.

Table 4.3

Percentage of fleet working in the reference cities.

While it is no surprise that cities in Australasia, Europe and North America report 100 per cent, they are not alone. Gorahi, Quezon, Belo Horizonte and Managua also achieve this, and Kunming, Bengaluru and Curepipe come close to it.

Notes: NA = not available; NR = not reported; N = no; Y = yes.
Figures in *italic* are estimates. Working fleet/facilities data: Bengaluru – 85% municipal fleet, 100% private fleet; Lusaka – 100% municipal fleet, 70–89% fleet.

Involving the community to plan for solid waste collection in their neighbourhood in Catia la Mar, Venezuela

© Jeroen IJgosse

communicate about what works – and what does not. For citizens, this might mean learning to place their waste in a container rather than throwing it in the street, or to separate materials for recycling. For the collection crews, it might mean learning not to mix everything together, and even being trained to explain new systems to citizens in the course of the route. User engagement, participation and good communications, therefore, are essential for the system to work.

A working collection service means that the professional providers and the household users relate to each other through daily, habitual, solid waste practices. Working together requires permanent and multidirectional communication channels. Some cities think of communication as a kind of advertising campaign that tells the users how to behave. This helps, but it places the users in the position of passive receivers. Active feedback systems and institutions that engage users have been proven to work better over the long term.

■ Creating effective channels of communication between users and providers

Communication fuels a good collection service, as the UK principles clearly suggest. Engaging users and facilitating their communication with the city and the providers is arguably the most important factor for effective waste collection.

The city government is responsible for ensuring that a service is provided, but also needs to ensure that it is a service that their 'customers' (i.e. households, businesses and institutions) will use. Any change in the type of service will probably require that both the users and the service providers change their ideas, modify trusted and established behaviours, and

Table 4.4

Feedback mechanisms and satisfaction levels

Notes: NA = not available; NR = not reported; N = no; Y = yes.
Figures in *italic* are estimates

City	Existence of feed-back systems	Satisfaction levels recorded
Adelaide	Y	Y
Bamako	Y	Y
Belo Horizonte	Y	Y
Bengaluru	Y	Y
Canete	Y	Y
Curepipe	Y	NR
Delhi	Y	Y
Dhaka	Y	N
Ghorahi	NR	N
Kunming	Y	N
Lusaka	Y	NR
Managua	N	Y
Moshi	NR	NR
Nairobi	Y	Y
Quezon City	Y	Y
Rotterdam	Y	Y
San Francisco	Y	Y
Sousse	Y	N
Tompkins County	Y	Y
Varna	N	N

Users cooperate better if they understand why solid waste services are set up in a particular way, and they are in a good position to monitor effectiveness and serve as a source of information as to how the system is actually working. Feedback systems include telephone lines for complaints, continuous or community monitoring of satisfaction and payment rates, and creating collaborative relationships between inspectors and the community.

Compliance and payment behaviour are also forms of communication. People communicate their satisfaction or discontent by obeying or violating the rules for disposal or recycling. They show approval by paying on time, and signal dissatisfaction with the system or the providers by withholding payment or paying too little, too late.

The providers of the service are what make the system work, and communication is also important for them. The people in provider organizations tend to be overworked and underpaid, and they suffer from a low status of their work: there is a tendency to assume that anyone who does 'dirty work' is somehow a 'dirty person'.[14] Under such circumstances, contact with users may seem unwelcome, or a luxury. In the midst of this stress, providers and their staff may forget why they are working and for whom.

A working ISWM collection system thus depends on a high degree of cooperation and trust between users and providers, so any attempt to improve and modernize solid waste services requires that both users and providers change habitual attitudes and learn new behaviours. In order for this process of innovation and mutual adaptation to work smoothly and effectively, there is a need for clear and continued communication, and the information channels need to be maintained.

A specific example of this is related to building recycling programmes, the technical success of which is dependent on users changing their behaviour and following the rules for recycling. When users understand the instructions and separate materials as requested by the

Intermediary's shop buying materials from itinerant buyers, pickers and the general public in Pakistan

© Monsoor Ali

providers, the amount of cross-contamination is reduced and the materials have a higher market value. When the communication system is incomplete or there is a lack of trust between users and providers, the result is often poor levels of separation at the household level. This translates to contaminated materials, which in turn result either in high post-collection sorting costs (the situation in Rotterdam), or low market value when the materials are marketed. Both of these impacts reduce the economic effectiveness of the recycling system and make it vulnerable to political critique or elimination. High degrees of separation reduce this contamination and increase the market value of the materials. These trade-offs are important to consider in the design of recycling programmes.

Box 4.11 Public engagement for enhanced recycling in the UK

Situated on England's south coast, Rother is a typically British municipality. In 2007, it launched a kerbside recycling service to around 35,000 homes, rolling out across the entire district over one month. A local recycling brand was developed and supported by a campaign, carefully choreographed to deliver the right information, in the right way at the right time.

The result was a rapid jump in recycling from 16 to 38 per cent within four months of introducing the new service, exceeding their 2010 target of 32 per cent recycling almost immediately. By the early summer of 2009, that had increased to nearly 50 per cent.

WASTE TREATMENT AND DISPOSAL: FRONT LINES OF ENVIRONMENTAL PROTECTION

Basic issues

Removal of waste from houses and city streets was the main priority of cities' waste management systems for nearly a century, with little or no attention to what was then done with it. The edge of town was usually far enough away, and better still if there was a swamp to be filled. Dumping waste into rivers or the sea was an acceptable strategy, where available. Finding its way into nature, waste enters the food chain and adversely affects ecosystems. The Great Pacific Garbage Patch, a vortex of estimated 3.5 million tonnes of plastic and other waste, covering an area the size of France or larger, is swirling in the north Pacific, causing 'birds and mammals to die of starvation and dehydration with bellies full of plastics; where fish are ingesting toxins at such a rate that soon they will no longer be safe to eat'.[15]

> **Box 4.12 Dandora among the dirty 30**
>
> The Nairobi, Kenya, dumpsite of Dandora was included in the list of the world's 30 most polluted places in a survey by the Blacksmith Institute in 2007. Originally located outside the city, it is now surrounded by heavily populated low-income estates, such as Dandora, Korogocho, Baba Dogo and Huruma, thus affecting the health of large number of residents of these ever-expanding settlements.

Since the emergence of the environmental movement in the 1960s, there is much broader understanding of the health and environmental risks of open dumping and burning, which pollute air and water resources, contaminate soils, and pose health risks to those living near such uncontrolled sites.

In countries where there is a low level of control and a lack of infrastructure, hazardous wastes from hospitals and industry often become mixed with the municipal or household wastes. This dramatically increases the health and environmental impacts from uncontrolled disposal; uncontrolled hazardous waste dumpsites were, indeed, a key driver behind 1970s waste legislation in developed countries, and drums of hazardous waste illegally exported to Western Africa – or dumped overboard in the Atlantic Ocean before even reaching the destination –

A typical controlled dump in South America – only the birds differ per continent

© Jeroen IJgosse

during the 1980s prompted the development of the Basel Convention, which regulates transboundary movements of hazardous wastes.

The environmental impacts of uncontrolled dumping are most acutely felt at the local level. Dumpsites are usually located in or adjacent to poorer communities, where the land costs are lower, and it is politically and socially easier to locate and continue to use these facilities.

In terms of health impacts, the informal- and formal-sector workers on waste disposal sites are on the front line – they are exposed to dangerous substances and face significant health risks. Waste disposal sites can attract dogs and rats, and sometimes also cows, goats and pigs, and these can be a mechanism for spreading disease as well.

Current environmental policy is generally founded on the principles of the 'waste management hierarchy'. The hierarchy is represented in many different ways; however, the general principle is to move waste management 'up the hierarchy', towards reduce, reuse, recycle (the '3Rs') nearer the 'top', diverting waste away from disposal, which is situated at the 'bottom'. The version of the hierarchy in Figure 4.2 emphasizes that a necessary first step is to get on the hierarchy in the first place by phasing out uncontrolled disposal in the form of open dumping.

Even in many developed countries, this first

Box 4.13 A crisis stimulates change

In many countries, a crisis and the political debate that it prompted were responsible for kick-starting the modernization process, which had been under way for years, but moving very slowly.

In 1971, drums of cyanide waste were dumped at an abandoned brick kiln near Nuneaton, UK, leading to a huge public outcry. The ensuing upheaval, along with press coverage of waste disposal drivers taking bribes to dump hazardous waste illegally and a report by the Royal Commission on Environmental Pollution on toxic wastes provided a catalyst for the first ever legislation to control hazardous waste. The consequent Deposit of Poisonous Waste Act 1972 was drafted in only ten days and passed through parliament within a month.[16]

In a similar way, the collapse of the Payatas dumpsite in Quezon City, the Philippines, in 2000, the associated loss of life of 200 waste-pickers, and the outrage in the press and media gave a big impetus to the process of creating a national waste management law, and this event is linked to the passage of Republic Law 9003, the Ecological Waste Management Act.

step was only taken during the 1970s or 1980s. Official statistics for 1990 show that 6 of the then 12 member states of the European Union (EU) were still using uncontrolled landfills, with 3 countries disposing of more than half their municipal solid waste by this route.[17]

The Japanese Ministry of Environment, working with the United Nations Regional Development Commission for Asia, is attempting to move the policy emphasis up the hierarchy in 12 low- and middle-income Asian countries.

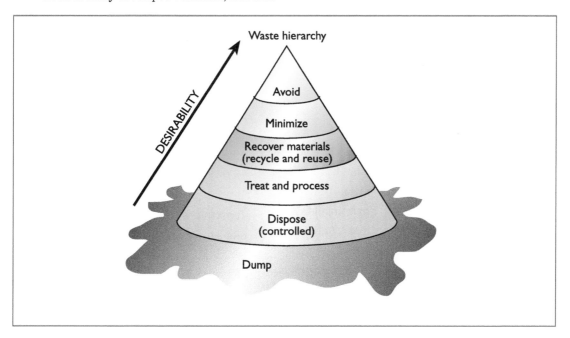

Figure 4.2

Waste management hierarchy

Source: Wilson et al (2001)

KEY SHEET 7
DRAFT REPORT OF THE INAUGURAL MEETING ON THE REGIONAL 3R FORUM IN ASIA, 11–12 NOVEMBER 2009

Choudhury R. C. Mohanty (UNCRD)

1 INTRODUCTION

1   The Inaugural Meeting on the Regional 3R Forum in Asia was organized in Tokyo, Japan, on 11 and 12 November 2009 by Ministry of the Environment of Japan (MOEJ) and the United Nations Centre for Regional Development (UNCRD) with support from the Institute for Global Environmental Strategies (IGES), with the participation of representatives of Asian countries (Bangladesh, Brunei Darussalam, Cambodia, People's Republic of China, Indonesia, Japan, Republic of Korea, Lao PDR, Malaysia, Mongolia, Myanmar, the Philippines, Singapore, Thailand, and Vietnam), international organizations and aid agencies: Asian Development Bank (ADB), Asian Institute of Technology (AIT), Asia-Pacific Forum for Environment and Development (APFED), Asian Productivity Organization (APO), Basel Convention Regional Coordinating Centre for Asia and the Pacific (BCRC China), Global Environment Facility (GEF), German Agency for Technical Cooperation (GTZ), Institute for Global Environmental Strategies (IGES), International Labour Organization (ILO), Organisation for Economic Co-operation and Development (OECD), Japan International Cooperation Agency (JICA), Secretariat of the Basel Convention (SBC), United Nations Centre for Regional Development (UNCRD), United Nations Department of Economic and Social Affairs (UNDESA), United Nations Environment Programme (UNEP), United Nations Economic and Social Commission for Asia and the Pacific (UN ESCAP), and United Nations Industrial Development Organization (UNIDO), and experts in the area of 3R/waste management from around the world.

2   The main objective of the *Regional 3R Forum in Asia* is to facilitate high-level policy dialogues on 3R issues, challenges and opportunities, as well as to provide a strategic and knowledge platform for sharing experiences and disseminating among Asian countries best practices, tools, technologies and policy instruments on various aspects of the 3Rs.

3   Delivering the opening remarks at the Inaugural Meeting, Mr Sakihito Ozawa, Minister of the Environment of Japan, emphasized the elimination of waste – *mottainai* – through the 3R approach. Introducing Prime Minister Hatoyama's Initiative which seeks to reduce CO_2 emissions by 25 per cent below 1990 levels by 2020, Mr Ozawa expressed his hope that the Regional 3R Forum would produce important results to help achieve a low-carbon and sound material cycle society in

the region. From the perspective of addressing global warming, Japan is also promoting a number of policies for 3Rs and effective use of resources. As part of the Hatoyama Initiative aiming to support global warming countermeasures, Japan would further promote the co-benefit approach in the developing countries in Asia, which would achieve both 3Rs of waste and climate change mitigation.

4   Expressing concerns over the rapid urbanization in Asia that has resulted in inadequate urban services such as water supply, sanitation, waste water treatment, sewerage system, drainage and solid waste management, Mr Kazunobu Onogawa, Director of UNCRD, noted the region's significant increase in waste generation in recent years as well as the diversification of types of waste with the growing presence of hazardous and e-wastes in the waste stream. Underscoring the importance of the need to build a climate-resilient society and economy given the fact that the hardest hit from climate change would be the poorer sections of the society, he urged the developing countries in Asia to identify an alternative path of more resource-efficient economic development that would prevent economic decline and environmental degradation.

5   Recognizing the important linkage between the Millennium Development Goals (MDGs) and 3Rs, Mr Muhammad Aslam Chaudhry, Chief of Global Policy Branch, Division for Sustainable Development (DSD), UNDESA, mentioned that the concept of 3Rs was beyond just better waste management and called for the building of an economy based on the life-cycle approach, covering both sustainable production and consumption. The success of 3Rs approach would largely depend on the right mix of policies and programmes implemented at the local level. At the same time, partnerships with business, trade and industry could advance the implementation of 3R concept by:

- facilitating economic development and creating markets around 3R policies;
- providing resources (technology, finance and market);
- developing and disseminating leading-edge technologies and products; and
- supporting corporate 'green' trends.

Expressing hope that the forum and follow-up actions would pave the way for scaling up the implementation of the 3R approach towards sustainable development, he urged that the outcome of the forum should provide meaningful inputs to the discussions that would take place in the current cycle of the Commission on Sustainable Development (CSD) in May 2010.

6   As the overall Chair of the Inaugural Forum, Mr Nobumori Otani, Parliamentary Secretary of the Environment of Japan, delivered a keynote address explaining Japan's policies for establishing a sustainable society by integrating approaches towards low-carbon, natural symbiosis and a sound material cycle society, as well as promotion of the 3Rs in Asia. He urged Asian countries to decouple economic development and environmental impact and shift towards the sound material cycle society by the integration of environment, economy and society through the promotion of 3Rs. The ultimate goal of the forum is to achieve low-carbon and sound material cycle societies in Asia. This will be realized through facilitating bilateral and multilateral cooperation aiming to increase resource and energy efficiency through the 3Rs, to promote environmentally sound management of wastes, and for capacity-building and institutional development in the countries.

Insights from the reference cities and global good practices in waste disposal

The state and status of waste disposal sites that take waste from a city tells a lot about environmental protection there and, consequently, about the state of modernization in the direction of controlling disposal.

In line with general tendencies, information from the reference cities confirms the orientation of US and Australian cities towards landfilling rather than incineration. The Philippines is one of the few countries that has institutionalized this orientation in their Ecological Waste Management Act, banning the incineration of municipal waste.

It is perhaps not coincidental that it is the three reference cities in these countries, Adelaide, San Francisco and Quezon City, which have adopted strong zero waste policies in order to divert as much waste from disposal as possible and to recover and valorize waste materials.

Incineration, in contrast, is the preferred final disposal for Rotterdam and many other European cities. Rotterdam, The Netherlands, incinerates waste that cannot easily be recycled in accordance with The Netherlands legal requirements, which prohibit landfilling any waste for which 'beneficial reuse' options – including reuse, recycling or energy recovery – are available. In this they are using their own interpretation of this law and are currently also opting to incinerate many plastics.

Developments, dilemmas and policy issues that can be observed in Kunming, Varna, Belo Horizonte and the two Indian cities today are similar to those that drove the modernization process in North America and North-Western Europe during the 1980s and early 1990s. Kunming is one representative of Chinese cities that are at the stage of the modernization process where they address environmental issues of adequate waste disposal, before addressing material recovery and waste valorization within their waste management system.

Other reference cities give a mixed picture. While some of them already have state-of-the-art landfills, others are struggling to get on the controlled disposal ladder by closing or upgrading their open dumps; moving from dumping to the hierarchy seems still to be a challenge.

Large cities such as Kunming, China, Delhi and Bengaluru, India, as well as Sousse, Tunisia,

Table 4.5

Waste disposal in the reference cities.

The 'controlled disposal rate' is more or less a parabolic function that goes up as GDP goes from low to high, then peaks, then drops as cities modernize and put increasing emphasis on diverting materials from disposal.

Note: ? = data uncertain. Figures in italic are estimates

City country	GDP, per capita, state-of-the-art (US$) (UNDP, 2007)	Disposed at simple controlled landfills (tonnes per year)	Disposed at dumped disposal sites (tonnes per year)	Generated to controlled disposal (including incineration) of total generated (%)	Lost or illegally dumped (tonnes per year)
Adelaide	39,066	341,691	0	46%	0
Bamako	6855	0	0	0%	198,757
Belo Horizonte	1046	1,136,246	0	88%	1405
Bengaluru	556	1,364,188	*350,000*	65%	209,875
Canete	3846	0	8490	0%	2040
Curepipe	5383	23,764	0	100%	0
Delhi	1046	1,810,035	0	71%	611,317
Dhaka	431	511,000	0	44%	509,248
Ghorahi	367	2200	0	67%	394
Kunming	2432	1,121,463	0	88%	0
Lusaka	953	77,298	0	26%	112,918
Managua	1022	0	376,878	90%	10,950
Moshi	400	0	46,538	0%	6205
Nairobi	645	22,776	*370,110*	3%	262,800
Quezon City	1639	450,020	0	61%	9221
Rotterdam	46,750	245	0	70%	0
San Francisco	45,592	142,330	0	28%	0
Sousse	3425	64,000	0	94%	0
Tompkins County	45,592	22,507	0	39%	0
Varna	5163	74,378	0	54%	*610*
Average				52%	
Median				58%	

have succeeded in attracting capital financing for disposal from multilateral development banks or private investors, either domestic or foreign.

Other cities are on the way to environmentally sound waste disposal due to a political commitment to clean cities and quality of urban living, as is the case in Quezon City, the Philippines, Ghorahi, Nepal, and Belo Horizonte, Brazil. Several of the reference cities have markedly improved – or are in the process of improving – their former waste dumpsites through strong donor support, typically in the form of bilateral cooperation. These cities and donors include:

- Dhaka, Bangladesh, and the Japan International Cooperation Agency (JICA);
- Lusaka, Zambia, and the Danish International Development Agency (DANIDA);
- Managua, Nicaragua, and the Spanish Agency for International Cooperation for Development (AECID).

Nairobi, Kenya, and Bamako, Mali, are still struggling to resolve their waste disposal problems but do not yet have controlled disposal. The Nairobi solid waste system is in an uphill battle to close the open dump of Dandora, currently the city's only option to dispose of its waste. Due to the scale and the extent of pollution, closure of Dandora requires serious political and financial commitment for technical measures to be implemented. Making the situation more complicated is an alternative site identified in Ruai, outside the city limits, which will incur increased costs associated with the longer hauling distance, which system users may not be willing to pay. In the meantime, Dandora continues to spread pollution and fill the headlines.

The situation is Bamako is also similar. There is a site designated for a controlled landfill approximately 30km outside the city limits in Noumoubougou. But the largely uncontrolled dumpsite in the middle of the city is still operating. Development of Noumoubougou, which now seems likely in the coming years, has been

delayed by a lack of clear financing for transfer and disposal operating costs, in spite of the availability of capital investment support. In a sense, the delay in developing this landfill is an indicator of financial sustainability because until operating funds are ensured, the risks are high to build and open it.

Some of the approaches and solutions being adopted in the cities are discussed here:

- phasing out or upgrading open dumps;
- adapting technologies to local conditions;
- reducing greenhouse gas (GHG) emissions through gas capture or methane avoidance.

■ **Phasing out or upgrading open dumps**

Although attention in high-income countries may now be moving on to the activities higher up in the waste management hierarchy by restricting landfilling of untreated municipal solid waste, many cities in low- and middle-income countries are currently working hard on phasing out open dumps and establishing controlled disposal. This is a first step towards good waste management and is designed to pave the way for a sanitary landfill, seen to be an essential part of any modern waste management system.

Over the last 30 to 40 years, development of environmental controls over waste disposal has come to be seen as a series of steps, as represented by the 'stepladder' in Figure 4.3. This

Figure 4.3

Stepwise progression controlling disposal

Source: Wilson (1993)

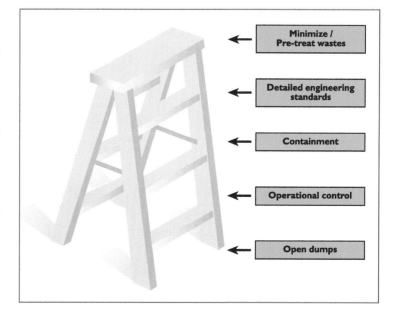

Daily operation of a landfill requires heavy machinery to place and compact the incoming waste

© Jeroen Ijgosse

Placing liners at the bottom of the landfill to prevent leachate infiltration into the groundwater is one of the essential characteristics of a modern (sanitary) landfill, as is the case in Lusaka, Zambia

© LCC-WMU Photo Library, Jan G. Tesink

progression represents the history of the development of waste disposal in many high- and middle-income countries. Step 1, for example, has focused on operational control of the site by organizing the receiving function, dividing the site into cells, compacting and covering the waste, and restricting access, so there is a fence and a gate. Step 2 has focused more on containment, restricting the migration of contaminating substances from the landfill site via leachate or landfill gas. Control features such as bottom liners, drains for surface runoff, leachate collection pipes in drainage layers, and some form of gas vents became common practice.

Step 3 has established detailed engineering standards and gradually increased their stringency, including hydraulic permeability and chemical resistance requirements for liners; drainage and filter functions of leachate collection and removal systems; gas extraction and utilization measures; and others. Step 4 is now moving beyond the landfill itself, diverting wastes up the hierarchy and restricting the range of wastes that can be legally landfilled.

There is extensive experience that intermediate steps can bring about some substantial improvements. The intermediary steps include either upgrading the operations at existing sites or developing new sites that are appropriate and affordable under local conditions. These steps mitigate current and future environmental risks, develop expertise, and provide valuable operational experience.[18,19]

The stepladder is important because it presents incremental measures that can be taken to significantly improve waste disposal at acceptable cost in low- or medium-income countries. Moshi, Tanzania, is an example of a small town in a developing country where decision-makers were not discouraged by the daunting task ahead of them, and undertook the first stage of modernization of their waste disposal. Since 2005, the site at Kaloleni is not a dump any more, but a controlled disposal site, where three Cs are successfully applied as locally affordable measures of environmental protection: confine, compact, cover.

KEY SHEET 8
PHASING OUT OPEN DUMPS

Jarrod Ball (Golder Associates) and Ljiljana Rodic-Wiersma
(Wageningen University and Research Centre)

Open-burning dumpsites are ubiquitous in developing countries. They are generally characterized by dumping of all kinds of municipal waste and by uncontrolled fires. Consequently, they have an adverse impact upon the environment (air, water and soil), causing pollution. They also adversely affect health and the quality of life of the people living in the general vicinity, and, more directly, of waste-pickers present on them. Finally, they constitute an unsightly feature in the landscape.

Based on the foregoing, it is important to phase out open-burning dumpsites in favour of controlled-disposal facilities, even if they do not meet the full engineering standards associated with landfills in developed countries. The internationally accepted approach in this regard is *progressive rehabilitation* to upgrade and phase out dumpsites.

Burning dumped waste next to a ravine, with houses in the background, Catia la Mar, Venezuela

© Jeroen IJgosse

**Waste dump in
Managua, Nicaragua**

© Jeroen IJgosse

The main steps in progressive rehabilitation are as follows:

- Recruit a facility manager responsible for managing the facility. The manager must understand the rehabilitation procedure, and must be able to handle all the stresses associated with the operation and dealing with waste-pickers.
- With the assistance of the landfill manager, initiate a public information and participation exercise to inform the public of the proposed dumpsite rehabilitation. This must include the waste-pickers and must be ongoing throughout the rehabilitation process.
- Provide the requisite resources (e.g. the people required for the above), as well as the machinery to accomplish the following.
- Establish control over vehicle access to the dumpsite (i.e. through only one access point).
- Establish a single working face in an area of the dumpsite that is not burning and establish a road suitable for traffic leading to it.

- Control waste dumping (i.e. allocate and control where loads are dumped and stop end-tipping – the pushing of waste over an extended slope, where the waste is un-compacted and can burn).
- Spread the waste in layers of a maximum 1m and compact as best possible with the machinery available.
- Extinguish fires in other parts of the dump-site by exposing smouldering areas and smothering them with soil (no water).
- Develop a draining system that prevents runoff water from entering the waste body.
- Create an operating plan (as simple as possible) that progressively levels areas of the landfill (always using a single working phase and some degree of compaction).
- Cover deposited waste as best possible with incoming soil, rubble or quenched ash. Vegetate if possible.
- Most important is to negotiate with the waste-pickers throughout the process. They will be most affected by the proposed reha-bilitation and are able to cause major problems on site if they feel that their liveli-hood is threatened. Consequently, they must be made part of the solution. This is achieved by:
    - recognizing the fact that they are on site and are there to stay;
    - formalizing the right for the regular or career waste-pickers to operate on site in a controlled manner;
    - developing a mutually acceptable working relationship that is facili-tated through negotiation between the landfill manager and the recog-nized leader of the scavenger community.

Even though it is a good governance issue, inclusivity can have a direct impact upon the technical performance of a waste disposal site. This point is reinforced by the example of Jam Chakro landfill in Karachi, Pakistan, which was built with donor funds in 1996. The site never really operated as a sanitary landfill and reverted to become an open dump, primarily due to failure to consult and take account of informal waste-pickers.[20] A similar situation is evolving in Sousse, Tunisia, one of the reference cities, which recently closed its dump and opened a new landfill.

Moving from open dumping to controlled disposal has many advantages for other parts of an ISWM system:

- Having managed and staffed *gate controls* enables the segregation of hazardous and non-hazardous waste, both through intervention of the staff and through the potential for direction of any difficult-to-manage wastes admitted to a remote part of the site. This may cost nothing in financial terms but may be the single most important measure to reduce pollution potential of the disposal site and improve occupational safety of workers and waste-pickers at the site.

- *Investing small amounts of money in a reasonable road* to the site will save much in terms of collection and transfer vehicle maintenance and will prolong vehicles' useful life.

- *Diverting waste from disposal through materials recycling and valorization of organic waste*, as discussed further in 'Resource management: Valorizing recyclables and organic materials and conserving resources', will prolong the useful life of the disposal site as well as that of collection vehicles that haul the waste to disposal.

■ **Adapting technologies to local conditions**

A large proportion of the costs of developing waste treatment and disposal infrastructure in high-income countries – and those with advanced modernization processes – is now spent on various engineered controls for environmental pollution prevention. This is reflected in high investment and operating costs. Operating costs for landfills range between 10 and 50 Euros per tonne of municipal waste. Incineration of municipal non-hazardous waste costs between 80 and 200 Euros per tonne, partially due to very high investment costs, in the order of 100 million Euros for a modern incinerator that meets the strict emission standards of the European Union.

All of this poses a challenge for cities in low- and middle-income countries, many of which are already struggling to replace their open dump with a better-performing controlled waste disposal facility. Upgrading may appear to be an impossible and even hopeless undertaking, particularly if 'Western' legislation has been copied, requiring the same advanced technology features as those applied in, for example, Germany or Switzerland. An additional problem is that European donors generally require all new facilities that they support to immediately meet current EU emission standards, which took 40 years to evolve. This leaves no room for phased development and usually discourages cities in developing countries from undertaking any steps at all. Expecting poor countries to switch immediately can act as a barrier to working on the improvement at all. When the investment, operating and maintenance costs of new facilities are prohibitively high, this tends to result in continuing the status quo of open dumping, even after 40 years of focus on environmental protection.

Sometimes, international equipment suppliers, often with funds from their home country, offer cities in other parts of the world subsidized equipment, which is combined with overoptimistic revenue projections – for example, high energy revenues, low operational costs, great market prices for recyclables, or dream-like technical promises. An extreme example is the incinerator that was marketed to a Nairobi private waste collector by a Swedish company in the 1990s, with the 'guarantee' that it would burn garbage and turn it into hundreds of litres of clean drinking water.[21] More plausible claims of energy from

Box 4.15 Failed treatment facilities in India[22]

In 1984, the Municipal Corporation of Delhi, India, built an incinerator to process 300 tonnes per day of solid waste and produce 3MW of power, with technical assistance from Denmark, at a cost of around US$3.5 million. The plant was designed for segregated waste as input, which was not practised by the households or promoted by the municipality. The plant had to be closed down within a week of its opening as the waste had a very low heating value and a high percentage of inert materials.

In 2003, Lucknow Municipal Corporation built an anaerobic digestion plant, as a 5MW waste-to-energy project, to process 500 to 600 tonnes of municipal waste per day at a cost of US$18 million. Private companies from Austria and Singapore provided the technical inputs, while Indian firms supplied the human resources for execution on a build–own–operate (BOO) basis. The plant was not able to operate even for a single day to its full capacity due to the high level of inert materials in the waste and was closed down. The operational difficulties and the ultimate failure were mainly due to the difference between the design assumptions that were based on European waste and waste management practices, and the actual field scenario in India.

Both facilities are landmarks to the failure of imported waste-to-energy technologies in India.

waste merit close examination because there are frequent hidden costs for supplementary fuel, maintenance or parts. Looking back to the Bamako example, one can say that if your city or your private operator can't manage to feed the donkeys, you won't be able to afford fuel for a tractor either. A donation of physical infrastructure does not change the financial and institutional conditions of your city; only focused modernization efforts, capacity strengthening and political commitment can do that.

There are examples in several continents of donor-funded incinerators that have never operated, but have sat for years as a kind of dinosaur in the landscape.[23] There are sanitary landfills built to meet EU environmental standards, which revert to being operated as an open dump because energy costs of the leachate collection system or fuel costs of operations are too high. Another example is the fleet of donor-provided collection vehicles, fitted with tyres of an uncommon size, which were not available locally, so that when the tyres needed replacing, the vehicles could no longer be used.[24] Inappropriate donations and investments are not only a waste of resources, they may also accrue debt to national governments or may break a solid waste organization by loading it up with debt.

Technologies developed in the industrialized countries are designed for their own local circumstances, characterized by high labour costs, high technical capacities and waste rich in packaging materials. An example of such technologies includes collection systems based on mechanical compaction during collection. These 'high-tech' approaches are associated with high investment costs, and depend on skilled maintenance personnel and expensive spare parts to keep them operational and to maintain compliance with the pertinent environmental standards. This is true for state-of-the-art compaction collection vehicles and sanitary landfills; it tends to be even truer for waste-to-energy incinerators, vehicles with advanced electronic control systems, and the many new processing technologies on the market.

A better approach is the other way around, when the characteristics of the waste stream and a good understanding of local conditions form the basis for choosing management strategies and technologies. The high moisture and organic content that make waste in low- and middle-income countries difficult to burn also make it an ideal material for composting, anaerobic digestion, animal feed or direct application to the land. Specific socio-economic, demographic and cultural circumstances are extremely important: in Africa where houses or household compounds often have dirt floors, street sweepings add such a volume of inert materials that direct land application may be more feasible – and more cost effective – than composting. The raw waste benefits agriculture without incurring the cost of managed composting.[25]

Clearly, simply importing European, American or Japanese disposal or incineration technologies to a low- or middle-income country, without considering how they will work under

Box 4.16 Understanding the function of technology

In Ghorahi, Nepal, based on the investigation by the Department of Mines and Geology, which was paid by the municipality, a suitable site was identified where thick natural clay deposits provided the necessary level of environmental protection, probably better than any engineered liner could do. Leachate collection and removal system was installed, as well a natural system for its treatment. At a safe distance from human settlements, the site incorporates a buffer zone consisting of forests and gardens.

local conditions, can be a recipe for environmental and economic disaster. Waste composition is of paramount importance. The developing world is littered by donor-funded Western incinerators that have never worked or require supplementary fuel to burn waste because they had been designed and developed for Western European or North American waste – waste with less moisture and more packaging waste such as plastic and paper than present in developing countries' municipal waste.

One way to approach this challenge is through understanding the properties and functions of the technology currently applied in developed countries, instead of copying their technical specifications. This is particularly relevant for landfills. Part of landfill technology is the aim of reducing emissions to groundwater via leachate. But if a site is available with 20m of naturally consolidated clay, it is better in terms of environmental protection of any groundwater underneath than any engineered liner.

In addition, various tools have become available to assess a possible impact of individual technologies or systems. The relatively new tool of Environmental Technology Assessment (EnTA) developed by the United Nations Environment Programme (UNEP) provides a valuable framework for assessing technology impacts not only on the physical environment, but also on the local social and economic circumstances.[26] ISWM is one of a number of tools and frameworks; but many others are equally useful.

■ Reducing GHG emissions through gas capture or methane avoidance

Improper disposal harms the global as well as local environment. Typical global environmental and climate issues associated with open dumping include:

- air pollution and release of particulates, fine particles and carbon dioxide (CO_2) from open or low-temperature burning;
- release of methane from anaerobic decomposition of organic materials under the surface; and

---

Box 4.17 Beware the 'magic solution' salesman

The Western market for novel waste treatment technologies is proving to be limited, and salesmen, both legitimate and unscrupulous, often target developing and transitional country cities desperate to find an easy answer to a difficult problem. A key message of this Global Report, however, is that there is no 'magic bullet'. The checklist below provides some questions to ask such salesmen and yourself in order to help you evaluate if their technology really is appropriate for your city:

- Is this technology suitable for your waste (e.g. is the heating value of your waste high enough to burn without support fuel)?
- Is the technology being proposed proven elsewhere? If yes, what documentation is there to prove this (i.e. do you wish to be a 'guinea pig' for a new technology)?
- Would the contract proposed require you to meet a specified minimum tonnage of waste? Is this realistic in your current situation? Would it discourage the city's recycling efforts in the future?
- Does the technology meet international emission standards (this is essential for waste-to-energy facilities in order to ensure that air emissions, including carcinogens such as dioxins, do not pose a risk to your citizens)?
- Are the costs both realistic and affordable? Are local markets available for the heat or other products from the facility? If yes, how do you know? If not, are there plans to develop the markets? Who will finance market development?
- Can the plant be run and maintained locally, using local labour and local spare parts?
- Has a suitable site been identified? Which criteria have been used to assess suitability? Will the developer pay for full and independent environmental and social impact assessments to international standards?
- Does your country have the institutional capacity to permit and regulate facility operations?
- Have you sought independent advice, perhaps at your local university, before signing any contract?

- release of waste to surface water, and resulting water pollution that causes release of methane, associated with anaerobic decomposition under water.

The availability of carbon trading schemes makes it possible to finance the improvement of disposal that reduces greenhouse gas emissions. There are at least two accepted methodologies for doing this: composting and landfill gas extraction.

Several of the reference cities have been utilizing, or are in the approval process for, such schemes. Dhaka has an exemplary composting initiative, combined with separate collection of organic waste.

RESOURCE MANAGEMENT: VALORIZING RECYCLABLES AND ORGANIC MATERIALS AND CONSERVING RESOURCES

Resource management is the third physical element of an ISWM system. The term represents a collection of public and private, formal and informal activities that result in *diverting* materials from disposal and *recovering them* in order to return them to productive use. Some products can be reused directly for their original or a similar use, while recyclables are returned to the industrial value chain and organic materials to the agricultural value chain. When there is enough value in these materials to produce an income stream, this document uses the European term *valorization* to cover both sets of activities and materials because it refers to extracting value.

Basic issues

Recycling is perhaps the most misunderstood element of ISWM and it has two 'faces': a commodities value 'face' and a service 'face'. The *commodities value face* is driven by the intrinsic economic value of materials to be found in waste. The origins of waste management are in rag-picking for its value, and the resource value left in waste remains a major driver for private-sector recycling activities. In some of the reference cities (e.g. Kunming and Lusaka), resource management is still a completely separate set of activities, institutions, actors and economic relations, and has little or no relationship to the solid waste system.

The basis for all private-sector recycling activity is 'valorization' of materials. Simply explained, when the owner of an item throws it away, it still has some retained value. For example, a cotton t-shirt includes cotton fibres, which were first grown, then harvested, then processed through a cotton gin to make them pliable, then wound into yarn. The yarn was woven into cotton cloth, which was then coloured, cut, sewed into a t-shirt shape, and finished.

When the t-shirt is no longer useful to its owner, it may be given to a younger or poorer member of the family or the community. If no one wants it, someone else may be willing to reuse it for work clothing and the global trade in used clothing takes advantage of this reuse value. When it isn't useful as a shirt, it can be used as a rag or cut down to make diaper for a baby, which recovers, at a minimum, the value in the materials used to make it. If the cloth is too old or too dirty to be worn or recut, the yarn still has value for making new textiles or for pressing into felt or industrial rags. When the yarn isn't recoverable, the rest can be burned for heat – it was recently reported that poor-quality clothing exported to Romania from Germany was being sold for fuel in wood stoves because the price – 5 Euros per kilo – was less than the locally available wood for burning. Thus, the commodities face of recycling drives most recycling before the onset of solid waste modernization and has since the 19th century.

Similarly, China has a long tradition of materials recovery, both organic nutrients and inorganic materials, as demonstrated by a thriving and effective network of small shops and large 'recycle markets' for metal recovery in the reference city of Kunming. The Chinese government commitment to promoting materials recovery, especially in industrial processes, is again becoming a priority in China, as is confirmed by the passage of their Circular Economy Law in August 2008.[27]

KEY SHEET 9

WASTE CONCERN AND WORLD WIDE RECYCLING: FINANCING DHAKA MARKET COMPOSTING WITH PUBLIC–PRIVATE PARTNERSHIPS AND CARBON CREDITS

A. H. M. Maqsood Sinha (co-founder Waste Concern) and Iftekhar Enayetullah (co-founder Waste Concern)

THE DHAKA MARKET COMPOSTING SYSTEM

Promoting the concept of waste as a resource and putting a market value on organic waste are primary interests of Waste Concern. Working in partnership with communities, Waste Concern operates a waste management system in Dhaka, Bangladesh, that implements a house-to-house waste collection system and collection of waste from vegetable markets. Household and market refuse are taken to a community-based composting plant where it is turned into organic fertilizer. In order to ensure utilization of the fertilizer and sustain the system, it assists communities to market the product by contacting and negotiating with fertilizer companies to purchase and nationally market the compost by-product or 'bio-fertilizers'.

The system introduced by Waste Concern has created a chain reaction among many sectors in Bangladesh. It has expanded the fertilizer industry and has created new entrepreneurs. It is providing jobs to urban poor residents who are hired to do the job of waste collection and processing. It has stimulated behaviour changes in urban communities who have begun to appreciate the value of waste and among professionals to learn how to harness communities in waste management and to experience the impact of converting waste into a resource. Amidst these changes, Waste Concern has helped to address the environmental problems of diminishing topsoil fertility (due to the use of synthetic fertilizers and pesticides) and greenhouse gas (GHG) emissions. A good indicator of the success of Waste Concern is the government's inclusion of composting and recycling in the National Safe Water and Sanitation Policy.

THE CHALLENGES THAT CREATED THE SYSTEM

The management of an increasing volume of solid waste in urban areas has become a serious prob-

Compost plant located at Bulta

©Waste Concern

lem in Bangladesh. Intensifying economic activities due to increasing urbanization and rapid population growth are contributing to the generation of 17,000 tonnes of urban waste per day nationwide. The World Bank predicts that in 2025 Bangladesh will generate 47,000 tonnes of waste daily in urban areas.

In Dhaka, 3500 tonnes of waste are generated per day, of which 80 per cent is organic. But Dhaka City Corporation (DCC) can collect only 50 per cent of the waste, which is disproportionate to the amount of budget used for collection, transportation and disposal using tax payers' money. At a collection rate of 50 per cent, the city is unable to take care of the additional increases in the city's waste. As a result, more uncollected waste is piled up on the roadsides or dumped in open drains and low-lying areas, further deteriorating the environment and the quality of life in the city. However, while almost 80 per cent of the waste is organic and can be converted to compost or soil conditioner, this potential of waste as a resource is unseen and the new resource is unutilized. Opportunities for developing partnerships between the government and other stakeholders in waste management who will engage in composting or recycling to reduce waste are not explored because of the absence of a waste management policy. Practices of waste segregation at source or at the household level, and waste reduction, reuse and recycling at source are unknown.

THE INNOVATION IN FINANCING

Recently, Waste Concern helped Bangladesh to harness a new opportunity of foreign direct investment using the Clean Development Mechanism (CDM) of the Kyoto Protocol by successfully developing a city-scale composting project to reduce GHG emissions while improving the environmental condition of the disposal site.

OBJECTIVES OF THE CLEAN DEVELOPMENT MECHANISM (CDM) PROJECT

The CDM project has been designed to:

- develop a sustainable model for solid waste treatment based on recycling;
- establish a large-scale composting plant for the resource recovery of organic wastes from households and vegetable wholesale markets of Dhaka City;
- develop an alternative solid waste management system to reduce the burden on the municipality, especially on landfills;
- create job opportunities for the urban poor, especially women and waste-pickers; and
- save hard currency at the national level and strengthen the trade balance by substituting, in part, chemical fertilizer with locally produced compost and enriched compost.

BRIEF DESCRIPTION OF THE PROJECT

The project as submitted to the United Nations Framework Convention on Climate Change (UNFCCC) is called Harnessing CDM for Composting using Organic Waste. It is a joint venture project of WWR Bio Fertilizer Bangladesh, Ltd and Waste Concern, which is the first compost project registered successfully with UNFCCC and the first organic waste recycling project in the world to earn carbon credits. The project is anchored on a 15-year concession agreement between Dhaka City Corporation (DCC) and WWR Bio Fertilizer Bangladesh, Ltd that was signed on 24 January 2006. (WWR Bio Fertilizer is a joint venture company of Waste Concern and World Wide Recycling BV, a Dutch company).

The significant features of the project as described in the terms of the concession agreement are as follows:

- WWR Bio Fertilizer Bangladesh, Ltd has the exclusive right to collect 700 tonnes of organic waste every day from different points of Dhaka City.
- Three compost plants will be established around the city. The first plant, which commenced construction on 25 November 2008, has a 130-tonne-per-day capacity. It is located in Bulta, Narayanganj (25km south-east of Dhaka City).
- Vegetable waste from the market is collected using the project's own transport networks and taken to a compost plant that is built on land owned by the project.
- The composting system will be capable of reducing 47,000 tonnes of urban waste that will be produced by Dhaka by 2025, as predicted by the World Bank.

THE PUBLIC–PRIVATE PARTNERSHIP (PPP) ELEMENTS OF THE PROJECT

The project is not a conventional public–private partnership (PPP) type because it does not involve a government agency as partner that shares the profits as well as the risks. It may be categorized better as a public–private cooperation project. The participation of the government is through the DCC, which has granted a concession to the private company WWR Bio Fertilizer Bangladesh, Ltd to collect and process waste. WWR Bio Fertilizer Bangladesh, Ltd will self-finance its collection and processing activities. It will procure vehicles to transport waste and build composting plants. There is no investment on the part of the DCC. On the other hand, WWR Bio Fertilizer Bangladesh, Ltd has Waste Concern and its Dutch partners – World Wide Recycling BV, FMO Bank and Tridos Bank – from The Netherlands as joint venture partners.

The CDM Project is a 700-tonne-per-day capacity compost plant that will produce compost fertilizer and improve soil in Bangladesh. At the same time, it will earn carbon credits from its capacity to reduce methane, a GHG, from the landfill. It was initiated by Waste Concern together with its Dutch

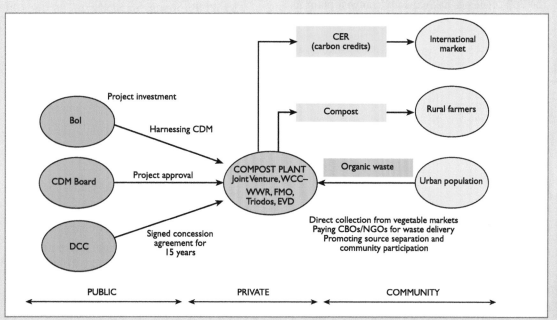

Figure K9.1

How the Clean Development Mechanism project uses public, private and community cooperation

Note: The Designated National Authority (DNA) of Bangladesh was established in 2004. Waste Concern submitted its PIN and PDD to the DNA on 29 February 2004. This project was approved by the DNA of Bangladesh on 8 August 2004 and registered with the UNFCCC on 18 May 2006.

partners who developed the methodology to account for methane reduction using aerobic composting technology, which was approved of by the Expert Methodology Panel of the Intergovernmental Panel on Climate Change (IPCC) of the UNFCCC. This methodology has opened a new channel to bring carbon financing to waste composting projects in developing countries. Using this methodology, Waste Concern and its Dutch partners were able to get the project approved by the CDM board of the Designated National Authority (DNA) of the government of Bangladesh, and later had it successfully registered with UNFCCC. As a joint venture project of Waste Concern and its Dutch partners, it was also registered with the Board of Investment (BoI) in Bangladesh. The project resulted in a 12 million Euro investment in Bangladesh.

THE CARBON FINANCING SET-UP

WWR Bio Fertilizer Bangladesh, Ltd, a joint venture company, has four major equity share holders: Waste Concern (Bangladesh) and World Wide Recycling BV, FMO Bank and Triodos Bank (The Netherlands). Total investment in this joint venture is 12 million Euros in a few phases for the 700 tonne per day capacity compost plant for

Temperature monitoring at Bulta

© Waste Concern

Dhaka City. Out of a total 12 million Euros, the mode of finance is as follows:

- 38 per cent of project cost: 4.6 million Euros as equity;
- 45 per cent of project cost: 5.4 million Euros as a soft loan from FMO Bank and Triodos Bank;
- 17 per cent of project cost: 2 million Euros as a local loan from a local bank in Bangladesh.

CURRENT STATE OF AFFAIRS AND EXPECTED RESULTS

WWR Bio Fertilizer Bangladesh, Ltd took the initiative to establish three large decentralized compost plants around the city. Total capacity of the composting plants is 700 tonnes per day of organic waste from the DCC area. The agreement between WWR Bio Fertilizer Bangladesh, Ltd and DCC mentions that the capacity of the compost plants will be increased gradually. The first 130-tonne-per-day compost plant was in operation by 25 November 2009.

The stages of increase are as follows:

- first year from the date of commencement, by 2009: reach capacity of up to 100 tonnes per day;
- second year from the date of commencement, by 2010: increase capacity up to 300 tonnes per day;
- third year from the date of commencement, by 2011: increase capacity up to 700 tonnes per day.

This project is an innovative model for waste recycling. WWR Bio Fertilizer Bangladesh, Ltd collects organic waste free of charge from the markets, using its own transportation network. The project makes profit from the sale of compost fertilizer, which is sold to farmers. In this way, the waste is no longer left behind in landfills

Box K9.1 Compost plant in Bulta, Narayanganj – a capacity of 130 tonnes per day

Basic information

- total plant area: 14,744 square metres;
- employment creation: 90 people;
- organic waste recycling capacity: 130 tonnes per day;
- production capacity: 32 to 39 tonnes per day;
- greenhouse gas (GHG) emissions reduction: 15,600 tonnes CO_2e per year;
- landfilling avoided by Dhaka City Corporation (DCC): 52,195 cubic metres per year.

Special features

- 100 per cent on-site wastewater recycling;
- rainwater harvesting from total roof and hard surface area;
- daycare centre for female staff;
- free meal for workers;
- health insurance for workers.

where it causes methane emissions and water pollution. It provides a solution to the increasing amount of waste in the city and at the same time receives carbon credits for its ability to reduce methane emissions at the landfill site.

The salient facts of the project are that it:

- collects waste from the DCC area at 700 tonnes per day in 3 phases;
- produces compost at 50,000 tonnes per year;
- reduces methane gas emissions at 89,000 tonnes of CO_2e per year;
- provides jobs to 800 urban poor residents;
- has a total project cost of 12 million Euros.

The first compost plant located at Bulta, Narayanganj, has a capacity of 130 tonnes per day. Some basic facts on this plant are highlighted in Box K9.1.

This project has pro-poor elements. It is not fully mechanized so that it allows opportunity to employ more people from the informal sector. A salary rate that is comparative with government rates and good working conditions is offered to the workers along with health insurance, a daycare facility and free meals. Compost produced from the initiative, which is cheaper

than chemical fertilizers, will help poor farmers to improve the health of their cultivable soil and can expect higher production. Finally, apart from improvement of the environment, the project helps municipalities to reduce their budget for waste management.

Project period considered 10 years (2009–2019)	
Total investment	12 million Euros
Equity	4.6 million Euros
Total compost sale	27.24 million Euros at 60 Euros/tonne
Total certified emissions reduction (CER) sale	9.76 million Euros at 13 Euros/tonne
IRR	16% with CER
IRR (without CER)	10.4%
Payback period	7 years

Table K9.1

Impact of carbon financing on total investment

Note: IRR = internal rate of return

In a CDM-based compost project, there are two major sources of cash flow. One is sales proceeds from compost and the other is certified emissions reduction (CER). From 1 tonne of organic waste, 20 to 25 per cent of compost can be produced and the price per tonne at the factory gate is 60 Euros per tonne. The financial implications of carbon financing on the total investment and cash flow of the system is illustrated in Table K9.1.

WHAT HAS GONE WELL? WHAT IS CHALLENGING OR STILL TO BE ACCOMPLISHED?

As of November 2009, the compost plant at Bulta was processing 70 tonnes of organic waste every day, planning to reach processing capacity of 90 tonnes per day in December 2009.

The project has several success factors and good examples:

- It is able to create a demand for compost in Bangladesh, which can improve soil conditions and ensure food security.
- The national agriculture policy of the country considers compost as necessary to improve soils.

Compost bagged to be marketed by private sector

© Waste Concern

- One of the world's largest marketing companies has signed a contract with WWR Bio Fertilizer Bangladesh, Ltd to market compost all over Bangladesh.
- Local banks are now interested in investing in similar waste-related projects, which is a good indicator of the success of the project.
- The project has shown that even without support from the government, the private sector can invest in solid waste-related projects and harness carbon funding to make the project attractive for financers.
- The government of Bangladesh, convinced of this project's good example, has taken the initiative to replicate this model all over the country with its own funding and by using the CDM approach.
- In contrast to the municipal system, the CDM approach promotes transparency and good governance since each step of its process is documented and properly monitored. In this project, for instance, monitoring equipment is installed in the compost plant, such as an electronic weigh bridge to keep a record of incoming waste, a gas meter to gauge oxygen, a thermometer to record temperature, a moisture meter to record moisture content.

EARLY CHALLENGES IN THE PROJECT

The project faced a number of challenges during its initial stages. The first was the anxiety-filled challenge of convincing policy-makers, engineers and bureaucrats about the benefits of the CDM and the opportunities represented by carbon trading. It was challenging to prove that aerobic composting of organic waste reduces methane emissions, a greenhouse gas. At this time, there was no methodology available from the UNFCCC to prove that aerobic composting can mitigate methane emissions. So Waste Concern, along with its Dutch partner World Wide Recycling BV, developed a methodology (AM0025) which showed that aerobic processing of composting does not generate methane gas.[1]

Second, a Designated National Authority (DNA) for CDM, which is necessary for project approval, did not exist in Bangladesh. To facilitate the establishment of the DNA, Waste Concern approached the United Nations Development Programme (UNDP) to assist the Ministry of Environment and Forests (MoEF) to set up the DNA.

Third, a most challenging part was getting the DCC to sign the concession agreement. Although the project had been approved by the DNA, and the DCC actively participated in the approval process, some DCC staff and officials who had acquired a vested interest in the management of the city's solid waste were against the project and openly opposed it. Prior to the agreement, the operation of the municipal waste management system was not transparent and not properly documented. There were issues related to 'ghost' labour, false trips and pilferage of gasoline for trucks used in waste collection and other acts of malfeasance that were not reported. But under the current agreement, these petty acts of graft and corruption have been eliminated since WWR Bio Fertilizer Bangladesh, Ltd undertakes the collection and recycling of waste every day without any cost to the DCC.

Fourth, due to a new regulation on compost standards and field trials introduced by the Ministry of Agriculture in 2008, the project had to wait for 12 months to get a licence from the government to market compost after the commencement of its first compost plant at Bulta, Narayanganj. Due to this delay, the project could not reach a capacity of 100 tonnes per day on 25 November 2009, as targeted, Starting with a capacity of 50 tonnes per day, instead, by 28 November 2009 the project was able to process 70 tonnes per day of organic waste and aimed to reach 90 tonnes per day in December 2009. By November 2011, the project is expected to run with full capacity of 700 tonnes per day.

And, fifth, 56 permits were required from different government agencies and departments, causing further unnecessary delay. For example, under the new regulations, before marketing, the compost has to satisfy the prescribed standard of the government and has to be tested in different crops for field trials.

LESSONS LEARNED AND REPLICATION

Organic waste, which is commonly generated by towns and cities of developing countries, can be converted to compost without any form of investment from the government by using the carbon financing scheme or the Kyoto Protocol's CDM approach. The scheme helps to overcome technological and financial barriers in waste management.

Carbon financing can open a new window of opportunity for poor cities to attract investment in waste management and promote public–private partnership or cooperation. The CDM allows the private sector to invest in collection, transport and disposal of waste, saving the government considerable overhead and management costs. It gives confidence to investors since

the project is endorsed by the government and the UNFCCC. It makes a waste-based project attractive to investors because it reduces the payback period.

CDM projects can be pro-poor. In small- and medium-sized towns, programmatic CDM will be appropriate for small-scale projects. In addition, there is an opportunity for bundling small-scale compost projects through the CDM.

However, a 'one-stop' approval process for CDM projects in Bangladesh will be necessary to reduce, if not eliminate, delay in project implementation. Furthermore, the CDM process being implemented by the UNFCCC has to be simplified for developing countries so that the transaction cost will be affordable.

There is also a need to inform government as well as private-sector officers and staff on CDM and carbon-trading initiatives.

IS THE PROJECT REPLICABLE?

Yes, the project is replicable. Currently, 47 replications of this model have been carried out by other groups (government, NGOs and the private sector) in 26 Bangladeshi towns. Recently, the United Nations Economic and Social Commission for Asia and the Pacific (UN ESCAP) has begun promoting Waste Concern's community-based composting model in Matale City in Sri Lanka and Quay Nhon City in Vietnam, which is now planned to be scaled up using carbon trading. Still more recently, a number of South Asian countries (such as India and Pakistan) are adopting the methodology (AM0025) developed by Waste Concern and its Dutch partner.

NOTE

1 Link to AM0025: http://cdm.unfccc.int/Projects/DB/SGS-UKL1134142761.05/view.

KEY SHEET 10
BUILDING THE PRIVATE SECTOR AND REDUCING POVERTY THROUGH SUSTAINABLE RECYCLING IN SOUTH-EASTERN EUROPE

Valentina Popovska (International Finance Corporation)

In 2005, the International Finance Corporation (IFC) initiated a three-year long Recycling Linkages Programme (RLP) in the area of recycling (paper, plastics and metal) and solid waste management across the economies of the Western Balkans (Albania, FYR Macedonia, Serbia, Bosnia and Herzegovina, Montenegro and Kosovo). The recycling industry was selected due to its ability to foster strong economic growth of the private sector while creating sustainable livelihoods among marginalized groups working in the sector. The RLP was jointly funded by the IFC and the Republic of Austria, Federal Ministry of Finance.

The ultimate objective of the programme was to improve the performance of the recycling industry in the region by creating positive economic, social and environmental impacts, resulting in significant increases in the volume of collected and recycled scrap across the region. The programme activities focused on:

- improving the regulatory environment by addressing the recycling industry in the region;
- strengthening operational capacity and access to finance of the private sector involved in recycling operations, including individual waste-pickers; and
- increasing public awareness on the benefits of recycling.

As a result, the RLP has:

- introduced waste packaging and waste electrical and electronic equipment (WEEE) regulations to local governments;
- trained 260 companies;
- assisted 80 companies in increasing their recycling operations, on average, by 20 per cent; and
- positively affected over 5000 suppliers of recyclable raw material, including over 4000 individual waste-pickers.

Poster for collection of electronic appliances as part of implementing the WEEE regulation in Galati, Romania

©WASTE,
Anne Scheinberg

By working with the private sector during the past three years, the RLP has influenced 37 per cent of the regional recycling industry, or the equivalent of 660,000 tonnes out of 1.8 million tonnes of recycled scrap material in the Western Balkans. The overall volume of recycling in the Western Balkans (only metal, plastic and paper streams) has grown by 72 per cent since 2005. The RLP's advisory programme has led to 30 million Euros in equity and loan investments, including 180,000 Euros in micro-loans to over 100 micro-businesses in the region. Finally, the programme has contributed to the reduction of current and future greenhouse (GHG) emissions achieved through increased recycled volumes in the region.

Throughout the implementation, the Recycling Linkages Programme has evolved in activities or pillars that addressed a wide range of issues, working from the street to the policy level. Operating through the lens of the private sector leads to strong development impacts upon a wide spectrum of key development target areas.

Although the RLP cannot influence the market dynamics for the different recyclable materials, or be the sole reason for increases in the percentage of recycling, it brings significant impacts to the fundamental growth drivers – namely, policy framework, business confidence, investment and professional capacity. The programme enabled an increase in quantities of materials extracted from the waste stream and a rise in the level of recycling, while reducing the amounts of waste requiring disposal and consequent emissions of greenhouse gases. This led towards the achievement of Millennium Developmental Goals (MDGs) and encouraged

Carton collector in Serbia
©WASTE, A. Scheinberg

corporate social responsibility and environmental protection, provided livelihood opportunities and created jobs for poor and marginalized people.

The IFC has continued its recycling advisory initiative in Southern Europe under the Integrated Solid Waste Management Programme (ISWMP), a two-year programme (2009 to 2010) jointly funded with the Republic of Austria, Federal Ministry of Finance.

The objective of the ISWMP is to encourage processes driving the improvement of solid waste management practices in the Western Balkan region by applying integrated solid waste management concepts based on sustainable waste collection, waste treatment (waste reuse, recovery and recycling), and reduction of waste disposed of at the landfill sites, resulting in abatement of GHG emissions.

The ISWMP focus is on improving waste management practices at municipal level, promoting private-sector participation in the waste management sector and supporting enforcement of the necessary waste management regulations and policies.

Individuals and micro-enterprises extract materials from waste and sell them to junk shops, which sell them to dealers or exporters, who, in turn, sell them to factories where they are inputs for new production. Recycling thus feeds many of the world's industrial supply chains. In most cities the recycling industry includes individual entrepreneurs, as well as micro-, small, medium, large and multinational private businesses, all of which upgrade, trade and further upgrade recovered materials.

The *service face* of recycling is relatively young: since the 1970s, *municipal recycling* has evolved and come of age as a series of strategies to combine waste collection with selective recovery of some of the materials in the waste stream. Municipal recycling is motivated partly by the commercial value of the waste materials, but far more by the *sink value* of the waste absorption capacity offered by the private recycling industry.

The term *sink* is used to explain the economic and environmental value of absorption capacity in nature, and the English name tells us something about its origin. A typical sink is a wetland or low-lying area, which can receive and biologically process waste materials. Sinks work by breaking down complex substances into simple elements. Forests are a sink for CO_2 because the trees and other plants use it in the process of photosynthesis. In this sense a modern disposal facility is an engineered sink, which is designed to safely absorb waste materials and control their release to – and contact with – the natural world.

Increasingly, cities in developed countries look at the recycling and agricultural supply chain as an economic sink: materials that go to an e-waste dismantler or a paper stock dealer don't have to go to the landfill or incinerator, and that saves money. This motivation can be so important that cities are willing to pay the recycling industry to take and process materials that can technically be used but have little or no economic value. When cities and towns upgrade disposal, the increasing costs make the sink function of recycling seem attractive, combined with the potential to earn income on high-value materials. A growing focus on recycling actually shifts the focus of waste management.

Cities and towns want to recycle because it keeps the waste out of their newly upgraded disposal facilities. This *diversion* has a value to the entire waste system. Modern recycling serves the entire waste management system by becoming a 'sink'.

While recycling is a win–win approach, there are also sources of conflict. Local authorities and users interested in the sink value work on recycling with the goal of improving environmental performance and conserving resources. This often brings them into competition with informal and formal private recyclers, who are focused on the commodity face and the value of traded materials. Current conflicts in Delhi between the informal recyclers and the newly privatized formal waste management system illustrate many of the issues associated with modernization and recycling.

Quantifying the contribution of the informal sector to waste management has been difficult[28] – few developing and transitional country cities have good statistics on the formal sector, and there is even less about the informal sector – which by definition tends not to keep written records. A recent Deutsche Gesellschaft für

Box 4.19 *Economic Aspects of the Informal Sector in Solid Waste,* **2007**[29]

In 2006, the Deutsche Gesellschaft für Technische Zusammenarbeit (the German Technical Cooperation, or GTZ) financed a study entitled *Economic Aspects of the Informal Sector in Solid Waste*, co-financed by the Collaborative Working Group on Solid Waste Management in Low- and Middle-Income Countries (CWG) (GTZ/CWG, 2007). Six cities formed the focus of the study on relationships between formal and informal solid waste activities. The cities, Cairo, Egypt; Cluj-Napoca, Romania; Lima, Peru; Lusaka, Zambia; Pune, India; and Quezon City (part of Metro Manila), the Philippines, represented five continents and ranged in size from 380,000 to 17 million people. In only these six cities:

- More than 75,000 individuals and their families are responsible for recycling about 3 million tonnes per year.
- The inputs from the informal private recycling sector to the recycling supply chain, and therefore to the economies in the six cities, have a value of more than US$120 million per year.

In addition, in Lusaka, more than 30 per cent of the city's waste collection service is provided by unregistered informal collection service providers.

Technische Zusammenarbeit (GTZ) project has changed that, providing for the first time specific data on both waste and money flows in the waste management (combined formal and informal) systems in six cities. This Global Report builds on that study, which is summarized in Box 4.19 and in more detail in Key Sheet 11 (overleaf).

Another problem can occur when a new landfill is designed for too much waste. This results in what is called in the US 'flow control'. It means that a local authority can require all waste to go to a specific disposal facility and disallow all forms of recycling to ensure that enough waste arrives at the disposal site. While flow control is more common in relation to waste-to-energy incinerators, Sousse shows this situation in relation to its new landfill. Much waste that was being informally recycled is not going to the new landfill.

A third issue at the intersection of resource management and financial sustainability occurs when the costs associated with using recycling as a sink appear to be too high. In some places, high recycling costs (often associated with poor design of collection routes or incomplete market-

ing strategies) can lead to a global discussion about whether it is 'worthwhile' for municipalities to invest in recycling or the separate management of organic waste. The answer is not so clear; but it is clear that local authorities have a stronger commitment to recycling when both 'faces' are represented and important.

In 2008 the global economic crisis reduced prices for many recyclable materials, especially globally traded materials such as paper and metals.[30] Confronted with a loss of their revenues, many cities and towns stopped their municipal recycling activities; the sink value wasn't enough to keep the activities going. Newspapers reported that the municipalities stopped recycling or decreased their efforts because they weren't seeing the economic benefits.

In contrast, studies by Women in Informal Employment: Globalizing and Organizing (WIEGO), Chintan-Environmental in India, Asociación Centroamericana para la Economía, la Salud y el Ambiente (Central American Association for Economy, Health and Environment, or ACEPESA) in Central America and CID Consulting in Egypt suggest that the informal recyclers tended to do the opposite. They recycled more because:

- less was taken by municipal recycling initiatives;
- they needed to recycle more kilos per day to make the same amount of money to support themselves and their families.

High costs point to another issue: labour costs also go down in crises, and this makes disposal less expensive as well. It may be that the use of recycling as a sink is, then, less attractive; as a result, during crises, local authorities may choose to spend money on healthcare or education, rather on the environment. These questions are asked more loudly in situations where disposal is relatively inexpensive, and paying the recycling industry to be a sink adds to total system costs.

KEY SHEET 11

KEY INSIGHTS ON RECYCLING IN LOW- AND MIDDLE-INCOME COUNTRIES, FROM THE GTZ/CWG (2007) INFORMAL-SECTOR STUDY

Sandra Spies (GTZ) and Anne Scheinberg (WASTE)

Until recently, there has been little attention paid to the similarities and differences in recycling performance in high-income countries as opposed to low- and middle-income countries now actively modernizing. Western European and North American countries saw rapid development of so-called 'municipal recycling,' driven by the high cost of landfilling and a need to divert waste from disposal, during the 1980s and 1990s, with the result that leading recycling cities are achieving between 50 and 70 per cent recycling.

Generally, the emphasis on improving waste management in most cities in low- and middle-income countries focuses first on increasing collection percentages and then, usually with donor or central government funding, on improving disposal. This is to be expected in the sense that the driver for public health is usually dominant in these countries; the driver for environment comes later and has less priority,

and resource management and related cost savings are interesting as solid waste strategies only after disposal has been modernized and regionalized. Before this, initiatives for separate collection systems for organics and recyclables are usually only included as an environmental project of a non-governmental organization (NGO) or community-based organization (CBO), or in response to donor or international NGO interest, and the informal recycling sector is seldom considered and rarely understood as part of such environmentally driven initiatives. A concrete example is the Philippines, where buying and selling of scrap metal goes back to World War II, when considerable metal waste materials from the American camps were being given away to locals who used them for all types of construction and for furniture-making. Enterprising locals then started to set up scrap-iron shops. The thriving junk shops of all types of recyclables today owe their origin to the scrap-iron shops. But the practice was never associated with recycling and resource management, only as simple 'buy-and-sell' livelihood until the mid 1990s when government began to recognize and regulate recycling as part of the solid waste management system. Some cities in low- and middle-income countries join zero waste initiatives and work on policies to support recycling; but without understanding that their situations and baseline conditions are somewhat different, these initiatives are not usually very successful.

Trading centre for sorted wastes, Philippines

© SWAPP

	Tonnes/year generated	Tonnes/year generated but not entering solid waste system	Percentage not entering solid waste system	Materials entering solid waste system (tonnes/year)	Tonnes initially entering formal waste and recycling	Tonnes initially entering informal waste and recycling	Estimated residential generation	Daily generation per capita (kg/capita/day)
Cairo	3,454,996	177,336	5	3,277,660	810,677	2,567,142	2,865,378	0.65
Cluj	194,458	33,981	17	160,477	145,779	14,575	163,085	0.70
Lima	2,725,424	37,349	1	2,688,074	1,839,711	848,364	1,956,228	0.72
Lusaka	301,840	91,437	30	188,890	90,720	98,170	245,996	0.52
Pune	544,215	17,885	3	526,330	394,200	132,130	369,745	0.33
Quezon City	623,380	10,135	2	613,245	489,606	141,831	380,261	0.25

Table K11.1

Citywide generation of solid waste in the six GTZ/CWG (2007) cities*,[1]

Note: * Amounts generated and entering both systems are not add to totals because of losses and rounding. In two cases, generation figures are established by utilizing coefficients for generation per household and per business or per capita multiplied by data regarding number of households, businesses and population within the cities. In other cities, the generation came from existing city data.

Source: GTZ/CWG (2007)

The specific situation in relation to recycling is one of the focus areas in the draft report *Economic Aspects of the Informal Sector in Solid Waste* during 2006 to 2007 (GTZ/CWG, 2007). The integrated sustainable waste management (ISWM)-structured research conducted offers a closer look at recycling and organic waste management in six low- and middle-income countries. Its focus on the interaction between formal and informal sectors suggests that understanding the activities of informal recyclers is perhaps the key ingredient for successful recycling and organics recovery in low- and middle-income countries. Some of the study's results that are relevant for conceptualizing, planning and implementing (new) formal recovery systems are presented here, supported with GTZ/CWG (2007) data.

The first set of insights relates to recycling and organic waste recovery in terms of the waste stream itself. Table K11.1 shows that the waste generated per capita is low, ranging from one quarter to three-quarters of a kilogram per person per day. Middle-income cities generate more, low-income cities less. This is not a surprise; but it is good to think about what it means: there is less waste and therefore less to be captured by recycling and organics recovery.

Looking at Table K11.2, it is possible to see how much of the waste could be recovered in the six cities. Low- and middle-income countries have relatively high percentages of organic wastes and varying percentages of recyclables in the household waste stream. In spite of their differences, the relative proportions of categories such as organics, glass and metal show significant similarities. The organic fraction is above 45 per cent (weight) in all cities and represents the best opportunity for diverting waste from disposal. In all of the cities, recovery of non-organic waste would reduce the total waste requiring disposal by at least 20 per cent and in Quezon City by as much as 40 per cent. Compared to two developed country cities, Rotterdam in The Netherlands and San Francisco in the US, percentages of paper, glass, metal and plastics are comparable; percentages of organics are significantly higher. This suggests that organics should be a priority for recovery in low- and middle-income cities.

The first thing to notice is that in these cities, the informal sector does most of the recycling and valorization of organic waste, and the

City	Paper %	Glass %	Metal %	Plastic %	Organics %	Other %	Total %
Cairo (reported 2006)	18	3	4	8	55	12	100
Cluj	20	5	3	8	50	14	100
Lima	13	2	2	11	52	20	100
Lusaka	9	2	2	7	40	40	100
Pune	15	1	9	13	55	7	100
Quezon City	17	3	3	16	48	13	100
Simple average	15	3	4	11	50	16	100
Rotterdam (2007)	7	2	1	0	4	85	100
San Francisco (2007)	24	3	4	11	34	21	100

Table K11.2

Composition of waste in the six GTZ/CWG (2007) cities and in cities in the US and The Netherlands

Source: GTZ/CWG (2007) project data, available from www.GTZ.de/recycling-partnerships

In many countries the informal collection plays a role in the material flow, here waste-pickers at work on a dumpsite in Romania

©WASTE, Ivo Haenen

Table K11.4 examines in more detail what amounts are recovered by different informal-sector occupational categories.

The informal recycling sector is diverse and has many occupations, some of which have parallels or competing occupations in the formal solid waste sector. In general, the informal recycling collection sector can be divided into five main types of activities and roles:

formal *solid waste* sector relatively little. In the IFC Recycling Linkages Programme in the Western Balkans, the estimate was more than 40 per cent of all recyclables entering the recycling supply chain via informal activity.[2] In three of the six cities in the study, the informal sector recovers about 20 per cent of everything that enters the waste stream. In Cairo, in 2006, this was substantially more because the Zabbaleen informal sector was at that time collecting almost all waste, and extracting large portions of organic wastes for swine feeding. The two cities in which the informal sector recycles less than 20 per cent are Lusaka and Cluj, both in regions with weak recycling markets and a relatively small informal recycling sector.

1 *Itinerant waste buyers* (IWBs) move along a route and collect recyclables from households (or businesses). In all Southern cities there is usually a payment made or something bartered for these materials, while in Cluj the household gives them as a 'donation'.

2 *Street-pickers collect* materials that have already been discarded by households. In some cases street-pickers extract materials from household waste set-outs, breaking bags, and/or picking up reusables or materials waiting for formal collection. In other cases street-pickers remove materials from dumpsters or community containers or secondary collection sites.

3 *Dump-pickers* work and often also live on the landfill or dumpsite, and sort through the waste as it is disposed of there.

4 *Truck-pickers* are informal members of formal-sector waste collection crews and ride with the trucks. They inspect the waste as it is loaded onto the truck and separate out valuable items for sale. Sometimes truck-pickers are actually paid members of the formal-sector work crew; but they may also be outsiders who have gained the right to work along with this crew.

Table K11.3

Material recovered by the formal and informal sector

Source: GTZ/CWG (2007) project baseline workbooks

	Material entering solid waste system (after household losses)	Tonnes recovered by formal recycling	Tonnes recovered by informal recycling	Percentage of total recycled, formal	Percentage of total recycled, informal
Cairo	3,277,660	365,724	2,161,534	11	66
Cluj	160,477	8879	14,575	6	9
Lima	2,688,074	9380	529,370	0	20
Lusaka	188,890	11,667	5419	6	3
Pune	526,330	0	117,895	0	22
Quezon City	613,245	15,555	141,831	3	23

Table K11.4

Informal-sector collection of recovered material in the six cities

Source: GTZ/CWG (2007) project baseline workbooks, based on secondary sources, experience and observation in the cities

	Total material recycled/recovered by the informal sector	Recovered by informal service providers (%)	Recovered by itinerant waste buyers (IWBs) (%)	Recovered by street collectors (%)	Recovered by dump-pickers (%)	Recovered via other routes (%)
Cairo	2,161,534	100	–	–	–	–
Cluj	14,575	–	2	40	58	–
Lima	529,370	7	27	30	6	30
Lusaka	5419	–	–	71	29	
Pune	117,895	32	34	–	10	24
Quezon City	141,831	–	72	16	8	4

Table K11.5

Informal-sector income and earnings from recycling in the study cities[3]

Notes: a represents earnings from between 50 days per year and full time. See Chapter 3 summary, and Cluj City Report (GTZ/CWG, 2007, Annex 6).

b 11,183 is the total number of individuals in the informal sector excluding the workers in the piggeries. The income earned in Lima includes the earnings of piggery workers, so they are included for calculating the income per worker. The total for six cities also includes these workers.

c Includes both recyclers and informal service providers.

Source: GTZ/CWG (2007) project socio-economic workbooks, Annex 5

	Cairo	Cluj[a]	Lima[b]	Lusaka[c]	Pune	Quezon City	For six cities
Total individuals working in the informal waste sector	40,000	3226	17,643 (11,183)	480	8850	10,105	80,304
Average earnings of an informal-sector worker (Euros/day)	4.3	6.28	5.4	2.03	3.29	6.26	4.3
Total estimated income (1000 Euros/year) earned by all people in the informal sector in this city	8979	1114	18,187	281	10,613	14,396	53,597
Total reported annual sales to recycling industry (1000 Euros/year)	26,337	2462	55,678	471	15,831	7077	107,856

5   *Junk shops* consist of small, medium or large traders of recyclables, usually with a commercial interest in only one or two classes of materials.

When a city aims to modernize its solid waste system and 'introduce' recycling, the consultants or planners often fail to analyse the performance of these informal recycling sub-sectors. There is an assumption that not much in recycling is happening, or that what is already occurring is not important. In fact, the informal sector may already be removing and recovering as much as 20 per cent of the waste at no cost to the local authority. This is a positive environmental externality which the municipality enjoys without having to pay for it because the environmental gain is a by-product of the economic interests of informal recyclers.

The informal sector has a significant economic footprint. While the estimates are difficult to verify, they are indicative and show the extent to which this sector affects the economic situation.

Table K11.6 considers the overall impacts of informal recycling (and service) activities on waste system costs. The formal-sector recycling infrastructure may well consist of second-hand imported equipment which may not fit the local context; the recycling collection may be poorly publicized or require 'heroic' measures from those agreeing to participate. As a result, formal systems have a high cost per tonne, associated with overcapitalization, the small volumes recovered, and most likely poor marketing performance associated with inexperience in valorizing recyclables.

In every city, informal-sector recovery represents a net benefit, and the formal sector (except in Cluj) represents a net cost. In all cities, the informal sector is responsible for substantially more benefits on a per tonne basis. This suggests that trying to eliminate the informal sector is a risky strategy that will increase overall costs. Once a city has disposal under control, they are likely to make legal or administrative claims on the recyclables. A recent court decision in Colombia sets a precedent that this may not be legal.[4] This competition over access to recyclables often leads to criminalization of waste-picking, police harassment or thug-like abuse of waste-pickers and their families.[5] When municipal administrations in low- and middle-income countries successfully prohibit informal activities and take over valorization, the resulting municipal recycling activities are generally

Table K11.6

Modelled total solid waste materials handling system costs per tonne, net costs per tonne (with revenues for material sales included), and total net costs for the whole sector for all tonnes handled by that sector

Note: parentheses indicate a net revenue.

Source: GTZ/CWG (2007), project data.

	Formal sector				Informal sector			
	Material handled (tonnes)	Total cost/ tonne (Euros)	Total net cost/tonne (Euros)	Total net sector cost (Euros)	Material handled	Total cost/ tonne (Euros)	Total net cost/tonne (Euros)	Total net sector cost (Euros)
Cairo	810,667	9.88	4.78	3,878,309	2,567,143	2.92	(31.51)	(80,898,939)
Cluj	145,779	25.45	(23.54)	(3,431,638)	14,575	5.29	(403.33)	(1,993,661)
Lima	1,839,711	41.64	41.14	75,689,381	848,364	90.84	(27.48)	(23,314,818)
Lusaka	90,720	127.53	76.34	6,925,113	98,170	5.05	(157.45)	(15,456,411)
Pune	394,200	22.74	22.74	8,964,662	132,130	11.67	(70.39)	(9,300,082)
Quezon City	489,606	31.40	25.36	12,415,362	123,639	36.19	(63.72)	(7,878,883)

more expensive, less efficient and more likely to be overcapitalized than parallel informal activities.

The most important conclusion for recycling planners is simple: the informal sector saves the city money and improves the environmental footprint of waste management activities at no cost to taxpayers or the city budget. Or, in other words, private-sector valorization activity has significant and quantifiable positive economic benefits as well as positive environmental externalities. The city authorities can increase these positive impacts – and work together with the informal sector to optimize and legitimize their activities, and institutionalize the benefits to citizens and informal enterprises – and in the process may even be able to harvest climate credits. This is a win–win approach, fitting into the models of inclusivity and sustainability. In contrast, the more common approach of establishing new public-sector exclusive rights to recycling and seeking to transfer valorization activities to an inexperienced public sector struggling to 'domesticate' expensive international environmental norms has very high risks, recovers small amounts of materials, and produces few economic or environmental benefits.

NOTES

1   The information in Table K11.1 is a summary of much more detailed information developed in GTZ/CWG, 2007, Annex 6, City Report and Workbook for Cluj-Napoca, Romania, which may have disaggregated generation figures by density of housing stock, such as in Lusaka, or by governmental sub-areas within the city, such as seen in Lima's data. See the City Reports for additional detail and specific sources.
2   Recycling Summit, 2008, closing phase 1 of the IFC Recycling Linkages Programme.
3   Earnings in the informal sector includes both earnings from recycling activities and from service fees for mixed or separate collection, street sweeping, etc. Sales to the recycling industry covers sales made to the recycling industry, but from both informal sector and formal sources.
4   GTZ/CWG, 2007.
5   Chaturvedi, 2006; GTZ/CWG, 2007.

KEY SHEET 12
THE DUTCH APPROACH TO PRODUCER RESPONSIBILITY

Frits Fransen (ex-ROTEB)

Since January 2006, Dutch producers and importers marketing packaging and packed products on the Dutch market, have been responsible for prevention, collection and recycling of used packaging materials, including associated costs. The overall aim is to reduce packaging materials in municipal waste and in street litter.

The Packaging, Paper and Cardboard Management Decision expanded upon European Union Directives 1994/92/EC and 2004/12/EC, aiming for 55 to 80 per cent recycling and at least 60 per cent recovery, or incineration plus energy recovery, from 31 December 2008. The municipalities take care of the actual collection and are financially compensated via a national waste fund financed by the producers and importers.

Since January 2008, packaging materials taxes have been levied ranging from 73 Euros per tonne for paper and cardboard, 434 Euros per tonne for plastics and 877 Euros per tonne for aluminium. The agreement deals with a wide variety of goods – for instance, cardboard boxes, wooden pallets, plastic carry bags, plastic shampoo bottles, aluminium and steel tins, glass bottles and jars, etc. 70 per cent of the packaging materials should have a useful application, including 65 per cent through recycling (70 per cent in 2010).

The recycling aim is further specified in Table K12.1.

The national agency Netvang coordinates developments on behalf of 15 national line-of-business organizations, monitors municipal efforts and performance, and pays financial compensation to the municipalities for new additional costs.

For instance, for plastics collection, which becomes mandatory on 1 January 2010 and is currently being piloted in approximately 300 municipalities, years of discussion between the business parties, the ministries of environmental affairs and economic affairs, and the Dutch municipalities have now been finalized, resulting in a financial compensation of 475 Euros per tonne for collection, 1.50 Euros per tonne per kilometre for transport, 100 Euros per tonne for sorting (if applicable), and a contribution to the start-up cost of the system.

Waste collection truck with a slogan on the side for awareness-raising

© Rotterdam

Table K12.1

Dutch national recycling goals in 2009 and their achievement levels in Rotterdam

Source: National Framework Agreement of 27 July 2007

Dutch national recycling goal	Achievement level in Rotterdam
75% for paper and cardboard	70% achieved in 2004
90% for glass	76% achieved in 2004
85% for metals	86% achieved in 2004
25% for wood	33% achieved in 2004
38% for plastics in 2009, 42% in 2012	19% achieved in 2004

Truck owned by an itinerant buyer of recyclables in Beijing, China

© Ljiljana Rodic

2   recyclable materials that can be separated, cleaned up, combined into marketable quantities, prepared for sale, and then sold into industrial value chains, where they strengthen local, regional and global production; and

3   bio-solids consisting of plant and animal wastes from kitchen, garden and agricultural production, together with safely managed and treated animal and human excreta. These materials are sources of key nutrients for the agricultural value chain, and have a major role to play in food security and sustainable development.

Insights from the reference cities and global good practices in resource recovery

The cities show a wide variety in approaches to recycling. The kinds of recycling and the completeness of the approach in the reference cities appear to grow in tandem with controlling disposal and increasing costs. This supports the hypothesis that modernization stimulates valorization and commercialization of three kinds of materials:

1   items, especially durable goods, which can be reused, repaired, refurbished or remanufactured to have longer useful lives;

The amount of material valorized in the cities varies substantially, as can be seen in Table 4.6.

The measure of successful recycling is simple: the *recovery rate* measures the effectiveness of diversion and records the percentage of total generated waste materials that are successfully redirected to productive use in industry or agriculture. The highest rate is shown by Adelaide. Bamako shows an even higher rate; but this is not strictly comparable as it refers to the widespread use of raw waste in agriculture. However, it does explain why

Table 4.6

Reference city recycling rates as a percentage of municipal solid waste.

This table begins to explore the extent to which the cities are recovering resources. 'All channels/sectors' means that this includes both formal and informal sectors – all types of recovery from high-tech recycling to swine feeding.

Notes: NA = not available.
*Includes recycling, reuse and recovery of organic waste.
Figures in *italic* are estimates. Adelaide: excludes commercial and industrial (C&I) waste. C&I waste is 1,251,935 tonnes.
Belo Horizonte: excludes construction and demolition (C&D) waste.
Tompkins County: excludes commercial waste.

City	Tonnes generated of MSW	Tonnes valorized, all channels/sectors	MSW valorized, all activities and sectors
Adelaide	742,807	401,116	54%
Bamako	462,227	392,893	85%
Belo Horizonte	1,296,566	12,200	1%
Bengaluru	2,098,750	524,688	25%
Canete	12,030	1412	12%
Curepipe	23,764	NA	NA
Delhi	2,547,153	841,070	33%
Dhaka	1,168,000	210,240	18%
Ghorahi	3285	365	11%
Kunming	*1,000,000*	NA	NA
Lusaka	301,840	17,446	6%
Managua	420,845	78,840	19%
Moshi	62,050	11,169	18%
Nairobi	876,000	210,240	24%
Quezon City	736,083	287,972	39%
Rotterdam	307,962	90,897	30%
San Francisco	508,323	366,762	72%
Sousse	68,168	4168	6%
Tompkins County	58,401	35,894	61%
Varna	136,532	37,414	27%
Average	641,539	195,821	30%
Median	441,536	84,869	25%

Bamako has survived so long without a controlled disposal facility. Table 4.6 examines the different kinds of reduce, reuse, recycle and organic waste recovery that are present in the reference cities. The table looks across the board at all types of recovery, not distinguishing whether the separation is at source, at the point of disposal, from waste-pickers taking recyclables from streets or secondary collection points, or retrieved by hand or machine from the disposal site.

Table 4.7 examines the total that is diverted from disposal and compares this to the goals for diversion and recovery. The destination is shown: reuse, percentage for recycling and percentage to the agricultural value chain. Table 4.8 provides a breakdown of what happens to organic waste. Organic waste is used in the cities to feed livestock, to make compost and for land spreading. Where organic waste has value, the value is related to residual nutrients that it can return to the soil and via plants to the food chain, or directly to the food chain via animals.

While the overall rate of recovery is important, it is also useful to know which types of valorization are the most effective in keeping

waste materials out of the dumpsite. Table 4.9 examines which two recovery operations are recovering the most materials. 'Operation' here means specific activity, such as informal street-picking, formal paper recycling and construction waste recovery. All activities are in the formal

City	Goal, diversion and recovery	Total recovered of total generated (%)	Total recycled of total generated (%)	Total to agricultural value chain of total generated (%)
Adelaide	25%	54%	28%	26%
Bamako	0%	85%	25%	31%
Belo Horizonte	16%	1%	1%	0.18%
Bengaluru	50%	25%	15%	10%
Canete	20%	12%	12%	0%
Curepipe	48%	NA	NA	NA
Delhi	33%	33%	27%	7%
Dhaka	0%	18%	18%	4%
Ghorahi	0%	11%	11%	NA
Kunming	0%	13%	13%	0.05%
Lusaka	0%	6%	6%	NA
Managua	0%	19%	17%	2%
Moshi	0%	18%	NA	20%
Nairobi	0%	24%	14%	5%
Quezon City	25%	39%	37%	2%
Rotterdam	43%	30%	28%	1%
San Francisco	75%	72%	46%	26%
Sousse	0%	6%	2%	4%
Tompkins County	50%	61%	61%	NA
Varna	50%	27%	27%	NA
Average	22%	29%	22%	9%
Median	18%	24%	18%	4%

Table 4.7

Actual recovery rates

Notes: NA = not available; NR = not reported
Figures in italic are estimates. Goals for diversion and recovery in italic have different goals for different waste fractions.

Table 4.8

Destinations of organic waste in the reference cities.

Solid waste master plans often target the large quantities of organic waste to go to composting or anaerobic digestion to produce biogas. Yet this table suggests that animal feeding, especially swine feeding, is important in many cities and represents an untapped potential.

Notes: NA = not available; NR = not reported. Figures in italic are estimates. Figures are calculated based on organic fraction in waste composition and total waste generated

City	Total organic waste generated in municipal solid waste	Tonnes to animal feeding	Generated to animal feeding of total organic waste (%)	Tonnes to composting or land application	Generated to composting or land application of total organic waste (%)
Adelaide	622,000	0	280,609	0	
Adelaide	424,158	0	0%	193,432	46%
Bamako	144,908	0	0%	144,908	100%
Bengaluru	1,500,606	0	0%	209,875	14%
Belo Horizonte	850,936	0	0%	2300	0.27%
Canete	5759	0	0%	0	0%
Curepipe	11,384	0	0%	0	0%
Delhi	2,066,250	NA	NA	165,565	8%
Dhaka	864,320	NA	NA	47,450	NA
Ghorahi	2595	0	0%	NA	NA
Kunming	582,000	500	0%	NA	NA
Lusaka	118,623	NA	NA	NA	NA
Managua	311,047	8395	3%	22	0.01%
Moshi	40,333	6205	15%	6205	15%
Nairobi	567,648	35,040	6%	5676	1%
Quezon City	370,201	12,723	3%	1592	0.43%
Rotterdam	80,070	0	0%	3255	4%
San Francisco	170,797	0	0%	132,000	77%
Sousse	43,968	2500	6%	0	0%
Tompkins County	17,112	0	0%	NA	NA
Varna	32,836	NA	NA	NA	NA
Average	410,278	4085	2%	60,819	19%
Median	157,852	0	0%	5676	3%

City	Tonnes recovered, all sectors	Percentage recoved by formal sector	Percentage recoved by informal sector	Operation recovering the most	Tonnes recovered by operation recovering the most	Percentage of total recovered by operation recovering the most
Adelaide	2,611,214	70%	0%	Formal recycling of C&D waste	1,257,182	48%
Bamako	392,893	0%	85%	Formal and informal land application (terreautage)	144,908	37%
Belo Horizonte	145,134	6%	0%	Formal recovery of C&D waste	132,934	92%
Bengaluru	524,688	10%	15%	Informal waste picking, IWBs of various materials	314,813	60%
Canete	1412	1%	11%	Informal IWBs collect various materials	548	39%
Curepipe	NA	NA	NA	Informal recovery of metals	NA	NA
Delhi	841,070	7%	27%	Informal recovery by waste pickers	1251	0%
Dhaka	210,240	0%	18%	Formal recycling of paper supplied by informal recovery	61,320	29%
Ghorahi	365	2%	9%	Informal IWBs collect various materials	300	82%
Kunming	NA	NA	NA	Informal recovery for recycling by IWBs and dump-pickers	NA	NA
Lusaka	17,446	4%	2%	Formal recovery organised by recycling industry	12,027	69%
Managua	78,840	3%	15%	Informal recovery by dump pickers of various materials	44,530	56%
Moshi	11,169	0%	18%	Informal animal feeding, land application of organic waste	11,169	100%
Nairobi	210,240	NA	NA	Formal and informal recovery of paper and plastic	NA	NA
Quezon City	287,972	8%	31%	Informal IWBs collect various materials	176,316	61%
Rotterdam	90,897	30%	0%	Formal recovery of paper and carton	21,125	23%
San Francisco	366,762	72%	0%	Formal recovery for recycling, various materials	254,101	69%
Sousse	4168	0%	6%	Informal recovery for animal feeding	1500	36%
Tompkins County	36,495	61%	0%	Formal recovery of metals for recycling	35,302	97%
Varna	37,414	2%	26%	Informal recovery by waste pickers	35,207	94%
Average	326,023	16%	15%		147,325	58%
Median	118,016	4%	11%		35,302	60%

Table 4.9

Optimizing recovery by building on successes

Notes: NA = not available; NR = not reported. Figures in italic are estimates. Tonnes recovered: Adelaide – includes C&D waste, and industrial waste for South Australia; Belo Horizonte – includes C&D waste; Tompkins County – includes tonnes prevented and reused.

sector unless it specifically says 'informal'. The reference cities also report formal activities in a more detailed way than 'informal recovery'.

One of the key principles of ISWM is to build on what you have. Therefore, studying the high-recovery parts of the system, and building on them, is a key means of arriving at good practices and making sound and sustainable plans. Of course, when making these kinds of choices, other criteria, such as efficiency, cost and inclusivity, are also important.

■ **Organic waste valorization**

Pre-industrial societies reused any edible leftover

Box 4.20 Plastic bag reduction

Delhi has been struggling with what to do with plastic bags for over a decade. During the late 1990s, a strong civil society movement against all plastic bags, driven primarily by schoolchildren and non-governmental organizations (NGOs) aimed to reduce the city's ecological footprint. However, this soon converted into a restriction on plastic bags based on their thickness. This was difficult to monitor. The city was more successful in phasing out flimsy plastic bags that contained dyes, suspected to contain toxins, mostly on account of the chief minister's personal interest in this issue. During recent years, recycled plastic bags have been targeted as being unsafe. Ironically, this has reduced the market for recycled plastic bags; but it is still easy to get plastic bags from virgin materials. In 2009, the courts banned plastic bags entirely. Although the ban has not been fully implemented, the virgin plastic industry has since formed an association to fight the ban and convince the public to start using plastic bags again. In these discussions, the plastic recyclers remain marginalized.

food as animal feed; other organic wastes, including the organic fraction of municipal solid waste, agricultural wastes and human excreta, were generally returned the soil. This organic resource loop is still important today, as was witnessed in UN-Habitat's bio-solids atlas, *Global Atlas of Excreta, Wastewater Sludge and Biosolids Management*.[31] The organic fraction is often between 50 and 70 per cent by weight of municipal waste in developing countries. Therefore, exploring possibilities of engaging citizens and service providers in 'recycling' organic waste through composting would alone significantly improve the situation and alleviate the problems of municipal solid waste. Composting organic waste, either at household level or on a more aggregated scale, can prove to be a worthwhile effort, as experiences from various countries testify.[32] A good example is Bangladesh (see Box 4.14).

Recycling or composting only makes sense if there is a market for the product. Whereas markets for recyclable materials exist in most countries in the recycling value chain, this is only partially true for organic wastes. Food waste may still have a market value as animal feed, and farmers used to accept or even pay for organic wastes in order to compost it themselves. This was the general disposal route in China before chemical fertilizers took over, and is still the case in parts of West Africa.

Organic waste in these circumstances may have a market value; but products made from it, such as compost, usually do not. For this reason, it is not reliable to assume that there is a 'market' for compost. Instead, a market for compost needs to be developed by building urban–rural linkages and by educating potential users and buyers of compost about its properties, nutrient value, and similarities or differences in relation to the fertilizers and mulches, which are better known. This process takes several years, and compost operations are generally not able to cover their own costs until compost itself becomes a commodity with a market price.

High-income countries have done much work on product specifications to give the buyer confidence in the quality of the compost product. An important consideration in this has been the presence of trace contaminants, particularly heavy metals, in the compost. This problem can be avoided by composting a source-separated organic fraction so that the possibility of cross-contamination by other waste components is avoided.

■ Prevention and reuse

Prevention and reuse are important because they improve the functioning of both the economic system and the waste management system. Low- and middle-income countries are generally better at these activities, which have to be formally promoted in high-income countries with higher material standards. Many developing and transitional country cities still have active systems of informal-sector recycling, reuse and repair, which often achieve recycling rates comparable to those in the West, at no cost to the formal waste management sector. One missed opportunity is that few countries either monitor or report the performance of prevention and reuse initiatives.

Prevention (sometimes called 'reduction') and reuse are the pillars of the new 3R Regional Forum in Asia, an initiative of the Japanese

Locally manufactured vehicle for collecting and transporting sorted recyclables in Vargas, Venezuela. The high sides increase the volume of the light recyclables that can be transported

© Jeroen IJgosse

Ministry of the Environment, the United Nations Centre for Regional Development (UNCRD), and 12 Asian countries. Two of these countries, the Philippines and Bangladesh, have cities represented in this Third Global Report. But while the progress of recycling in the Philippines and composting and climate credits in Bangladesh make them global good practice leaders, their record with prevention and reuse remains questionable. Particularly in the Philippines, where Republic Act 9003 encourages prevention and prohibits incineration, there is a need for strong prevention work. All countries could begin by tracking and reporting results of prevention initiatives, as well as monitoring the numbers of materials flowing through (private) reuse channels.

■ Applying lessons from high-diversion recycling successes

Current recycling initiatives in Europe, North America and Australasia were put into place during the mid 1980s. By now, they have between 20 and 25 years of monitoring, evaluation and experience that can be highly useful for decision-makers and practitioners. This Third Global Report includes three developed country cities with high recycling rates: Adelaide, Tompkins County and San Francisco, and one, Quezon City, whose solid waste law and current practices deliver higher recycling rates than the rest of the reference cities in low- and middle-income countries. In different ways, these four cities bear the burden to illustrate the principles

Box 4.21 Institutionalizing the sink value of recycling in The Netherlands

The Netherlands has a long tradition of favouring recycling over other forms of waste management, whether it is profitable or not. Three examples demonstrate this:

1   Extended producer responsibility (EPR) agreements in The Netherlands, called *covenants*, establish that certain industries must pay for recycling as part of their responsibility for safe end-of-life management of their products. The *paper covenant* creates an obligation for paper recycling companies to reimburse municipalities for collection and processing costs when the economic value of the paper is too low to cover these costs.

2   Municipalities pay non-governmental organizations (NGOs) and non-state actors, including recycling centres and second-hand shops, a *diversion credit* based on the number of tonnes that they collect and market. Many non-profit recycling shops claim these credits, which are usually a part of the cost of disposal.

3   Recycling in The Netherlands is such an important sink that it is fully paid for out of the waste management fee, the '*afvalheffing*,' which every household pays to their municipality. In Rotterdam the ROTEB, the waste management company, calculates this fee by adding together all of its capital and operating costs of sweeping, collection, recycling, composting and disposal, and spreading it across all households and businesses in the city.

of what works and what does not. A fifth city, Bamako, reports high recycling rates based on the fact that the raw waste is recovered by spreading on fields after decomposing on illegal dumps. This approach to high recovery is a kind of emergency measure, which, while it has lessons for other countries, is not considered to be a global good practice.

No recycling system based solely on 'awareness-raising' receives high recovery rates; but neither is high recovery possible without this. High diversion such as is demonstrated in Tompkins County or San Francisco or Quezon City is based both on providers modifying and adapting the system and on users adjusting their behaviour. A whole range of approaches to home composting, source separation at the office, separate collection of kitchen and garden waste, and prevention have been tried and there is information available on how they work.

In the same way, countries such as Brazil have more than 20 years of experience in building modern recycling initiatives, together with the informal sector, on the model of the pioneering work of Asmare in Belo Horizonte, franchising depots to informal-sector operators, and the work of Linis Ganda in Metro-Manila, licensing itinerant waste buyers (IWBs) to enter

certain areas and collect source-separated recyclables. Cañete is in the process of applying these experiences to its own modest waste stream, based in part on successful experience in the capital of Peru: the combined cities of Lima and Callao.

Whatever the specific idea, making use of this global body of experience and applying the lessons learned is a key element of good practice.

■ Working with the informal sector

Not only does the informal recycling sector provide livelihoods to huge numbers of the urban poor, thus contributing to the Millennium Development Goals (MDGs), but the information in the reference cities suggests that informal recyclers divert 15 to 20 per cent of the city's recyclables, which translates to savings in its waste management budget by reducing the amount of wastes that would otherwise have to be collected and disposed of by the city. In effect, the poor are subsidizing the rest of the city.

There is an opportunity for the city to build on these existing recycling systems in order to maximize the use of waste as a resource, to protect and develop people's livelihoods, and to reduce the costs to the city of managing the residual wastes. Bengaluru's approach to recycling in businesses is a good example of such cooperation: businesses hire waste-pickers to manage their waste and do basic cleaning. The waste-pickers get a modest salary and also get to keep the materials. The formal and informal sectors are working together, for the benefit of both.

With the exception of Quezon City, the reference cities still have some way to go in learning how to combine high recycling performance with integration of the informal sector into municipal recycling. In spite of their efforts in this direction, Belo Horizonte's recycling effort recovers only very small percentages of waste; as a result, too much is being landfilled and the landfill is reaching capacity more quickly than anticipated. San Francisco's way of incorporating the private sector provides some interesting ideas.

Box 4.22 A scrap of decency

AMONG those suffering from the global recession are millions of workers who are not even included in the official statistics: urban recyclers – the trash pickers, sorters, traders and re-processors who extricate paper, cardboard and plastics from garbage heaps and prepare them for reuse. Their work is both unrecorded and largely unrecognised, even though in some parts of the world they handle as much as 20 percent of all waste.

The world's 15 million informal recyclers clean up cities, prevent some trash from ending in landfills, and even reduce climate change by saving energy on waste disposal techniques like incineration.

They also recycle waste much more cheaply and efficiently than governments or corporations can, and in many cities in the developing world, they provide the only recycling services.

But as housing values and the cost of oil have fallen worldwide, so too has the price of scrap metal, paper and plastic. From India to Brazil to the Philippines, recyclers are experiencing a precipitous drop in income. Trash pickers and scrap dealers in Minas Gerais State in Brazil, for example, saw a decline of as much as 80 percent in the price of old magazines and 81 percent for newspapers, and a 77 percent drop in the price of cardboard from October 2007 to last December.

In the Philippines, many scrap dealers have shuttered so quickly that researchers at the Solid Waste Management Association of the Philippines didn't have a chance to record their losses.

In Delhi, some 80 percent of families in the informal recycling business surveyed by my organization said they had cut back on 'luxury foods,' which they defined as fruit, milk and meat. About 41 percent had stopped buying milk for their children. By this summer, most of these children, already malnourished, hadn't had a glass of milk in nine months. Many of these children have also cut down on hours spent in school to work alongside their parents.

Families have liquidated their most valuable assets – primarily copper from electrical wires – and have stopped sending remittances back to their rural villages. Many have also sold their emergency stores of grain. Their misery is not as familiar as that of the laid-off workers of imploding corporations, but it is often more tragic.

Few countries have adopted emergency measures to help trash pickers. Brazil, for one, is providing recyclers, or 'catadores,' with cheaper food, both through arrangements with local farmers and by offering food subsidies. Other countries, with the support of nongovernmental organizations and donor agencies, should follow Brazil's example. Unfortunately, most trash pickers operate outside official notice and end up falling through the cracks of programs like these.

A more efficient temporary solution would be for governments to buoy the buying price of scrap. To do this, they'd have to pay a small subsidy to waste dealers so they could purchase scrap from trash pickers at about 20 percent above the current price. This increase, if well advertised and broadly utilized, would bring recyclers back from the brink. In the long run, though, these invisible workers will remain especially vulnerable to economic slowdowns unless they are integrated into the formal business sector, where they can have insurance and reliable wages.

This is not hard to accomplish. Informal junk shops should have to apply for licenses, and governments should create or expand doorstep waste collection programs to employ trash pickers. Instead of sorting through haphazard trash heaps and landfills, the pickers would have access to the cleaner scrap that comes straight from households and often brings a higher price. Employing the trash pickers at this step would ensure that recyclables wouldn't have to be lugged to landfills in the first place.

Experienced trash pickers, once incorporated into the formal economy, would recycle as efficiently as they always have, but they'd gain access to information on global scrap prices and would be better able to bargain for fair compensation. Governments should charge households a service fee, which would also supplement the trash pickers' income, and provide them with an extra measure of insurance against future crises. Their labour makes our cities healthier and more liveable. We all stand to gain by making sure that the work of recycling remains sustainable for years to come.

Source: 'A scrap of decency', Bharati Chaturvedi, *The New York Times*, op-ed, 5 August 2009, www.nytimes.com/2009/08/05/opinion/05chaturvedi.html
Note: the author, Bharati Chaturvedi, is the founder and director of the Chintan Environmental Research and Action Group.

■ Designing for high diversion and valorization

During the last 25 years, it has become clear that there is a real difference between well- and poorly designed formal recycling activities. Studies, particularly in Canada where there has been substantial monitoring of recycling collection efficiency, demonstrate that separate collection has a much different economy of scale than waste collection: the amounts set out by households are smaller, the specific densities different, the participation rates more variable, and the types of set-out containers and logistics of loading vehicles makes it possible to have routes that are significantly longer and serve up to three times the number of households per route day as is possible when collecting mixed waste. A local authority who simply sends a truck to collect recyclables on the same route may be saving resources but throwing away money.

North American experience has also indicated that co-collecting some materials works better than collecting single materials or materials streams. Separate collections during the 1980s and 1990s *commingled* paper and cardboard, on the one hand, and glass, plastic and aluminium cans, on the other: the densities of the combined streams were more favourable for both collection and transport and made it relatively easy for high-performance post-collection sorting in materials recovery facilities (MRFs). Increasingly, at the time of writing, North American separate collection focuses on *single stream collection* of the main materials, which are collected separately from non-recyclable waste.

The high-performing recycling and organic waste systems tend to have some elements in common:

- Nothing is collected for which there is not an established market or a market development strategy, *even if the commodity value of the to-be-collected material is null.*
- They are based on the results of a highly participatory recycling planning process, either for the city itself, or at regional or national level (such as the county plan in Tompkins County).
- They are based on phased implementation and horizontal learning.
- They are designed to be convenient for ordinary people and do not rely on heroic measures or special levels of effort.
- They focus on an 'opportunity to recycle' that is as easy and convenient – or easier and more convenient – then the normal way of disposing of mixed waste.
- They give multiple options to households and businesses (e.g. collection at the household, plus a depot in the neighbourhood, plus collection at the market, plus a place to sell large quantities of materials).
- They rely on logistical interventions in the household that give users physical and visual reminders. Examples include containers of different colours and shapes, carrier bags for bringing recycling to drop-off centres, and the like.
- If there are financial incentives, they are modest and reward high-performing households with recognition, rather than direct cash payments for materials.

NOTES

1   UN-Habitat, 2009, p129.
2   Documentary film *La Vache Qui Ne Rit Pas*, CEK, Bamako, Mali.
3   'Nigeria: Country loses N10 billion annually to unhealthy environment – Rep', 29 July 2008, http://allafrica.com/stories/20080729031 1.html.
4   METAP Solid Waste Management Centre website: 'Highlights – Decision Makers', Support document, www.metap-solidwaste.org/index.php?id=12.
5   Hawley and Ward, 2008.
6   UN-Habitat, Global Urban Observatory 2009. Data compiled from national demographic and health surveys.
7   Scheinberg and Mol, in press.
8   See www.chinadaily.net/olympics/2007-02/11/content_6005068.htm.
9   WRAP (2009) *The Waste Collection Commitment*, www.wrap.org.uk/local_authorities/waste_commitment.html.
10 Coffey, 2009.
11 See UWEP City Case Study for Bamako, and UWEP Programme Reports, 1995–2002, WASTE, Gouda, The Netherlands.
12 See UWEP City Case Study for Bengaluru, and UWEP Programme Reports, 1995–2002, WASTE, Gouda, The Netherlands.
13 WASTE feasibility study for a PPP in Nairobi, 1999.
14 In many countries, waste work in both the formal and informal sectors is often reserved for members of disadvantaged classes and religious or ethnic minorities (e.g. in Egypt: Coptic Christians; in India: people from lower social strata; in the Balkans and Central Europe: Roma people; in the US: the dominant group of new immigrants).
15 Various sources, e.g. www.greatgarbage patch.org/.
16 ERL (1992) *Quantification, Characteristics and Disposal Methods of Municipal Waste in the EU – Technical and Economic Aspects*, Report for the European Commission.
17 See www.wasteonline.org.uk/resources/InformationSheets/HistoryofWaste.htm.
18 Rushbrook and Pugh, 1998.
19 Ali et al, 1999.
20 Rouse, 2006.
21 WASTE project for UNEP IETC for private waste collection company in Nairobi, 1999.
22 Joseph, 2007.
23 See examples given in 'The scale of the solid waste problem' in Chapter 2.
24 This specific example is from Zambia, provided by Manus Coffey, pers comm.
25 A study by the Dutch Agricultural Economics Institute (LEI) and WASTE in 2003, showed that, at that time, there are also few health or environmental problems with land-spreading.
26 Hay and Noonan, 2002.
27 See www.china environmentallaw.com/2008/08/30/china-adopts-circular-economy-law/.
28 Wilson et al, 2009.
29 GTZ/CWG, 2007.
30 Chaturvedi, 2009.
31 Le Blanc et al, 2006.
32 Rothenberger et al, 2006.

CHAPTER 5

THE INTEGRATED SUSTAINABLE WASTE MANAGEMENT GOVERNANCE FEATURES IN THE REFERENCE CITIES

How do you manage a solid waste management system? Until the 1990s, the answer would have been framed around technology. But this does not allow us to answer many related questions (e.g. for whom is the waste managed; how can conflicting ideas and claims be dealt with; who is responsible for planning the system and creating it; who operates it; who maintains it; who pays for it; who uses it; and, above all, who *owns* it?). All of these questions can be addressed more easily within the framework of integrated sustainable waste management (ISWM). Integrated sustainable waste management is based on 'good garbage governance', and its goal includes inclusive, financially sustainable and institutionally responsive waste management, which functions well for users and providers.

This chapter focuses on three key governance features of ISWM, which together form the 'second triangle':

1 inclusivity and fairness, with a dual focus on inclusivity for users of the ISWM system and inclusivity for providers;
2 financial sustainability; and
3 sound institutions and proactive policies.

Successful ISWM interventions, represented by the 'first triangle' consisting of the three physical elements discussed in the previous chapter, rest on and are supported by this governance triangle.

 ## INCLUSIVITY

This section begins by looking at who are the stakeholders in a waste management system. It then looks at issues relating to equity for the system users in order to ensure that a fair and adequate service is provided to all citizens; and at equity among service providers – large and small, formal and informal – in terms of a fair distribution of economic opportunities for providing the service or valorizing materials.

Key issues and concepts

■ **Who are the stakeholders?**

This discussion starts with a basic assertion that solid waste management is a *public* service. Solid waste management is a public good in the sense that addressing public health and environment and resource management are tasks that are considered responsibilities of the governmental

Waste-picker in Belo Horizonte, Brazil, on her way to sell her collected plastic materials

© Sonia Diaz

sector. These are responsibilities which individual families or businesses cannot organize for their neighbours, just as they cannot arrange taxation or military protection.

But different responsibilities fall to different levels of government. *National government* sets the policy, financial and administrative framework within which the city needs to work, and national ministries and departments are influenced by an increasingly globalized vision of correctness, which comes into the discourse as 'best practice'. Many different ministries and departments may be involved, as may regional government, neighbouring jurisdictions, transnational institutions and the private sector. They will often listen to different stakeholders and therefore have diverging institutional positions on what is needed and who is responsible for it.

While national authorities create the boundary conditions, it is the *municipal authorities* who are responsible for solid waste management in a city – that is, for establishing the legal, regulatory and financial boundary conditions that make it possible to provide the service or extract materials for valorization. Historically, solid waste is a municipal responsibility because municipal authorities are the main stakeholder responsible for public health: they receive the blame if the service is not provided or falls below an accepted or agreed-upon standard. This does not mean that they have to provide the service on their own, especially when a range of other stakeholders are looking for opportunities to plan the system, make investments, provide the service, organize users, supply the economic actors with equipment, valorize materials, and have cleaner neighbourhoods.

The second key concept is therefore *stakeholders* – that is, the people and institutions who have a stake in waste management and recycling. Providers, users and the local authority will always be included. Some other stakeholders are not that obvious, nor do they think of themselves as involved in waste, although they are certainly users of a waste removal service. These unrecognized stakeholders include professional associations, churches and religious institutions; healthcare facilities, and sport and social clubs; schools, universities, educational and research institutions, engineering and planning departments, and the consultancies that serve them; and the many sub-groups of waste generators and service users, including market stallholders, kiosks, hotels, restaurants, sport clubs, hospitals, hydroelectric companies, transport operators, and schools and kindergartens.

Box 5.1 Public interest litigation in India: Sugar PIL or bitter PIL?

Public interest litigation (PIL) has been an important means of urban policy-making in Delhi during the last 12 to 15 years, from sealing 'polluting industries' to banning plastic bags. The response to solid waste in Delhi has also been shaped, in part, by two PILs: *B. L Wadhera versus the Union of India and Others* (1996) and *Almitra Patel versus the Union of India and Others* (1996).

Dr Wadhera's contentions were that 'the MCD [Municipal Corporation of Delhi] and the NDMC [New Delhi Municipal Council] had been totally remiss in the performance of their statutory duties to scavenge and clean Delhi City', violating the citizens' constitutional rights to a clean environment. This resulted in the 1998 Bio-Medical Waste (Management and Handling) Rules, as well as dramatic moments in court, when top municipal functionaries were summoned to explain their non-performance. Such public humiliation increased the need to find a new way out and privatization began to be considered seriously.

Almitra Patel filed her PIL 'to protect India's peri-urban soil and water, and the health of its urban citizens through hygienic practices for waste management, processing and disposal'. The petition hastened the creation of the influential 2000 Solid Waste (Management and Handling) Rules. A famous statement made by the judge related to the urban poor. He declared: 'Rewarding an encroacher on public land with an alternative free site is like giving a reward to a pickpocket for stealing.' This aptly reflected the trend in courts to criminalize the poor, legitimizing their exclusion from the gaze of policy.

Given the influence of the courts in Delhi, this trend filtered to several levels and was, in part, responsible for the institutionalized exclusion of informal-sector players in Delhi. The rules emphasize the need for recycling. However, they do not embed informal-sector rights.

The main *service users* are households and commercial and institutional waste generators, including most of those mentioned above. Non-governmental organizations (*NGOs*) and community-based organizations (*CBOs*) are important as representatives of wider sections of the user community; but they may also have a separate role as private-sector providers. Trade and professional associations and chambers of commerce have a similar representational function for businesses and institutions, as do labour unions and syndicates for workers.

Both the *informal* and *formal private-sector* actors are, or may wish to become, *service providers* in waste collection, *commodities traders* in recycling, and/or *agricultural producers* who process and sell compost and other materials in the agricultural value chain.

Private-sector recyclers in both formal and informal sectors are often recycling a significant proportion of the city's waste – probably as much as 20 per cent.[1] The private recycling sector also includes commodity traders and their industrial customers for recycled materials; they form the *industrial supply chain*, which also includes manufacturers, retailers and others who supply products that end up in the municipal waste stream.

■ Inclusivity and equity for users

Equity of service means that all users – or, said another way, all waste generators – need to have their waste removed regularly and reliably and disposed of safely. It also means that all residents have the streets swept and litter removed – or that there is an agreement that they do this themselves for their own communities. And providing a regular collection and sweeping service depends on there being a safe and agreed-upon place to put the collected waste.

Naples provides a case study to illustrate this last statement. The waste management service in Naples broke down in early 2008, with waste piling up in the streets because all of the region's landfills were full and the collectors had nowhere to take the waste. According to one press report: 'Naples has been choked by waste over and

Recycling facility of one of the waste-picker co-operatives in Belo Horizonte, Brazil

© SLU, Belo Horizonte

over in recent decades, partly due to mismanagement, corruption and mafia involvement in trash pickup, but also because of Neapolitans' refusal to sort their trash.'[2] Other contributing factors included authorities arguing with each other and not involving local citizens and other key stakeholders in the decision-making process. This resulted, among other things, in a situation where it became impossible to site new disposal facilities because the levels of trust between authorities and citizens were so low.

User inclusivity is roughly divided into several sub-categories, which include:

- consultation, communication and involvement of users, both in decision-making, and in *doing for themselves* in relation, for example, to home composting and waste prevention;
- participatory and inclusive planning and system design, which includes inclusivity in siting facilities;
- institutionalizing inclusivity, for example, feedback mechanisms, client surveys, and solid waste forums and 'platforms'.

■ Inclusivity for providers and economic actors

In some countries the local authorities have an automatic monopoly on providing waste services, and no private businesses are allowed to participate. In other countries, the private sector is the only actor providing service provision. In both of these cases it is the local authority – acting in the context of national legislation – that sets the institutional stage and has the end responsibility. An inclusivity focus challenges exclusive business models and questions the need for restricted access to the economic activities in waste management.

Inclusive business models allow and enable non-government stakeholders to initiate waste-related activities; they invite or legalize economic participation of formal private-sector firms, informal entrepreneurs, CBOs, NGOs and other non-state actors. Such models have labels such as private-sector participation (PSP); public–private partnerships (PPPs); pro-poor PPPs (5-Ps); joint ventures (JVs); para-statal organizations, including a range of municipally owned institutions and companies; informal integration or formalization initiatives; and the like.

Because of its prominence in the global discourse on entrepreneurship, sanitation and poverty alleviation, this Third Global Report has a strong focus on provider inclusivity, specifically in terms of opening or maintaining economic opportunities for informal entrepreneurs to provide waste services and participate in recycling. *Provider inclusivity* therefore covers:

- inclusivity in providing solid waste, sweeping and cleaning services;
- inclusivity and protection of livelihoods related to valorizing materials – specifically, formal and informal recycling and organic waste management.

Informal service providers (ISPs) are private entrepreneurs who collect and remove waste and excreta and who have private economic relations with waste generators. Informal-sector providers of services are responsible for a significant percentage of waste collection services in a wide range of cities and towns. In Lusaka, Zambia, for example, informal service providers are responsible for more than 30 per cent of all collection coverage;[3] in Bamako, Mali, they serve clients too poor to pay the city-regulated fees of the *Groupement d'Intérêt Économique* (GIE) micro-collectors. Think, for example, about a one-man operation in La Ceiba, Honduras, or Ithaca, New

York, who is collecting waste using the family donkey cart, tractor or truck, and being paid directly by clients; or a woman in Delhi, India, going door to door with a handcart collecting waste from 100 households per day, and who makes her living from the fee she charges and from the materials she separates. In sanitation, ISPs include 'frogs' who operate manual latrine- and *puisard*-emptying services, or day labourers who clean drains and gutters. In some countries there are informal street sweepers, usually women, who work directly for households.

Waste removal is not the only activity of informal or semi-formal enterprises. Working for themselves and their clients, but without recognition or protection or supervision from the city authorities, literally millions of people collect and recycle a significant proportion of the waste in the world's cities at no direct cost to taxpayers. Yet, all too frequently, the city authorities reject this activity, labelling it as illegitimate, illegal or even as a crime.

Informal recyclers represent a large and growing stakeholder group in most low- and middle-income countries – which is present, as well, in high-income countries and world-class cities and is reported to comprise as much as 1 per cent of the world's population.[4] The 75,000 informal-sector recyclers in the six Deutsche Gesellschaft für Technische Zusammenarbeit (GTZ) study cities[5] recycle about 3 million tonnes per year.[6] Informal private recycling activity is based on extraction of valuable materials from various waste streams (or their separate collection from generators), followed by upgrading and trading the materials to industry or agriculture. In all of the reference cities, there is at least some recycling of metal or paper or plastic by the private recycling sector, and in all cities except Adelaide and Rotterdam the informal sector is responsible for a significant amount of extraction. In many cities in low- and middle-income countries there is also substantial valorization of organic waste for animal feed, and in high-income countries, for production of compost.

KEY SHEET 13

WIEGO, ITS WORK ON WASTE-PICKERS AND THE FIRST WORLD ENCOUNTER OF WASTE-PICKERS IN COLOMBIA IN 2008

Lucia Fernandez and Chris Bonner (WIEGO)

Women in Informal Employment: Globalizing and Organizing (WIEGO) is a global action–research–policy network. It is committed to helping strengthen democratic member-based organizations of informal workers – especially women – and to build solidarity and organization at an international level. It has supported the development of StreetNet, an international organization of street vendors, and the regional networks of home-based workers, HomeNet South and HomeNet South-East Asia. It is currently supporting the International Union of Food, Agricultural, Hotel, Restaurant, Catering, Tobacco and Allied Workers (IUF) in its efforts to build an international network of domestic workers.

WIEGO's first contact with waste-pickers was through its member organizations in India, the Self-Employed Women's Association (SEWA) and the Trade Union of Waste-Pickers in Pune (KKPKP). WIEGO knew little about the situation of waste-pickers in other countries and continents. Thus, the first step made was to identify and map organizations of waste-pickers and supportive non-governmental organizations (NGOs) and individuals. Through this process, WIEGO found that waste-pickers in many Latin American countries had made great progress in organizing themselves into local co-operatives and national cooperative movements, and were engaged in building a network across Latin

An waste buyer, in Siddhipur, Nepal, who pays for recycled materials with onions

America. With the assistance of the AVINA Foundation and researchers and activists belonging to the Collaborative Working Group on Solid Waste Management in Low- and Middle-Income Countries (CWG), WIEGO was able to forge links between waste-picker organizations in Asia and Latin America. This collaboration resulted in the jointly organized and highly successful First World Conference and Third Latin American Conference of Waste-Pickers, held in Bogotá, Colombia, in March 2008. Since the conference, waste-pickers have continued to build their connections and raise their voices nationally, across regions, and globally.

KEY SHEET 14

FIRST WORLD CONFERENCE AND THIRD LATIN AMERICAN CONFERENCE OF WASTE-PICKERS, BOGOTÁ, COLOMBIA, 1–4 MARCH 2008

Lucia Fernandez and Chris Bonner (WIEGO)

During a sunny morning on 1 March 2008 in Bogotá, Colombia, the First World Conference of Waste-Pickers began. This was a special day for waste-pickers in Colombia, 1 March being the annual Day of the Waste-Picker. Participants were full of energy and were highly motivated to share experiences, to promote networking, to develop global strategies addressing their problems and to plan future actions. There were around 700 people from 34 countries present on this day: waste-pickers (leading the proceedings), non-governmental organizations (NGOs), governmental institutions, development agencies and private enterprises.

During the course of the four-day event, participants had the opportunity to take part in one of five parallel sessions: the experiences of waste-pickers, the progress of waste-pickers in the recycling value chain, opportunities and risks posed by technological changes and privatization, corporate social responsibility, and the role of waste-pickers in the public system of solid waste management. They also had the opportunity to present and discuss local and global networks, and the difficulties and opportunities faced when forging links and maintaining networks. The conference concluded with two declarations prepared by waste-pickers: Declaration of the Third Regional Conference of Latin American Waste-Pickers and the Global Declaration of the First World Conference of Waste-Pickers.

Both declarations focus on increasing recognition of the work done by waste-pickers, improving their position within the waste management system and value chain, transforming waste management systems to make them more environmentally sustainable and socially inclusive, and increasing the organizing and networking capabilities of waste-pickers around the world.

DECLARATION OF THE THIRD REGIONAL CONFERENCE OF LATIN AMERICAN WASTE-PICKERS

In Bogotá, between 1 and 4 March 2008, the delegates of 15 Latin American countries – Argentina, Chile, Peru, Brazil, Bolivia, Mexico, Puerto Rico, Costa Rica, Guatemala, Ecuador, Paraguay, Venezuela, Nicaragua, Haiti and Colombia – gathered as members of grassroots organizations of waste-pickers, also known as *pepenadores, cartoneros, cirujas, clasifi cadores, buceadores, guajeros, minadores, catadores, thawis, barequeros* and countless other denominations according to where they come from.

We declare the following commitments in the framework of the Third Latin American Conference of Waste-Pickers to the public, governments, communities, society in general, cooperation agencies and our own organizations:

- Promote global recognition of the profession of waste-pickers and our organizations through creating spaces for discussion and develop strategies for having an active presence in those spaces.
- Generate actions and strategies for the recognition of the Latin American network of waste-picker organizations and certify the work and the profession of waste-pickers and our organizations.
- Commit to sharing knowledge with waste-pickers and their national organizations, their local structures and the members of the different movements.
- Promote the progress of waste-pickers and their organizations in the value chain to gain access and share in the benefits generated by the activity.
- Contribute to a world mobilization from each country in a connected effort that seeks to have a World Day of the Waste-Picker aimed at recognizing the activity of the people who work as such.
- Demand from governments, by the congress participants, that they prioritize waste-pickers' organizations in the solid waste management system, giving the required conditions for their inclusion through the development of social, financial and environmental affirmative actions.
- Review laws and public policies so that they include waste-pickers' organizations in their formulation, considering them as actors in decision-making.
- Commit to generating the capacity-building, training and knowledge to professionalize the activity (a commitment of the participant organizations).
- Commit globally to promoting contact with as many waste-pickers and waste-picker organizations as possible.
- Advance, together with the world, regional and local committees with the aim of controlling the value chain and its income

through networks and production centres.
- Work to implement the commitments of the Declaration of the Second Latin American Congress of Waste-Pickers.

March 2008

GLOBAL DECLARATION OF THE FIRST WORLD CONFERENCE OF WASTE-PICKERS

At the First World Conference of Waste-Pickers, grassroots organizations of waste-pickers from around the world gathered in Bogotá, Colombia, from 1 to 4 March 2008, representing waste-pickers from Asia, Africa, Europe and Latin America, to make a declaration to the public, governments, support organizations, society in general, and their own organizations, joined by technical adviser delegates, technical support organizations, government representatives, NGOs, universities, enterprises, micro-enterprises and other civil society groups.

We declare:

- Our commitment to work for the social and economic inclusion of the waste-picker population, to promote and strengthen their organizations, to help them move forward

Logo for Wastepickers without Frontiers

in the value chain, and to link with the formal solid waste management systems, which should give priority to waste-pickers and their organizations.

- Our agreement to reject incineration and burial-based processing technologies and to demand and work on schemes of maximum utilization of waste, as activities of reuse, recycling and composting represent popular economy alternatives for informal and marginalized sectors of the world population.

- Our commitment to continue sharing knowledge, experience and technology, as these actions will promote and accelerate contact with the greatest possible number of waste-pickers and their organizations across the world, making visible their living and working conditions and their contributions to sustainable development.

- Our commitment to advocate for improved laws and public policies so that their formulation effectively involves waste-picker organizations. Waste-pickers should become actors in decision-making, searching for improved common conditions, and for capacity-building activities and knowledge for the recognition and professionalization of their work.

March 2008

REFERENCES

www.wiego.org

www.inclusivecities.org

Inclusivity in the reference cities and global good practices

Both in the cities and in good practices from around the world, there are two main areas where inclusivity is important. 'User inclusivity' refers to the active involvement of households, waste generators and other system users in making the ISWM system work. 'Provider inclusivity' refers to broadness of access to the economic opportunities in waste and recycling. One way of understanding the difference between users and providers in an ISWM system is that users are often the ones who pay for and get the service or the benefits, and providers are usually the ones who do the work and earn income from creating those benefits. Of course, in real life it's not so simple; but this helps to keep straight the differences. Inclusivity can be divided into a number of sub-categories:

- *Information inclusivity* loosely translates to building on what is already working. Two specific areas of good practice include:
  - assessing existing systems before making plans;
  - designing interventions and changes based on building on what is already functioning.
- *User inclusivity* has a focus on all types of consultation, communication and involvement of users, both in decision-making and in *doing for themselves*. The sub-categories include:
  - participatory and inclusive planning and system design;
  - inclusivity in siting facilities;
  - self-provisioning, for example, home composting and waste prevention.
- *Provider inclusivity* creates and maintains equity of access to livelihoods and economic niches. Sub-categories include:
  - PPPs and PSPs;
  - working together with informal service providers;
  - inclusive access to valorizing materials.

- *Institutional support for inclusivity* makes it more than just the good will of one city administration. In this, the two main institutions considered are:
  - solid waste platforms and forums;
  - feedback mechanisms and satisfaction surveys.

User inclusivity: Consultation, communication and involvement of users

Civil society participation in decision-making processes on urban environmental issues is mandatory by national or state law in a number of Latin American countries. In Peru, the General Law of Solid Wastes 27314 states that the process of making the Integrated Solid Waste Management Plan must include the main stakeholders. Based on this, an Environmental Municipal Commission (EMC) was created in Cañete, Peru, that developed the plan.

The certification programme *Selo Verde* in north-east Brazil (see Box 5.2) shows how certifying stakeholder inclusivity and environmental policy development can form a driving force for change.

City authorities are essential in giving life and legitimacy to any policies for inclusivity: they need not only to be seen to engage with other stakeholder groups, but to do so with *one voice* representing the view of the municipality.

Box 5.2 *Programa de Selo Verde*, Ceará State, Brazil

In 2003, by state law an environmental certification and award programme for municipalities, called *Programa de Selo Verde*, was introduced by the Ceará State Council for Environment Policy and Management (CONPAM). The programme seeks to improve public policy related to urban environmental management and to stimulate community participation in the process. Subscription to the certification is voluntary and training is provided to municipalities on a range of environmental management and public policy issues. Municipalities who subscribe to this programme have to create a multi-stakeholder environmental commission CONDEMA. By 2008, 146 of the 184 municipalities have applied for certification, which has to be renewed annually and is reviewed yearly on a progressive basis. Municipalities have to demonstrate long-term progressive institutional improvements in terms of environmental legislation, infrastructure planning, environmental education, creation of dedicated funds, and public consultation mechanisms. Solid waste is one of the main topics addressed and the certification programme has stimulated municipalities to introduce improvement measures related to waste management.

KEY SHEET 15

THE EVIDENCE BASE FOR HOUSEHOLD WASTE PREVENTION: HOW BEST TO PROMOTE VOLUNTARY ACTIONS BY HOUSEHOLDS

David C. Wilson (Imperial College London and Research Managing Agent, Defra Waste and Resources Evidence Programme, UK)

In 2005, the UK Department for Environment, Food and Rural Affairs' (Defra's) Waste and Resources Evidence Programme (WREP) commissioned a comprehensive series of research projects on *household waste prevention*. This pioneering body of research sought to understand consumer behaviour in relation to household waste prevention through investigating how different initiatives work in practice. This key sheet presents the findings from a comprehensive synthesis review commissioned by Defra WREP in 2008 (from Brook Lyndhurst, the Social Marketing Practice and the Resource Recovery Forum), which draws together not only the findings from this research, but also the international evidence base.

The review defined waste prevention as including *strict avoidance, reduction at source* (e.g. through home composting) and *reuse* (where products are reused for their original purpose). Recycling was excluded. The review set out to answer questions about the extent to which

waste prevention is practised at the household level; what the barriers and motivations are; and what options and measures exist to encourage waste prevention behaviour, either by engaging directly with households or through the products and services provided to them (including waste collection services).

Over 800 literature sources were identified, of which 88 were selected for detailed review and 48 others for more summary review. An international review drew on 106 sources. Target documents included reports to Defra and the UK Waste and Resources Action Programme (WRAP), together with academic papers and key pieces of practitioner research. The desk element was complemented by further evidence gathered from stakeholders (through web surveys, telephone interviews and workshops).

The key finding – given the breadth and complexity of waste prevention behaviour in the light of an extensive literature review and in the light of considerable international experience – is that a *coherent basket of measures* will be required if waste prevention activity is to increase.

From a *householder point of view*, the review identified that there is no single activity involved in 'waste prevention' since it involves not one but many behaviours. On the basis of reported surveys, these behaviours have very different levels of participation, with one source estimating that up to 60 per cent of the public does at least one of them, at least some of the time.

People attending a home composting clinic organized by the municipality in Dorset

© Dorset County Council

Detailed data is provided for each behaviour in the technical report and related modules.[1]

The literature reveals a general hierarchy in the popularity of performing waste prevention behaviours, from donating goods to charity at the top, through small reuse behaviours around the home, to activities that involve changes in consumption habits at the bottom.

Barriers to engaging householders in waste prevention behaviour operate at both a societal and individual level. At the societal level, modern consumer culture is antipathetic to many of the behaviours required for household waste prevention to occur, particularly by conferring status through the acquisition of 'stuff'. Stemming from this, the public seems genuinely confused about what 'waste prevention' means, and there is a general tendency to think that it is equivalent to recycling, and no more. The recycling norm is now so well developed that it is often hard for people to think beyond this.

A key opportunity to engage the public in waste prevention activity is through campaigns at both local and national level. Campaigns can comprise a mix of interventions and are not restricted merely to communications. Local authorities and stakeholders think that more consolidated evidence is needed on two fronts: what interventions and approaches work; and what communication messages should be used. This evidence will need to be collected from future campaigns, not from past work, because historic data is often weak.

On the products and services side, the review focused on three aspects: reuse infrastructure and the role of third-sector organizations within it; retail development of product refills; and the provision of a service to substitute for appliance ownership (e.g. provision of laundry, garden, DIY services, etc.) on new housing developments (referred to as 'product service systems'). The first of these would appear to have particular potential to contribute to greater prevention of waste.

Many barriers were identified in relation to the further development of a reuse infrastructure. These included operational difficulties (funding, capacity, logistics), difficulties on the consumer side (attitudes towards second-hand goods, lack of knowledge), regulatory issues (concerning, for example, the relative price of waste treatment options) and institutional issues (notably the often poor relationship between stakeholders at the local level). The main opportunities for improving performance are to ensure more strategic planning for reuse in local authority services, and to foster better coordination and joint working between public bodies and the third sector. The review suggested that proper rewards and incentives for reuse activity (e.g. paying reuse credits or a higher cost of landfill compared to the costs involved in reuse) also need to be in place.

The review examined literature that reported the impacts or potentials (based on scenarios) of various policy measures designed to influence either household behaviour directly or the products and services provided to them. The international review suggested that it is the overall package of policy measures and how they complement each other that is the key to successful waste prevention. The package tends

Organizing activities around recycling for children

© Dorset County Council

Promoting smart shopping outside local shops

© Dorset County Council

to combine the following: prevention targets; producer responsibility; householder charging; public-sector funding for pilot projects; collaboration between public-, private- and third-sector organizations; and intense public awareness/ communications campaigns.

The review used the available evidence to estimate the potential of these and other policy measures in the UK (in terms of million tonnes per annum). For local-level interventions, evidence from WRAP in the UK suggests that if the target is tonnage reduction through waste prevention, then the priorities should be to focus on food waste, home composting and bulky waste.

In addition to identifying options for enhancing waste prevention, the review was charged with identifying gaps in the evidence. The principal gap – identified in both the literature and by stakeholders – is robust and comprehensive quantitative data. This applies to almost all aspects of household waste prevention and the situation is probably not dissimilar to that which existed in the early days of recycling. The challenge, now, will be to put in place systems that can capture evidence as it is generated from new local interventions; to develop best practice evidence from leading local authorities; and to continue to investigate key topics at strategic level (e.g. extended product warranties; product life-cycle impacts; waste arising and collection arrangements; consumer attitudes and behaviours to particular waste prevention activities; and so on).

Stakeholders, in particular, want evidence of what actually works on the ground to promote waste prevention and what outcomes (weight, carbon and costs) can be expected from different measures. Work is still required, too, in order to ensure that sensitive and effective monitoring and evaluation mechanisms are in place to gather the evidence that will be needed for the development of the required basket of future policy measures at local and national level.

NOTE

1   See http://randd.defra.gov.uk/.

It helps if there is only one department or section responsible for solid waste within the municipality. In Belo Horizonte, Brazil, the development of the Waste and Citizenship Movement was strengthened by having a designated waste management agency Superintendência de Limpeza Urbana (SLU) representing the municipality in the construction of a new waste management agenda for the city. This was further consolidated when by municipal law a Department of Social Mobilization was created with the SLU whose prime responsibilities are environmental education and provision of technical advice to waste-pickers' organizations and other community-based organizations partnered with the SLU.

■ Information inclusivity

One of the most common failures in modernizing waste management systems is a failure to understand how the system is already working. Understanding how the system works will increase the chances of success in introducing sustainable changes in waste management, improving performance and satisfying users.

Time pressure on planning or consulting may result in too little time spent on investigation; foreign consultants may use a standard approach with no time for understanding the local situation, and the local members of a team, who know something, may not know everything. For this reason it is critical to include a baseline that looks at the behaviours and relationships of providers and users, at strengths and weaknesses of what is already occurring, and at opportunities for improvement that build on strengths and fix weaknesses.

A key focus of the city research is on the role of the informal sector in service provision and recycling. A key good practice is examining, first, what the situation is before making changes and even before planning. So the data collection instruments asked the cities to provide a process flow diagram, and to use this to report on the number of people in both formal and informal waste and recycling sectors, the tonnes that they collected, and the average daily wages that

both earn. The largest number of cities could not or did not report this. From the cities that reported, it is notable that Bengaluru and Delhi in India describe a large contingent of people working in the informal sector. Quezon City in the Philippines also has many informal recyclers; but they have been partially formalized. Other cities, which report large numbers of informals but do not have an estimate of the numbers, include, for the informal recycling sector, Dhaka, Kunming, Varna and Managua, and for informal service providers, Lusaka and Bamako. Discussions with city profilers suggest that there are informal recyclers active as well in San Francisco, Curepipe and Sousse.

■ User inclusivity: Consultation, communication and involvement of users

Changing or modernizing the waste system implies changing people's habitual behaviours. In a city in a low- or middle-income country seeking to expand service coverage, this may mean persuading people to put their waste in a household or communal container rather than dumping it in the street. In Northern Europe, it's about bringing e-waste to the EPR depot rather than storing it in the closet or putting it in the residual waste bin. They seem different, but the structure is the same: the users and providers have to agree on new practices that change business as usual. It is for this reason that involving the service users is critical to the success of any waste management system.

One of the key reasons for authorities to endorse an inclusive approach, involving all stakeholders from the beginning, is to avoid problems later on. Establishing a platform and/or developing the strategic plan in a participatory manner provide a solid foundation of socio-political dialogue. With a clear and transparent approach, and the political commitment to open up the decision-making process, the final results are left open, giving the decisions back to the people. It is thus possible to move from argument to implementation.

A good service is one that people will use and be willing to pay for, something that is more likely if they have been involved in its design and have had something to say about the planning. Moreover, providing a good collection service to slum areas as well as middle-class districts is more than just an equity issue – infectious diseases or an economic blight will affect the whole city.

The Waste and Citizenship Movement in Brazil is a grassroots approach to inclusivity that started in Belo Horizonte. Brazil has received recognition as a global leader in good practices promoting inclusivity both for users and providers. In the reference cities, user inclusivity, also called community participation, is a key feature in Adelaide, Tompkins County, Ghorahi, Bamako, Bengaluru, Cañete and, to a lesser extent, in Delhi, Rotterdam and San Francisco. In all the cities, it is related to consultation with citizens, to involvement of stakeholders in planning and to the presence of user feedback mechanisms.

Table 5.1 shows two complementary indicators of user participation and inclusivity in the reference cities. The cities are grouped according to the number of platforms, forums, etc. (many; few or one; none) and as to whether there are legal measures in place to ensure participation in the (waste) planning process.

First, in relation to participation in planning processes, an assessment was made of indicators as a presence of stakeholder platforms in the reference cities, as well as the existence of legal measures for planning and participation.

In Bamako, Quezon City, Belo Horizonte, Managua and Delhi, specifically, there are legal measures that obligate, establish, promote or support participation, and at the same time these cities have a rich diversity of participation mechanisms in the form of platforms and forums. Cities such as Dhaka and Kunming have reported

neither having legal measures related to participation in planning processes or the presence of participation platforms.

In some cities, there is strong stakeholder participation without the presence of supporting legal measures. Moshi, Tanzania, is a good example of stakeholder participation, and dialogue and communication between the municipality and citizens, as well as CBOs. A stakeholder platform has been active since 1999. In Moshi the following aspects of user inclusivity are reported:

- commitment by the municipal council and dialogue with citizens and other stakeholders;
- learning by doing: practising stakeholder dialogue and a pilot project with a private contractor; initiatives to engage CBOs; recognition of their status; and
- active stakeholder participation in the form of discussion forums, stakeholder platforms and involvement in decision-making.

■ Inclusivity in planning and siting

Policy-makers and practitioners around the world know how difficult it is to develop the ISWM sector. An inclusive planning process has proven to be a key to success. The most comprehensive guidance for inclusive strategic waste management planning was published by the World Bank in 2001,[7] and has since been used in many locations. The *Strategic Planning Guide for Municipal Solid Waste Management* (SPG) was one of the outputs of an international donor and practitioner platform called the CWG, which stands for Collaborative Working Group on Solid Waste Management in Low- and Middle-Income Countries.[8] The SPG was written by Environmental Resources Management (ERM)-UK with contributions from many international organizations and practitioners working on the issue around the world.

The SPG is based on a stepwise planning process, where stakeholder working groups are entrusted to research options and propose specific aspects of the planned new system. The

Table 5.1

Participation and inclusivity in the reference cities

| | Legal measures for planning and participation | |
Platforms, forums, etc.	Yes	No
Many	Belo Horizonte, Bamako, Delhi, Quezon City, Managua	Bengaluru
Few, one	Adelaide, Cañete, Ghorahi, Rotterdam	Moshi
None	Lusaka	Dhaka, Kunming

SPG emphasizes the need for 'facilitators' to manage a complex process of discussion and debate between stakeholders. The outputs from the process are a 'strategy' and 'action plan', with the strategy focusing on those issues which stakeholders can agree on, and the action plan dealing with the often more contentious measures required to implement the strategy, such as specific technologies and sites for waste infrastructure.

The assessment steps in the SPG were used as the basis for strategic planning in Bengaluru, where it served to unite the activities of the Swabhimana platform with the high-profile activities of the Bangalore Agenda Task Force (BATF), as well as in Bamako. In both places it contributed to and strengthened technical outcomes and the institutionalization of inclusivity in their respective platform structures.

Further testing of the strategic planning guide was funded by the UK Department for International Development (DFID), and resulted in a series of practical *Waste Keysheets*[9] to assist in stakeholder consensus building, and with implementing the initial steps in the planning process. Training materials were prepared as part of a later World Bank–Mediterranean Environmental Technical Assistance Programme (METAP) project.[10]

■ Inclusivity in siting facilities

Modernization of waste management includes developing environmentally sound facilities. A key part of an ISWM solution, at least in the medium term, will be somewhere between a fully fledged sanitary landfill and a modest but well-operated controlled disposal facility. Upgrading an existing disposal site may be an appropriate first step; but at some point, most cities will have to take the decision as to where to site a new landfill. Economies of scale are likely to favour a regional facility, so issues of inter-municipal cooperation and equity between communities become important.

When the focus shifts from simple collection and removal to environmentally appropriate

> **Box 5.3 Participatory urban waste planning in Vietnam**
>
> In 1998, participatory planning in the urban waste management sector was tried out for the first time in Ha Long and Cam Pha, a beautiful United Nations Educational, Scientific and Cultural Organization (UNESCO) World Heritage location in the north-east of Vietnam.
>
> All of the major stakeholders involved in and influencing urban waste management were invited to join the process in a structured series of workshops in order to prepare a provincial waste management strategy. There were no prearranged results, and the final form of the plan depended purely on the results of the process.
>
> The initiative was a great success and resulted in the creation of a fresh approach to waste management planning. The World Bank Strategic Planning Guide methodology was born.

disposal, the stakes for participation increase because very few people want a large new landfill site next to their home. Inclusivity also means transparency in communicating about what the risks are and making agreements for sharing the risks as well as the benefits. Inclusivity in risk-sharing – especially in relation to siting disposal facilities – connects users, providers and the local authorities in one set of governance processes and social agreements.

The traditional top-down approach to siting has been for the city to hire consultants to find the 'best' site, which is usually based on technical criteria with a focus on geology and geography. People living near candidate sites organize themselves into protest groups, and 'battle lines' are drawn. The city is labelled as non-democratic, the protesters are written off as 'yet another example of NIMBY' – that is, 'not in my backyard'. Once that has happened, effective decision-making ends, rhetoric escalates and agreements become close to impossible. Even when siting is based on technically correct reasoning, and engineers carry out state-of-the-art environmental and social impact assessments, the results may still be perceived as biased and politically tainted.

One specific area related to user inclusivity in planning processes is inclusivity in siting processes for new disposal facilities, and the related risk-sharing among stakeholders. No one wants to live next to a waste management facility, but someone has to. How successful are authorities in avoiding NIMBY situations so that

The rural county of Tompkins, New York, with its population of approximately 100,000, had a full landfill in the mid 1980s. Like most other US counties, they accepted the challenge of regionalization and hired technical consultants to identify geologically appropriate sites for the landfill.

When they had the shortlist, the county decided to depart from the normal ways of doing business. They held community meetings in each community and asked one simple question: 'Suppose your community turns out to really be the best and the most environmentally sound location for our new regional landfill. What would your community need in order to make it acceptable that the landfill comes here?'

Several communities reacted positively to being involved in this way. The one that was eventually selected as technically the best asked for a new school and recreation centre; for house prices in the community to be benchmarked at current values; for the county to guarantee it would buy any house within a certain radius of the landfill for that (inflation-corrected) benchmark price, for an agreed-upon period of years during construction and after opening; and for the municipality to receive a 'host community fee' for each tonne of waste disposed of over the life of the landfill.

The total cost of all of these measures to the county was a fraction of what was usually spent at the time on legal fees in settling siting issues, and the host community was content.

The key was that the local authority asked stakeholders for their opinions, listened to their answers and respected their position.

resources. Within solid waste system upgrading, implementation relates to the same two basic fields of work:[11] services and commodities. In this sub-section the focus is on inclusivity in solid waste services, generally related to sweeping, cleaning or removing waste, litter, excreta or undesired materials from living areas and commercial districts and 'disposing' of them.

For there to be a service, there have to be service providers. Inclusivity for providers is used to refer to the attitudes and actions of the *public authorities* towards individual, micro-, small- and medium-sized *private economic actors*. The private sector in this case also includes civil society providers – that is, community-based organizations (CBOs), non-governmental organizations (NGOs), and other actors who want to earn livelihoods.

Waste collection *service* providers can be public or private, informal or formal, large or small, local or international. They can use their own muscles for energy and sweeping, or move waste with animal traction, or be small, medium, or large motorized vehicles of all types. There is already a large literature and experience with inclusivity in services under the terms (pro-poor) *public–private partnership (PPP or 5-Ps), private-sector participation (PSP)*, as well as a large body of experience on micro-privatization in East Africa.

citizens feel sufficiently included and the trust level is high? Tompkins County in New York State provides an interesting case study of a successful participative siting process.

■ **Provider inclusivity: Working with informal and formal entrepreneurs in providing services**

Provider inclusivity is about ease of entry for economic actors in relation to providing solid waste services, and extracting and valorizing

---

Profile of informal activities in solid waste.

Half of the cities have given information or estimates for the number of informal sector workers in their cities. These 10 cities together have a total of 350,000 informal workers, who collect an average of 32 tonnes per person per year, or just under 3 tonnes per person per month

Note: Figures in italic are estimated

City	Workers, informal sector	Tonnes collected per worker, per year, informal	Informal-sector workers as percentage of total population	Informal sector workers per km²
Belo Horizonte	421	24	0.0%	1
Bengaluru	*40,000*	6	0.5%	*50*
Canete	176	7	0.4%	0
Delhi	173,832	5	1.3%	117
Dhaka	*120,000*	2	1.7%	*329*
Ghorahi	39	8	0.1%	1
Lusaka	480	205	0.0%	1
Managua	3465	18	0.3%	12
Quezon City	14,028	17	0.5%	87
Sousse	*150*	27	0.1%	*3*
Total	352,591			
Average	35,259	32	0.5%	60
Median	1973	12	0.4%	8

KEY SHEET 16
THE INTERNATIONAL LABOUR ORGANIZATION AND THE MODEL OF MICRO-FRANCHISING IN EAST AFRICA

Alodia Ishengoma (International Labour Organization, Tanzania Office, Dar es Salaam, Tanzania)

One approach to address the need for jobs for the poor and better services is to involve the micro-private sector – small- and medium-sized enterprises (SMEs), community-based organizations (CBOs), community-based enterprises (CBEs) and non-governmental organizations (NGOs) in the provision of municipal services. In East Africa, Dar es Salaam city authorities were the first to adopt this method for solid waste collection. The idea was conceived in 1992 with the support of the Global Sustainable Cities Programme (SCP), which started as the Sustainable Dar es Salaam Project (SDP), financially supported by the United Nations Development Programme (UNDP) with the United Nations Centre for Human Settlements (Habitat), or UNCHS (now UN-Habitat) providing the steering technical assistance. The decision to privatize solid waste management (SWM) services in Dar es Salaam was made not only against a background of the dismal performance in collection at less than 4 per cent, but privatization fitted well in the overall government policy, which sought to reduce the state's involvement in the provision of services. Privatization of a franchise type was opted for and done gradually, undergoing significant changes in approaches, in part driven by rising costs and the behaviour of stakeholders.

Prior to the first privatization that was implemented in September 1994, some donor-funded emergency clean-up campaigns were conducted until August 1994, followed by piloting with only one contractor company, hired to provide collection services in ten city centre wards. In 1993, the city authorities enacted a refuse collection and disposal law, which also covered the same ten wards. However, this system underperformed. The situation deteriorated to the extent that in June 1996 the government decided to dissolve the then Dar es Salaam City Council (DCCl) and put in place the Dar es Salaam City Commission (DCCn) to address the problem of solid waste. Immediately, five companies were contracted to provide service to 24 wards. Collection increased to 20 per cent by 1997. Together with the International Labour Organization (ILO), UNCHS (Habitat), through their SDP, the DCCn, developed a strategy for increased privatization to cover the whole city. Conducted in August 1998, increased privatization attracted some 70 franchisees, most of which started operating by January

CBO members using simple working tools to transfer collected waste to the communal container, Dar es Salaam, Tanzania

© Alodia Ishengoma

1999. Collection rates rose to about 40 per cent by the end of 1999, but were collected in only 44 out of 73 wards. The DCCn term ended in January 2000 when the Dar es Salaam City was divided into three municipalities under local governance of one city council – the Dar es Salaam City Council and three municipal councils – namely, Ilala, Kinondoni and Temeke.The SWM service was decentralized into these three municipalities that continued to follow the same procedure of supervision and to support the operations of the franchisees. The DCCl retained its authority over cross-cutting issues, including running of the common dumpsite.

The ILO was invited in 1997 to give input in the field of job creation and the involvement of SMEs and CBOs. This marked the commencement of a series of public–private partnership (PPP)-based SWM projects, which were replicated in another 7 and 15 municipalities in Somalia and in East Africa (Kenya, Tanzania and Uganda), respectively, building on the previous experiences in the SWM project in Dar es Salaam (1997 to 2003), particularly ensuring that the PPP approach never caused loss of livelihoods when applied to urban service delivery systems. The PPP was accepted as a solution. Through this approach, franchisees were empowered by other stakeholders – UN-Habitat, the World Health Organization (WHO), the United Nations Development Programme (UNDP)/LIFE, the Danish International Development Agency (DANIDA), the United Nations Industrial Development Organization (UNIDO), CARE and the Swedish International Development Cooperation Agency (Sida) – to implement sustainable SWM. Some franchisees started with very little capital but were able to get basic working tools. Although recycling activities remained informal, recyclers were also supported technically to complement collection.

The involvement of the private sector is clearly a success. The cities became considerably cleaner than they previously were. Privatization created a great potential for improving people's lives, especially among the women and youths who did not have jobs. The earlier phase of limited privatization had created about 2300 jobs and involved some 52 franchisees. With increased privatization, over 4000 jobs were created in East Africa by the end of June 2006 with the use of funds from UNDP (1997 to 2001), the ILO Programme on Boosting Employment through Small Enterprise Development (IFP/SEED) and ILO–ASIST (2002 to 2003), and from the UK Department for International Development (DFID) (2004 to 2006). It resulted in other significant achievements, such as stopping child labour in waste management, and women and men becoming involved in activities they had not done before (e.g. women pushing waste handcarts and loading trucks and men doing road sweeping).

Despite these achievements, some issues were not resolved. The franchisee selection process turned out to be difficult. Applicants could not meet all of the requirements. Most low-income areas did not have franchisee applicants to serve them. Other issues were poor working conditions and unwillingness to pay for the collection service, which have both remained a big challenge.

The lesson learned from this experience in Dar es Salaam is that most, if not all, municipal authorities in East Africa are in favour of privatizing solid waste collection and disposal services; but they lack the strategy and capacity for implementation and so they continue requesting support even after project implementation. The PPP strategy with proper support, especially in building capacity for business development, project management, supervision and procurement, appears to be among the best options for SWM. There are various PPP options or arrangements depending on the situation; hence, the best is yet to be identified. The commitment and active cooperation of various stakeholders, including policy-makers, is a prerequisite for effective private solid waste collection service delivery. Up-scaling requires appropriate public–private partnership arrangements built on trust, business principles and legal reinforcement. The SWM PPPs are in line with the poverty reduction strategy processes and the Millennium Development Goals (MDGs). These strategies make it possible to create employment and generate income from waste. But making these strategies work in cities will require expanding the skills of the city planning and management staff.

Still more remains to be done in PPPs. Time and resources are required for the PPP programmes to register impact upon the ground. Systems and behaviour changes require time and patience. It would be helpful to conduct a study on poverty relative to willingness to pay for PPPs in SWM and to determine a proper enabling environment where people earn a living and become willing to pay a SWM fee. Another good study would be on cross-subsidies (i.e. to determine appropriate cross-subsidization schemes for different types of service recipients – for example, where the businesses pay more and households pay less, or according to income and per solid waste volume/weight). The bottom line

is that there is a need to continue support for micro-private and informal-sector participation in pro-poor PPP SWM services delivery in order to:

- consolidate what has already been achieved; and
- improve the sustainability of the systems that have been developed and produce a reliable and tested model that can be recommended with confidence to other developing cities/municipalities.

CBO members emptying collected solid waste into the communal container, Dar es Salaam, Tanzania

© Alodia Ishengoma

Inclusivity in relation to providers means above all that the economic niches, which produce income, are available to a wide variety of sizes and types of enterprises, and that conditions of entry and access to contracts are transparent, fairly allocated and managed according to rule of law. In the cities, this has been reported in two main areas:

1   in relation to rules enabling and regulating participation of the private sector in PSPs, PPPs, contracts and the like, and whether these rules also allow, protect and regulate the participation of micro- and small enterprises (MSEs), informal service providers (ISPs); individual entrepreneurs; CBOs and NGOs;

2   with specific reference to recycling are the access and claims of micro- and small, especially informal, waste-pickers and recyclers to pick materials that are secured and protected.

In cities such as Adelaide, San Francisco, Varna, Kunming and Rotterdam, the informal sector is reported to play no role in the handling of waste; in cities such as Moshi, Quezon City, Delhi and Bengaluru, on the other hand, the informal

sector is responsible for 50 to 100 per cent of all ongoing waste activities.

■ Inclusive access to valorizing materials and working in municipal recycling

Provider inclusivity is, if anything, more important in relation to resource management. In many cities, at least as many – if not considerably more – people earn their livelihoods in recovery, valorization and recycling as are employed in the public services of waste collection and street sweeping. In fact, many formal employees of the waste system supplement their income or personal possessions by separating materials for repair, reuse, recycling and feeding animals.

The informal sector comprises recycling experts, working efficiently but under poor working conditions and without recognition; they are frequently ignored, denied or harassed. The new-found realization that they make such a significant economic contribution to reducing the burden of waste management of city authorities will hopefully secure informal recyclers the room to be treated as professionals and key stakeholders in ISWM in their cities.

A lively global discussion and associated advocacy in Brazil, India, and other middle-

Table 5.3

Formal and informal participation in waste system

Notes: NA = not available; NR = not reported. Figures in *italic* are estimates. Belo Horizonte: formal tonnage excludes C&D waste. Moshi: informal tonnage calculated based on tonnes handled by formal sector and tonnes lost.

City	Tonnes handled, formal	Tonnes handled, informal
Adelaide, Australia	742,807	0
Bamako, Mali	263,469	198,757
Belo Horizonte, Brazil	1,286,666	9900
Bengaluru, India	*1,662,210*	226,665
Canete, Peru	8632	1270
Curepipe, Republic of Mauritius	23,764	0
Delhi, India	1,677,237	822,163
Dhaka, Bangladesh	1,018,000	*210,240*
Ghorahi, Nepal	2275	300
Kunming, China	803,000	NA
Lusaka, Zambia	90,720	98,170
Managua, Nicaragua	375,220	61,685
Moshi, Tanzania	40,150	*15,695*
Nairobi, Kenya	NR	NR
Quezon City, Philippines	494,984	231,878
Rotterdam, Netherlands	307,962	0
San Francisco, USA	508,323	NR
Sousse, Tunisia	64,168	4000
Tompkins County, USA	58,401	NR
Varna, Bulgaria	*100,715*	35,207
Average	501,511	119,746
Median	307,962	25,451

Box 5.5 The early bird gets the ... e-waste

Delhi is one of the hubs of electronic waste recycling in India, with approximately 25,000 individuals in the informal sector dismantling and extracting materials from all kinds of e-waste – computers, televisions and even washing machines. Repeated campaigning by global groups has drawn a great deal of attention to the toxic nature of some aspects of this recycling. However, the actual recyclers have been, until recently, sidelined in these debates and new legislation may divert waste away from them. Besides, several formal-sector players have also started entering and scoping the market. Now, some of these informal recyclers have worked with the non-governmental organization (NGO) Chintan and the Deutsche Gesellschaft für Technische Zusammenarbeit's (GTZ's) ASEM Programme, forming an association, 4R, with the intention of developing themselves into formal players, earning a legal and legitimate livelihood. While in the solid waste sector, informal actors reacted after the formal actors began their operations; here, the informal players have been able to react pre-emptively. This also points to the power of information at the right time.

income countries is gradually leading to a shift in power relations between the informal recyclers and service providers, on the one hand, and formal institutions of government, industry, the financial sector, the broader society and the recycling supply chain, on the other. The Philippines has recently published a national framework plan for the informal sector in solid waste management.[12] Experts, advocates and waste-pickers themselves are talking in global meetings such as the First World Conference of Waste-Pickers in Bogotá, Colombia, in March 2008,[13] and the Collaborative Working Group on Solid Waste Management in Low- and Middle-Income Countries (CWG) workshop on Waste Management in the Real World in Cluj-Napoca, Romania, one month earlier.

The activity, research and global discussions have produced a compelling body of evidence that the models for sustainable, affordable waste management and recycling outside of the developed world work best when they are built around the integration of waste-pickers and other informal recyclers and service providers into modernizing ISWM systems. When this happens, the resulting systems are robust, socially responsible and economically productive. Poor performance, strong stakeholder resistance, poor design and serious overcapitalization are some of the risks that the recycling systems might face if the integration principle is not respected.

In evaluating the attention that the reference cities give to provider inclusivity, one set of key indicators can be understood through the institutional context. Where there is inclusivity, there will usually be a clear and institutional space that allows individuals, and micro- and community, formal and informal, private-sector enterprises to participate as contractors or initiators or independent traders. A legal context that invites and protects this often means that individuals are recognized as having official occupations. 6 of the 20 cities report some form of institutional support to informal, micro- and small entrepreneurs, and 4 report that there is

City	Support organizations, movements, initiatives	Occupational recognition
Adelaide	NR	3
Bamako	5 or more	NR
Belo Horizonte	2	1
Bengaluru	1294	1
Cañete	1	None
Delhi	4	NR
Managua	3	NR
Quezon City	5 or more	1

Table 5.4

Institutional context for inclusive provider practices

Note: NR = not reported.

city or national occupational recognition of informal waste collection and/or recycling. Institutional support to the rights of the *catadores* in Belo Horizonte, Brazil, has contributed to 'waste-picker' being recognized as an official profession, as detailed further on.

Related to this is the key question of whether an inclusive enabling environment makes a difference, in practice, in terms of what happens on the ground. For example, does recognition and inclusion of informal recycling result in significant tonnages of waste, recyclables or organic materials moving through the informal system? Similarly, a mixed solid waste system gives opportunities to many stakeholders to earn livelihoods, conserve resources and keep the city clean and healthy. Internal diversity promotes sustainability.

From Tables 5.3 and 5.4, it is possible to see that Delhi, Bengaluru and Quezon City stand out by having a strong support structure for the informal sector combined with an important role played by that sector in handling the waste. Interestingly, in Belo Horizonte, known for its pro-inclusivity approach towards waste-pickers, the amounts actually processed by the informal sector are minimal.

By contributing to the diversion of waste from disposal, informal-sector recyclers are actually helping the city stretch its budget for waste management. One aspect of good practice is to encourage city authorities to recognize, engage with and cooperate with the informal recycling sector. If such efforts result in increasing the quantities of materials recycled, there is a case to be made for investing part of the savings in improving the working conditions of the informal recycling sector – for example, by financing equipment or paying for health insurance.

Such positive engagement is beginning to happen, particularly in Brazil, as Box 5.6 suggests. Cities in all kinds of countries may learn from Brazil's positive experiences.

Resource valorization, which involves extracting valuable materials and items from the waste stream, cleaning and upgrading them, and trading them as inputs to industry, commerce or agriculture, is the second area where inclusivity is critical.

A recent court case in Colombia has ruled in favour of waste-pickers who were being denied access and may represent a landmark ruling worldwide.

Efforts of local and international NGOs to help the pickers often focus on their social and

educational weaknesses, lack of identification papers and unhygienic work conditions. Such a focus tends to put donors on the side of pushing the informal waste sector to exit to work in other low(er)-paid occupations, which is also a general focus of the ongoing International Labour Organization (ILO) initiatives, working to eradicate child labour in scavenging.[16] There are other initiatives that aim to 'clean up the streets and get rid of waste-pickers' or to 'save people from this horrendous and undignified work'. Yet the millions of informal service providers and recyclers in Asia, Africa, the Americas or Eastern Europe are professionals in a legitimate, if not legalized, economic activity.

■ **Institutionalizing inclusivity: The solid waste platform**

There is increasing acceptance of the need for stakeholder mobilization during the planning and development process; but such active inclusivity is seldom maintained once the system is in place. Yet, a successful ISWM system also needs to stabilize and institutionalize practices and mechanisms for two-way communication between all stakeholders – in particular, the municipal authorities, the service users, both formal- and informal-sector service providers, and the wider community.

A platform for dialogue on solid waste issues is one example of how to increase ownership and to anchor institutional memory in ISWM. A platform is often created in solid waste management at the beginning of the modernization process, under the general rubric of 'stakeholder mobilization'. It is commonly initiated or convened by an NGO, and brings together a diverse group of individuals, businesses, organizations, municipal and government officials, and institutions. Bengaluru is a classic solid waste platform in the broadest sense of the word, and its characteristics serve as a general description of what a platform is and does.

A platform maintains open channels of communication between actors who are normally isolated from, or actively antagonistic to, each other. What makes a platform more than just a

Box 5.8 The Swabhimana platform: Key to waste planning in Bengaluru[17]

The Swabhimana platform in Bengaluru, India:

- provides representatives for planning or evaluation teams or meetings;
- sponsors, promotes, organizes and attends events, ranging from promotional days to study tours to training events to working meetings;
- organizes individuals into working groups for specific purposes;
- mobilizes technical expertise to complement or balance the expertise offered by the formal authorities;
- shares information among the members and also with other platforms; and
- prepares or commissions key knowledge products, such as handbooks and brochures.

Another example comes from Bamako in Mali, where each commune has its own multi-stakeholder platform. Unusually, the Bamako platforms have a role in operations: they represent a forum for users and providers and the local authorities to talk to each other and resolve issues.

series of meetings is its continuity over time, the fact that it does not depend on the results of elections, and the fact that it provides a safe social space for discussing differences, resolving conflicts and arriving at a common way of looking at the situation. A key feature of platforms is that they have permeable boundaries; 'members' are self-selecting and represent themselves as much as their organizations; and the local authorities neither own them nor control their activities. This last feature, in particular, makes platforms an important host organization for long-term processes and a repository of institutional memory. Unlike elected government, platforms survive elections and make a bridge to new administrations.

Box 5.9 COGEVAD: A stakeholder platform in Mali[18]

COGEVAD, the Committee for the Management and Recycling of Waste in Commune VI (one of six cities in the Bamako district of Mali), was a platform in the Urban Waste Expertise Programme (UWEP). COGEVAD and the corresponding platform in Commune IV, COPIDUC, were the focus of the UWEP programme's exit strategy from Bamako. Each platform became the formal owner of the physical, social and information infrastructure in its city, and the institutional home for further developments in waste management.

■ Feedback mechanisms

The presence of feedback mechanisms and surveys is an expression of satisfaction level of households that was examined in the reference cities (see Table 4.4, page 102). Cities such as Adelaide, Belo Horizonte, Managua and Rotterdam indicated that they have feedback mechanisms. However, when feedback systems are in place, the satisfaction levels of the service may not actually be recorded – and in some of the reference cities, it is unclear whether there is satisfaction monitoring. Bamako and Quezon City are two of the few examples where satisfaction levels has have been reported. The example of Quezon City clearly shows the importance of the relation between implementing a new waste management system and receiving feedback on how the new system is perceived by the users.

Campaigns or competitions for the cleanest city in a country or at state level can strengthen feedback mechanisms and foster the sense of all

Box 5.10 Integrating user feedback mechanisms with municipal monitoring and supervising tasks in Quezon City, the Philippines

In 2002, Quezon City introduced the package clean-up system of waste collection to replace the old system where city-contracted hauliers are paid on a 'per trip' basis. Under the new system, the hauliers are paid based on the computed hauling requirements of their assigned routes.

Under the terms of reference of the city-contracted hauliers, a clear set of guidelines was formulated prescribing the truck's and its personnel's equipment requirements and the proper conduct of garbage collection, enumerating any corresponding fines and penalties.

The most common types of violations committed by the city-contracted hauliers are the following: backlogs (no or incomplete garbage collection in a cell), incomplete equipment, improper collection, spillage while in transit, scavenging while on route, unauthorized crew, and soliciting money during collection, among others.

In 2008, the patrolling monitors of the City's Garbage Collection Section reported violations translating to almost 2.5 million Philippine pesos in deductions to payments due to the city-contracted hauliers. Because of these monitoring efforts, garbage collection efficiency is maintained at near perfect rates (99 per cent) and problems in waste collection are quickly reported and resolved.

Most recently, in the results of the 2007 Centre for Health Development of the City's Department of Health on Waste Collection Services Survey, an impressive 104 per cent of households in the city report having satisfactory waste collection services – the highest in all municipalities included in the survey, the rate even exceeding the total number of registered households in the city, a result of the effort placed in covering the waste collection needs of even those in depressed areas inaccessible to dump trucks. As of 2006, there are 239 identified inaccessible areas in the city, 88 per cent of which are now serviced by pushcarts and pedi-cabs under the package clean-up system. This process led Quezon City to being considered the second cleanest city in Metro Manila in 2004.

stakeholders contributing to the cleanliness of the city. The inclusivity extends beyond the city boundaries and the city sets an example for its neighbours. In Tanzania, Moshi has won the official title of the country's cleanest city, several years in a row. The city council is committed to achieving higher levels of cleanliness and maintaining the status and the good image of the cleanest municipality in Tanzania. Other stakeholders from the grassroots to the municipal level are also involved, including the Chaga and Pare tribes, who hold cleanliness in high esteem in their culture, regardless of income.

The institutionalization of feedback mechanisms is therefore also important. Strategies include creating different windows for receiving feedback, providing accurate and timely follow-up to the user, and incorporating user ratings in employee evaluation and reward systems so that the workers also modify their behaviour.

Much feedback is related to complaints, such as reports of illegal dumping or non-compliance with the expected collection frequency and time schedule. Here it is essential for the municipality to provide rapid follow-up in order to establish the nature of the non-compliance and to correct it. This requires involving the private service provider if collection is outsourced.

From the provider side, feedback needs to go beyond recording and immediate response. Patterns of complaints need to be analysed and corrected, resulting perhaps in adjustment of collection routes and established times and locations. This avoids the situation mentioned in Byala, Bulgaria, where lack of user feedback contributed to the municipality spending money to collect nearly empty containers in the off-season. Involving the user in this process from the start can be fruitful, as was seen in Catia La Mar, Venezuela.

FINANCIAL SUSTAINABILITY

An economist labels an activity as financially sustainable if it earns more than it costs and supply meets demand. For a purely commercial activity, in this case both supplier and provider are happy and nothing needs to be done. Financial sustainability in waste management is a more complex issue because waste management, from an economist's point of view, is all of the following:

- a demand-driven business;
- a policy-driven activity; and
- a public good.

Financial sustainability is important, in different ways for all of the three ISWM physical elements: collection, disposal and resource management.

Collection

Collection of waste is a public good for which there is a need in all cities and a matching demand in all cities. It is associated with the public health driver, and as such provides both a public and a strong private benefit. Collection is a public service that benefits all city residents by cleaning the city and protecting public health and the environment. Those who benefit but do not pay, the so-called free-riders, make the system more expensive, and put pressure on the providers to either enforce fee payment or create incentives for payment. As with any public good, access to benefits is easy, and it is costly to exclude free-

Box 5.11 Involving the community in assessing collection routes in Catia La Mar, Venezuela

In Catia La Mar, Venezuela, community representatives were trained by local NGOs in coordination with municipal field inspectors. They were asked to assess the collection routes, to observe set-out practices from the users, to record types of containers employed, to observe the impact of including bulky waste on punctuality, and to evaluate loading practices. This led to an expected improvement in the collection routes, but an unexpected bonus in that the *users* saw the importance of modifying their waste management practices. This included using receptacles that could accommodate storage of waste between collection days without causing odour or nuisance; and calling the municipality to make use of the specific service already offered for collecting bulky waste rather than mixing it with the domestic waste.

The municipality institutionalized the mutual behaviour change in publishing and distributing annual collection calendars with information of the routes, schedules, expected rules related to solid waste collection, and contact details where the citizens can place their queries and give their complaints.

riders. In Bamako, Mali, for example, the definition of 'household' creates an opportunity for free-riders. There is a clear solid waste fee, which is assessed per household – and most households pay. The problem arises because 'household' often means more than one family, either because a man has more than one wife, and each has their own household within the compound, or because different generations live together. The result is that some 'households' of 20 people pay the same fee as a household of 5 people. Free-riders also exist in high-income countries in the form of 'garbage tourists' who carry their waste to an area where disposal is free in order to avoid paying the per-bag or per-can fee in their own home area. The role of the public authorities is to ensure that levels of cleanliness are maintained in spite of individual behaviour.

Economic demand means that someone – in our case, citizens and businesses in the city – chooses and values these benefits enough to be willing to pay for them. Therefore, the operational cost to provide levels of collection is affordable for cities as a whole, and can often be recovered from users.

In the reference cities, if there is an issue with providing and paying for waste collection services, it is generally a matter of managing free-riders and cross-subsidies, rather than a lack of demand or insufficient willingness to pay. For example in Moshi, Tanzania, citizens already think that they are paying for waste management: there is no transparent communication about what is being paid for, and what is the value for the money. Non-payment of fees or poor cost recovery may also occur in instances when collection service exceeds demand or is simply inefficient. This was the case in Byala, Bulgaria, during the early 1990s, where collection trucks came by too frequently and travelled half empty, raising the costs of collection.

In low- and middle-income cities, providing waste collection services in slum areas or poor neighbourhoods can be a challenge. Such is the case in Dhaka, Bangladesh, where this is due to complicated legal issues and the lack of cross-

Raising awareness among citizens to pay for waste collection in Maputo, Mozambique goes hand in hand with collection service improvement

© Joachim Stretz

subsidies. In contrast, in some cases low-income neighbourhoods actually do pay to private (informal) waste collectors, as happens in many cities in Latin America; but the service provided is not adequate. Normally this occurs in those areas where there is no official waste collection service either provided by the municipality or a formal private contractor. As such, the informal sector provides an alternative *private solution*, where prices are set according to market rules. This is often on a fee-per-bag basis. Citizens could at times actually be paying up to five times the collection fee that the municipality would be charging. However, they are willing to do so because their demand of their waste being removed from their household is serviced, even though the informal waste collector might just bring the waste to the next vacant lot or river a couple of blocks away, or leave it at the official communal waste collection point, where the collection truck of the official system would pass the next day.

Disposal

Disposal is associated with the environmental protection driver. Disposal cost rises in tandem with increasing environmental standards for waste management, and the investment and operating costs of meeting the standards and reducing emissions to the air, water and soil. Increased awareness and scientific knowledge of the environmental impacts of waste management drive both the demand for improvements and the cost of implementing them. While environmental groups

Box 5.12 Where there is a will, there is a way

Ghorahi, Nepal, is a small economically undeveloped town far away from the capital city of Kathmandu. Nevertheless, the municipality has done an impressive job in addressing the environmental issues related to waste disposal through its own initiative and active participation of local stakeholders, without externally funded projects or technical assistance. The municipality developed its own Karauti Danda sanitary landfill – one of only three landfills in the country, and all with its own financing.

and those living close to the dumpsite may be vocal in demanding improvements, 'demand' for controlled disposal cannot be understood or analysed as market demand in the same way that demand for collection can: most of the users are not direct beneficiaries (those living close to the dumpsite are the only ones that are). For this reason, availability of investment funds and full cost recovery for users of engineered landfills with high environmental protection standards are more likely in high-income countries, where income is higher and fees generate surpluses, than in low- and middle-income countries.

The ability to plan, develop, and provide long-term disposal capacity depends, in many cases, on the availability of capital financing on the one hand, and the ability to raise operational funds on the other. Such is the case in Bamako, which has no controlled disposal and no landfill. Ghorahi and Dhaka are both exceptions: although Nepal and Bangladesh are both poor low-GDP countries, both cities have managed to build a disposal facility that protects the environment, Dhaka with funds from the Japan International Cooperation Agency (JICA), and Ghorahi without external support. Sousse, Tunisia, also has a new landfill with donor financing.

Even though direct-cost recovery is a challenge, upgrading landfills and increasing environmental standards are valuable processes to cities and their people. They may also become a local and national development priority, as well as a priority on the agendas of donors and development organizations. Kunming is a good example: in the city and surrounding region, there are tens of new landfills being developed.

In some instances, cities might need a bit of a push to start doing something about the environmental standards of waste management either from the national government or the international development organizations. The risk of a mismatch between local needs and external financing priorities in these circumstances is high, and often comes about where there is international pressure without a strong locally recognized driver, such as is the case in Bamako – which has never built its landfill – and in Managua as well.

As a result, in some cases investment plans are based on internationally recognized standards and technologies that may be too expensive or a misfit for the specific local situations. The issue is not whether environmental protection is a value to cities and their residents or not, but rather whether investments are based on local needs, build on the existing resources, and do not put too much of a financial burden on users and municipal budgets.

In some cases, the investment is carried out but is not used to full capacity because of the high operation costs. For example, the city of Arad in Romania (in 2000) and Gotce Delchev in south-west Bulgaria (in 1998) both built engineered landfills that stood empty for years. The Bulgarian one was publicly financed and the Romanian one privately financed; but in both cases running the landfill was beyond the capacity of municipal governments – and almost impossible to recover from users. Another example is Manila, where there is overinvestment in developing a landfill without a proper collection system, landfill costs are high, and illegal dumping is high.

In other cases, investment funds are available; but the financial carrying capacity for operating costs is missing and investments may not be carried out even though a lot of time and effort might have been spent in their preparation. It is not uncommon that international organizations such as development banks invest in a range of studies, and the development of frameworks and instruments that remain on the shelf and do not translate to actions on the ground.

KEY SHEET 17
CLOSURE AND UPGRADING OF THE OPEN DUMPSITE AT PUNE, INDIA

Arun Purandhare and Sanjay K. Gupta

Pune is a city of 3.5 million people in the western Indian state of Maharashtra, generating 1000 to 1200 tonnes of solid waste per day. From 1992 to 2002, the city's waste was indiscriminately dumped in a former stone quarry near Urali-Devachi village. Consequently, water wells in the vicinity were contaminated by leachate. The ambient air also suffered due to odour, high suspended particulate matter, sulphur dioxide (SO_2) and nitrogen oxide (NO_x), along with flying litter, flies and birds.

The site had an area of 17.4ha with a sloping terrain. The waste height of 18m was present on an area of 3.2ha. It was decided that this area

Fencing and levelling of the dumpsite for upgrading

© LCC-WMU Photo Library, Jan G. Tesink and Michael Kabungo

Preparation of the site after closure

© LCC-WMU Photo Library, Jan G. Tesink and Michael Kabungo

Laying HDPE liner

Covering of the old dumpsite while dumping is continuing in adjacent area

Gas venting pipe on waste dump

should be closed in accordance with the 2000 Municipal Solid Waste Rules. Due to an acute shortage of land, upgrading the existing dumpsite was chosen in lieu of constructing a new landfill. Since no alternative site was available even for the duration of the closure works, the tipping continued throughout. The job of closing the dumpsite was awarded to M/s Eco Designs India Private Limited in 2002 and the design and construction of the closure was completed by 2003. The waste in the dumpsite was dressed and brought to the required levels. The capping comprised of layers provided in accordance with the 2000 Municipal Solid Waste Rules, which included a high-density polyethylene (HDPE) geomembrane to prevent the ingress of rainwater. Passive vents were provided to release any remaining landfill gas, as most of the organic waste had been burned due to frequent fires. The cost of the closure amounted to approximately 10 million rupees. A 'piggy-back' landfill of 1.2ha was built over this closure as a temporary measure, at a cost of 8.8 million rupees.

Resource management and sustainable finance[19]

Recycling happens for two economic reasons, either for the market value of the secondary materials as a business and/or as a policy-driven activity related to avoiding the cost of disposal. For most cities the valuable materials – which comprise about 15 to 20 per cent of waste generated and include metals and high-grade waste paper – are already being recycled by private actors who depend on the value.

The next 20 to 60 per cent of the waste stream may be technically recyclable in some countries under some conditions; but they generally cost more to recycle than they are worth in the marketplace. Table 5.5 presents a classification of recyclable materials according to their commodity, or economic, value.

Recycling type 3 and type 4 materials (see Table 5.5) are policy-driven activities that require some intervention in the market, either by subsidies or market development. It is not easy to achieve high recycling rates through market intervention. Some key issues are as follows:

- Municipalities, which believe that they can earn revenues from recycling, usually focus on type 1 materials, which brings them into conflict with informal and formal recycling businesses. Delhi offers us an example of how problematic this is: a successful recycling programme in terms of the municipal solid waste budget often has negative consequences for inclusivity and the efficiency of disposal. Recycling is important, but it should not be seen as a revenue-generating strategy.

- Sometimes there is an emphasis on collecting low-intrinsic-value materials that burdens collection prices without considering the market for recyclables, as is the case in Ghorahi and Curepipe.

- Even some developed countries struggle with reaching high recycling rates, when final elimination options are readily available at low costs, such as incineration in Rotterdam or too much landfill space in Adelaide.

- Last, but not least, recycling takes a lot of know-how: retrieving valuable materials efficiently, at the right point, and processing it in the right way takes skills. If not done properly it may be inefficient and costly, as is the case in Curepipe, where materials are extracted only at the end of the chain at the disposal site.

Type	Examples: incidence in waste	Pre-modernization approach	Economic value
Type 1: high intrinsic value, globally traded commodities	High grades of waste paper, aluminium used beverage containers (UBCs), ferrous and other non-ferrous metals, about 10 or 15% of household waste. In recent years, also polyethylene terephthalate (PET) to China.	Recycled by individuals or enterprises through private initiatives and very rarely end up at dumping sites, except in extreme circumstances, such as the global paper market crash at the end of the 1980s.	Price paid for the materials covers or exceeds the cost of labour and equipment involved in extracting or collecting them.
Type 2: moderate intrinsic commodity value, locally traded commodities	Glass, tin, steel cans, rubber, non-PET polyolefines (polypropylene (PP), low-density polyethylene (LDPE), high-density polyethylene (HDPE)), textiles, low-grade paper. Total is about 10 or 15% of household waste.	Recycled by private enterprises only when there are local markets or a temporary shortage raises the price.	Have some value but not enough to cover the cost of extraction, processing and marketing. Recycling is not 'profitable' or even able to cover costs on its own.
Type 3: non-commodity materials with local options for 'beneficial reuse', subject to 'market development'	Kitchen, garden and small livestock waste for composting (about 40% of household waste); mixed waste with high organic content combined with high-density floor sweepings (as found in some West African countries, about 60% of household waste).	Small-scale private arrangements for removal by swine farmers or by cultivators who use the waste in composting or as a soil conditioner, with or without payments to generator or to remover.	Not a commodity, and so no intrinsic commodity value, but with some use value and some environmental value. May depend on willingness of government to 'purchase' compost for public uses: cemeteries, parks, sports fields, mine reclamation, erosion control, landfill cover and highways.
Type 4: negative value materials that damage the environment ('highly negative environmental externalities')	Healthcare waste, hazardous wastes, chemicals, fluorescent light bulbs, used engine oil, end-of-life e-waste, automobiles, accumulators, batteries, white-/brown goods, less than 5% of household waste.	Illegally dumped or traded for parts or residual use value, often partially burned to more easily extract metals.	Some residual value added, but not enough to cover cost of safe management with or without recovery.

Table 5.5

Understanding revenues and values of different types of materials to be recycled

Note: All percentages are estimates and are based on weight, not volume.

Source: VNGI (2008)

Insights from the reference cities and global good practices in financial sustainability

The good practice principles of financial sustainability in waste management are the same as for any business or household trying to function within a limited budget: know your costs, know your revenues and live within your means. However, organizing waste management strictly as a business would be limiting since, as discussed earlier, it is both a public service and a policy-driven activity. Therefore, good practices include the setting-up and fine-tuning of market interventions, such as subsidies and financial incentives.

■ Costs

This sub-section looks at how the reference cities are counting costs and revenues, and how they are raising investments and managing their budgets. Some attention will be paid to the management tools and the financial incentives that they use.

Cities report two different costs: their total system costs and their total solid waste budget. Using the system costs as the basis, it is possible to see that cost recovery per household ranges from none (two cities don't ask users to pay) to

100 per cent. Generally, countries in higher-income countries recover more of their costs; but the picture is not so clear.

Tables 5.6 and 5.7 are as close as the data will support to looking at costs across cities. None of the cities report a clear cost per tonne. However, they do report a total budget and a total number of tonnes handled by the formal sector, and this is used as a proxy for cost per tonne. It is important to note that the cities themselves might not recognize this number.

The financial information from the reference cities suggests that cost and budgeting mechanisms are generally fragmented and difficult to analyse. The cities are either unable or perhaps unwilling to share their costs, or they are not actively using them in setting fees.

In contrast, Rotterdam is an example of a city that knows its costs and uses them precisely in setting fees. This is an example of good practice: *knowing your financial costs*, you can control your solid waste improvements, both financially and technically. Financial management systems in waste management could often be helped if the following issues were considered:

- Aggregate all costs incurred for the waste management system in order to know your system costs and be able to plan for improvements in the future. This helps transparency and accountability when choosing investments.
- You have calculated total cost and the costs are too high? Practise activity-based costing in order to understand the cost of each activity. This will help you to spot the expensive activities and to make decisions about cutting costs or allocating resources. For example, it would make it easy to understand the cost of recycling materials versus disposing them, or the cost of street sweeping versus collection.
- Make sure to look at all operating costs and not only regular cash expenditures needed to operate and maintain the service. These include routine provision for financing vehi-

Table 5.6

Waste management costs and budgeting in the reference cities

Notes: NA = not available; NR = not reported. Figures in *italic* are estimates. Formal costs reported often include only municipal costs.

City	Solid waste management costs, formal (US$)	Total waste management municipal budget (US$)	Percentage municipal budget to solid waste management
Adelaide	NR	43,285,119	10%
Bamako	NR	1,443,308	NR
Belo Horizonte	115,500,000	115,500,000	5%
Bengaluru	42,295,420	57,830,211	NR
Canete	116,847	269,927	NR
Curepipe	1,158,043	1,468,164	NR
Delhi	99,726,833	99,726,833	3%
Dhaka	15,755,620	15,755,620	NR
Ghorahi	43,291	66,272	15%
Kunming	NR	NR	NR
Lusaka	NR	NR	3%
Managua	12,469,780	12,469,780	NR
Moshi	NR	NR	NR
Nairobi	NR	NR	4%
Quezon City	7,876,745	21,026,248	9%
Rotterdam	62,396,252	108,875,999	NR
San Francisco	NR	11,139,005	0%
Sousse	2,524,661	2,366,870	NR
Tompkins County	5,867,560	6,028,845	NR
Varna	19,336,142	19,336,142	5%
Average	29,620,553	32,286,771	5.89%
Median	12,469,780	14,112,700	4.50%

City	Total waste management budget (US$)	Solid waste budget per capita (US$)	Solid waste budget per household (US$)	Solid waste budget per capita as pecentage of GDP per capita
Adelaide	43,285,119	40	95	0.10%
Bamako	1,443,308	1	5	0.14%
Bengaluru	57,830,211	7	26	0.71%
Belo Horizonte	115,500,000	47	146	0.69%
Canete	269,927	6	24	0.14%
Curepipe	1,468,164	18	70	0.33%
Delhi	99,726,833	7	37	0.69%
Dhaka	15,755,620	2	10	0.52%
Ghorahi	66,272	1	5	0.31%
Kunming	NR	NR	NR	NR
Lusaka	NR	NR	NR	NR
Managua	12,469,780	12	65	1.22%
Moshi	NR	NR	NR	NR
Nairobi	NR	NR	NR	NR
Quezon City	21,026,248	7	37	0.45%
Rotterdam	108,875,999	187	364	0.40%
San Francisco	11,139,005	13	31	0.03%
Sousse	2,366,870	14	55	0.40%
Tompkins County	6,028,845	60	138	0.13%
Varna	19,336,142	62	160	1.19%
Average	32,286,771	30	79	0.47%
Median	14,112,700	13	46	0.40%

Table 5.7

Annual municipal waste management budget calculations

Note: NR = not reported. Figures in *italic* are estimates. Total waste management budget calculations: Adelaide – calculated based on costs per household and number of households; Bamako – complete budget DSUVA reported, which non-solid waste management activities; Bengaluru – based on total sources of funds received by the municipality; Belo Horizonte, Delhi, Dhaka, Managua, Rotterdam, Varna – based on total costs, budget not reported.

cle or equipment maintenance or upgrading associated with changing norms and regulations, as well as the human resource costs associated with planning and management of waste in municipal departments.

- Understand the cost impact of investments made to improve the service. For example, if you buy compactor trucks to replace donkey carts, you don't need to feed the donkeys anymore, but you will need to spend on diesel, spare parts, maintenance, etc. Likewise, if you open a landfill you need to be able to pay for the closure of the old disposal site in order to avoid the trap of having two open sites and, thus, more harm to the environment. Investments are not single expenditures paid from the great budget in the sky.
- The amount of waste is often either under or (more often) overestimated in cities where there is no weighbridge at the disposal facility. This may be because the cost effectiveness of a desired and planned landfill seems higher when there is more waste because the cost per tonne goes down.

Box 5.13 Making getting to know your costs a priority, from Hong Kong to Nicaragua

In their Municipal Waste Management Strategic Plan prepared during the late 1990s, the City of Hong Kong formulated two main objectives that would guide its chosen path in waste management for the next decade:

1   reduce the amount of waste going to the landfill through a number of recycling, composting and incinerating activities; and
2   create transparency in knowing, controlling and reporting on all the costs related to solid waste management.

Implementing the latter resulted in an intensive internal learning process for the municipality in accounting and a major reorganization of tasks and responsibilities.

In Managua, Nicaragua, the municipality has learned that the fragmentation of solid waste management responsibilities within the organizational chart has contributed to a misunderstanding of what the real costs are related to solid waste management. Currently, the waste costs budgeted and reported are only those designated to the Public Cleansing Department at the central municipal level. Unravelling the internal administrative lines and rules forms part of the internal capacity-building process to understand how much waste actually does costs. This process of tracing where, throughout the municipality, costs are incurred on waste management includes accounting for:

- fuel and lubricants bought and controlled by a central municipal depot;
- spare parts and tyres bought and controlled by the procurement unit;
- protective clothing and implements bought and controlled by the Human Resource Department;
- repairs done by a central municipal mechanical workshop;
- sweeping and illegal dumping clean-up activities delegated to five municipal district offices.

Adding all of the separate parts together reveals that the costs were actually underestimated by more than 50 per cent.

Box 5.15 Belo Horizonte, Brazil

The waste management agency Superintendência de Limpeza Urbana (SLU) annually prepares an activity report and a separate financial report, disclosing in detail the activities undertaken and the related costs occurred. These are submitted to the city council and are public knowledge. The SLU also uses the information for monitoring, evaluation and future planning. Comparative unit costs of activities conducted by the in-house services and those outsourced to private contractors are reported every year and allow for assessment of trends over a number of years. Similarly, unit costs for different composting, recycling activities, waste collection and sweeping activities are calculated per trimester.

Box 5.16 Why the donkeys of Bamako, Mali, were dying in 1999[20]

In Bamako, Mali, the donkeys and the owners of the *Groupements d'Intérêt Économique* (GIEs) or micro- and small enterprise (MSE) collection enterprises were, until recently, the casualties of a poorly functioning cost-recovery system. The city council set the tariffs per household for waste removal, based on what they thought was politically acceptable, without considering the real service costs to the GIEs.

A law prohibiting the donkey carts from using paved roads made the situation worse: the collectors began to overload the carts and underfeed the donkeys in order to make ends meet. The result was that the donkeys did not take in enough calories to replace the energy they used for pulling the carts and usually died within a year. This is a good example of poor financial practice: where operations costs are not covered, the system will not be sustainable.

---

overestimated. As a result, when cities struggle to secure financial resources to sustain and improve service coverage and quality, it is important to seek the right amounts of money for the right investments for the right amounts of waste. Extensive guidance on setting up such systems, and on other aspects of finance and cost recovery for solid waste management, is available.[21]

■ Revenues

Concerning revenues used for financing solid waste, the information from the reference cities shows that a variety of sources are tapped into for financing waste costs. All cities from the high-GDP countries (Adelaide, Rotterdam, San Francisco and Tompkins County), as well as a number of other cities, including Kunming, Moshi and Nairobi, use one bill, either a direct waste bill or through the utility company. Seven of the cities, headed by Delhi, Belo Horizonte and Bengaluru, complement the direct waste fee with revenues collected through property tax, municipal income tax or national transfers.

Quezon City and Ghorahi are the only reference cities where no fee is charged to the citizens for waste collection services. Important indica-

The costs of providing municipal waste management services are thus commonly underestimated, while the amount of waste is often

---

Table 5.8

Sources of operational funds: Revenues.

This table combines several types of information relating to sources of funds. In column 2 are the costs per household per month converted to USD. Column 3 describes how the fee is assessed, and column 4 whether it is differentiated and, if so, along what parameters. The rightmost column indicates whether the user fees contain an incentive to participate in diversion from disposal through recycling or other forms of recovery.

Notes: N = no; NR = not reported; Y = yes. Monthly fee per household: Adelaide – based on average annual fee; Delhi – only for daily collection, door-to-door, by informal sector; Varna –based on annual fee per capita.

City	Monthly fee per household (US$)	Fee collection type, per household	Fee differentiation or waiver description	Fee incentives for diversion
Adelaide	8	Direct bill for property owners	Differentiated by amount of waste bins	Y
Bamako	2.40–4.80	Direct bill (mostly informal), and formal tax bill	Differentiated by HH agreements with informal sector	NR
Belo Horizonte	3.90–7.90	Fee is part of property tax	Differentiated by property tax	N
Bengaluru	0	Direct bill	None	N
Canete	3.00–3.90	Direct bill	None	N
Curepipe	0	Fee is part of property tax	None	NR
Delhi	0.45–1.15	Fee is part of property tax, fee for daily collection informal sector	Differentiated by HH agreement with informal sector	N
Dhaka	1	Fee is part of property tax (7%)	Differentiated by income group	NR
Ghorahi	0	No fee, developing / starting system	None	NR
Kunming	0.35–1.45	Direct bill	Differentiated by category 'permanent residents' and 'other residents'	N
Lusaka	18.80	Direct bill collected by franchisers	Differentiated by service provider	N
Managua	0.50–5.00	Direct bill	Differentiated by type of residential zone and 'frontage'	N
Moshi	1	Direct bill	Low Income HH are not required to pay a fee	N
Nairobi	0.15–0.30	Utility (water) bill	None	N
Quezon City	0	Direct bill in some barangays (neighborhoods), no household bill on municipal level	Fee differentation at Barangay level, no municipal fee for HH	NR
Rotterdam	33	Utility bill	Waiver for low income housholds	N
San Francisco	22	Direct bill	None	Y
Sousse	NR	Municipal tax	None	N
Tompkins County	15	Direct bill, volume-based	Pay as you throw' for HH in combination with annual fixed fee	Y
Varna	4	Direct bill	None	N
Average	8			
Median	1			

tors in relation to payment of waste fees include the payment rate, the willingness to penalize non-payment and the updating of the fees. Examples from the reference cities give an insight into how politics often play a key role in these issues.

In Bengaluru, the payment rates are at a low of 40 per cent and the fees are minimal; but the reason for non-payment is attributed to a political choice, which doesn't penalize users for non-payment. There is no action to improve the rate of fee collection because the politicians want votes. There is willingness to pay; but there is a lack of willingness to charge a cost-covering fee to users. Bamako had this problem in the 1990s, but reduced the problems by creating communal platforms, such as COGEVAD in Commune VI and COPIDUC in Commune IV, so that now their payment rate is 50 to 60 per cent. Kunming is another example where the payment rate is 45 to 50 per cent, but the fees have not been raised. In all of these three cities, the collection system works rather well, even though cost recovery does not.

In Managua, for the last 10 to 15 years the municipality has only collected between 20 to 30 per cent of the billed waste collection fees to citizens. Collection fees, which have remained unchanged since 1993, do not include all costs incurred related to solid waste management and from which all very low-income households are exempt by law.

Knowing your revenues and how far they could potentially be increased is also critical. In waste management, revenues either come from a payment for removal or through the recovery of valuable materials from the waste stream, and these are collected in many different ways – for example, through:

- a specific solid waste management levy, which may be collected separately or via electricity or water bills;
- a charge, tariff or fee, levied by the service provider – this may be a flat rate, or be related to the quantity of non-recyclable or

City	Fee collection type, per household	Fee differentiation or waiver	Fee incentives	Cross-subsidization
Adelaide	Direct bill	Graduated	Y	NR
Bamako	Other: direct bill (informal) and formal tax bill	Informal verbal agreements	NR	NR
Belo Horizonte	Other: part of property tax	Levels	N	N
Bengaluru	Mix: direct bill Other: part of property tax	None	N	N
Cañete	Direct bill	None	N	N
Curepipe	Other: part of property tax	None	NA	N
Delhi	Other: part of property tax	Levels	Y	N
Dhaka	Other: part of property tax (7%)	Levels	NR	Maybe?
Ghorahi	None	None	NR	N
Kunming	Direct bill	Levels	N	Maybe?
Lusaka	Direct bill	Levels	N	Y
Managua	Other: part of property tax	Levels	N	Maybe?
Moshi	Direct bill	None	N	Maybe?
Nairobi	Utility (water) bill	Levels	N	N
Quezon City	None	NA	NA	N
Rotterdam	Utility bill	Waiver	N	Y
San Francisco	Direct bill	Graduated	Y	NR
Sousse	Other: tax bill	NR	0	NR
Tompkins County	Direct bill + volume-based	Levels	Y	NR
Varna	NR	NR	NR	NR

total waste collected, or both;
- indirectly, through their general local fundraising, which may be through local taxes of some kind, commonly on property;
- earned from sale of depreciated equipment or materials, or through the marketing of valorized recyclables or compost.

The source of funds for the solid waste system in Cañete is 30 per cent supported by solid waste fees and 70 per cent by taxes or other public sources, including the municipal general fund. Even though the quality of the service is generally recognized as good, only 40 per cent of households, 60 per cent of institutions and 90 per cent of commercial users pay the fee, whereas 100 per cent of industrial users pay the fee. These characteristics are the same in almost all cities in Peru.

Payment rates and cost recovery for waste management, especially for collection, depend largely on the efficiency of the fee collection system and the level of trust between users and providers, rather than financial incentives. Belo Horizonte is an eloquent example of a middle-income country city reaching 95 per cent collection and payment rates due to an efficient

Table 5.9

Fees and incentives

Notes: NA = not available; NR = not reported; N = no; Y = yes

and transparent fee collection system through a city tax, called the Urban Cleansing Tax. The system has a long-standing tradition and is trusted by the citizens. However, this only covers 40 per cent of the solid waste budget, which is known by the municipality and a deliberate strategy. It has been a policy (and political choice) to finance the remainder of the costs through other municipal taxes (primarily property taxes) for which it has been easier to apply yearly inflation correct rates.

For the local authority, a possible alternative is to collect the waste fees with other utility charges; but this may not be feasible when sanitation or water have been privatized. In Russian cities, for example, and prior to 1989 in Bulgaria as well, housing management companies collect one fee, which covers water supply, heat, sewage, waste, gas, television antenna, etc. Collection with electricity charges has also been used in Jordan and Egypt,[22] and is frequently proposed for other cities.

In general, a well-designed and transparently functioning tariff or fee system can recover some of the costs of operating the service, especially those associated with direct benefits of removal. Recovering costs for controlled disposal is less feasible, so at least in the short term, much of the full cost will continue to be paid for by the local, regional or national authorities from general revenues as part of the government's public health and environmental protection responsibilities. Kunming and Ghorahi are both cities where disposal is well organized but the costs are not recovered from users – in Kunming because fees are kept deliberately affordable, and in Ghorahi because there is no fee to users at all. Moshi struggles with this issue because users feel that they have already paid for disposal and don't want to pay again.

This may be unavoidable: a lack of electricity or water is a life-and-death situation; but too much waste is only a nuisance, and a solvable one at that. Thus, people may choose to make private legal and illegal arrangements to get rid of their waste rather than (agree to) pay, and this has public health consequences for the city as a whole. And because protecting public health remains a municipality responsibility, the municipality is not in a position to actually decide not to provide a service to those who do not pay. It cannot 'cut off' the collection service as a water company can disconnect water supply, or a telephone company can discontinue the telephone service to non-payers.

In some of the newly independent states of the former Soviet Union, the municipal waste collection company contracts directly with the management company for each housing block.

City	National or state government	Donor funding	Carbon financing	Landfill / Gate fee	License, permit, franchise fees	Fees from users	EPR / CSR	Revenues from valorization
Adelaide	Y	NR	Y	Y	Y	Y	Y	Y
Bamako	N	Y	N	NR	NR	Y	Y	Y
Belo Horizonte	N	N	Planned	Y	NR	Y	NR	Y
Bengaluru	Y	N	Planned	Y	Y	Y	NR	NR
Canete	Y	N	N	N	N	Y	N	N
Curepipe	Y	NR	N	NR	NR	N	N	N
Delhi	Y	N	Y	NR	Y	N	NR	Y
Dhaka	Y	Y	Y	NR	NR	Y	NR	N
Ghorahi	N	N	N	N	NR	N	NR	N
Kunming	Y	N	Y	Y	NR	Y	NR	N
Lusaka	Y	Y	Planned	Y	Y	Y	NR	N
Managua	N	Y	Planned	N	NR	Y	NR	N
Moshi	Y	Y	NR	NR	Y	Y	NR	NR
Nairobi	NR	Y	NR	NR	Y	Y	NR	NR
Quezon City	NR	N	Y	Y	NR	Y	NR	N
Rotterdam	Y	N	N	NR	NR	Y	Y	Y
San Francisco	Y	N	Y	Y	Y	Y	Y	Y
Sousse	NR	Y	Y	Y	NR	Y	Y	NR
Tompkins County	Y	N	Y	Y	Y	Y	Y	Y
Varna	Y	N	NR	NR	Y	Y	Y	Y

Table 5.10

Sources of funds.

This table looks at the range of sources of financing in use in the reference cities. Most cities use more than one source, and some, like Adelaide, San Francisco and Tompkins County, use virtually all of them.

Notes: N = no; NR = not reported; Y = yes.

Most sign up and pay, but some don't, preferring instead to make their own arrangements, such as burning, dumping, using street bins or neighbours' bins, or taking the waste to work. So the definition of the term 'collection coverage' changes – the service may be available to 100 per cent of the city, but perhaps only 80 per cent actually sign up to and use the service.

In most of the cities, charges are levied at a 'flat rate', depending on the size or value of the property, rather than directly on the quantity of waste generated. True 'pay-as-you-throw' systems are relatively uncommon: more are being introduced in high-income countries, at least partly to provide an incentive to householders to segregate their wastes for separate collection and recycling, for which a charge is not made. Nairobi is an exception. The private waste collectors sell their own garbage bags for a fee, which includes the cost of collecting and disposing the amount of waste that fits in the bag.

While pay-as-you-throw systems have been presented as *the* solution and a key waste prevention incentive in the US, they are hardly used in The Netherlands and are generally less popular in Europe, where governments have a somewhat broader social vision of what they should be paying for. And it is true that one negative side effect of the pay-as-you-throw system has been so-called waste-tourism, where citizens employ evasive measures for not paying for all the waste that they generate, either adding their waste to their neighbours' waste or bringing their waste to the next city, where the pay-as-you-throw system is not applied.

In many low- and middle-income cities, itinerant waste buyers are already collecting – and paying or bartering for – source-separated materials door to door, often making a small payment based on weight; in this sense, an 'incentive system' to encourage separate collection already exists and could be built upon.

■ Sources of funds for investment

The landscape of sources of investment funds and what seems to work in the cities in terms of actual use of capacities built and delivery of the system is very varied. Table 5.10 shows that investments made up by any combination of sources are able and likely to finance installations and fleet that work at full capacity or nearly full capacity. The only outlier is Moshi, where the municipality is not involved in financing and the investments are used at 55 per cent capacity.

The reference cities fund their investments in solid waste service improvements from a range of sources – for example, through:

Table 5.11

Benchmark indicators for sustainable financing.

The purpose of this table is to test some benchmarks based on the data points provided by the reference cities. Tables 5.9, 5.10 and 5.11 provide a wealth of possibilities for cross-comparison.

Note: Figures in *italic* are estimates. Curepipe, Delhi, Ghorahi and Quezon City do not have a central municipal fee. Cost recovery: Nairobi – from fees ranging between 25 and 20 per cent; Sousse –: only based on household fee; Tompkins County and Varna – calculated based on total collected fees reported and total reported costs.

City	SWM percentage of Municipal Budget	Population using and paying for collection as percentage of total population obligated to pay	Percentage of population that pays for collection	Reported cost recovery percentage collected via fees	Solid waste annual fee as percentage of average annual household income	Solid waste budget per capita as percentage of GDP per capita
Adelaide	10%	100%	100%	90%	0.21%	0.10%
Bamako	NR	95%	54%	NR	2.00%	0.14%
Belo Horizonte	5%	85%	81%	36%	3.60%	0.69%
Bengaluru	NR	40%	28%	NR	0.15%	0.71%
Canete	NR	40%	29%	30%	0.90%	0.14%
Curepipe	NR	0%	0%	0%	0.00%	0.33%
Delhi	3%	0%	0%	58%	0.00%	0.69%
Dhaka	NR	80%	44%	30%	2.00%	0.52%
Ghorahi	15%	0%	0%	0%	0.00%	0.31%
Kunming	NR	50%	50%	NR	1.00%	NR
Lusaka	3%	100%	45%	NR	NR	NR
Managua	NR	10%	8%	50%	0.14%	1.22%
Moshi	NR	35%	21%	20%	0.30%	NR
Nairobi	4%	45%	29%	38%	0.15%	NR
Quezon City	9%	20%	20%	0%	0.00%	0.45%
Rotterdam	NR	100%	100%	100%	0.00%	0.40%
San Francisco	0%	100%	100%	100%	1.43%	0.03%
Sousse	NR	50%	50%	0%	NR	0.40%
Tompkins County	NR	95%	95%	35%	0.11%	0.13%
Varna	5%	100%	100%	76%	0.90%	1.19%

- a direct grant from national or regional government;
- building up provisions for investments from regular revenues;
- donor grants or loans;
- franchise, permit, or concession fees;
- revenues from sale of real estate, equipment;
- revenues from valorizing recyclables or organic waste;
- central government transfers;
- engaging private investors in modernization in a public–private partnership (PPP) arrangement; or
- a credit line.

In order to successfully live within their means and deliver a good waste management system, cities need to think critically and make their own choices, and to accept donations when they have made sure that they understand what they are getting into and when local experts have ensured that technologies are suitable and plans meet local demand, policies and priorities. The donors will not be there to pay the fuel bills, order the tyres, or answer to the citizens at the next elec-

tion, so it should not be the donors who decide what equipment is needed to keep the city clean.[23]

A good practice example of financing a new landfill site is that of Ghorahi, where the municipality decided to set aside its own budget for its own waste management modernization agenda. The municipality was engaged in the process and did it at its own pace, paying attention to all needs and involving all stakeholders in the process. The site location is also a fortunate one, being only 5km away from the city and resting on a natural clay bed. The landfill is run by a management system that includes a committee where local community leaders and the business community is also involved.

Table 5.11 demonstrates that the cities which report solid waste as a percentage of the total range from 3 to 15 per cent. The payment rate is up to 100 per cent of households receiving the service in the four richest cities: Adelaide, with 90 per cent, and Tompkins, San Francisco and Rotterdam at 100 per cent. From the user side, Belo Horizonte households pay 3.6 per cent of family income on solid waste, and are the highest proportionally. Leaving aside Quezon City and Ghorahi, where no payment is asked,

Managua households pay the lowest percentage of their household income, with Nairobi, Bengaluru, Adelaide and Moshi in the same order of magnitude.

Belo Horizonte, Brazil, also used its own accumulated financial resources and national loan mechanisms to finance major infrastructure investment (upgrading of the landfill and construction of a large 1500 tonne per day capacity transfer station).

One of the difficulties with investment financing is that it is often tied to the priorities of the giver. Grants may look like 'free money,' but they very often have conditions and requirements that limit the receiver's scope for making independent decisions. In this, perhaps the most risky grants are those for specific technologies or equipment, and it is wise to take a long and critical look before accepting these – or indeed any – grants. Another issue to be mentioned in connection with financing is the extent to which planning and service reliability are hampered by the way that finances are transferred from central municipal funds to the SWM service.

In conclusion, cost recovery is part, but not all, of the sources of funds story, especially in low- and middle-income countries. Getting people to pay for primary collection, where they can see the benefit of keeping their neighbourhood clean, may be a realistic first step. Expecting them to be equally willing to pay for secondary collection and environmentally sound disposal is optimistic, because they don't immediately experience the impacts of problems of the status quo. The proposal of David Morris of the Institute for Local Self-Reliance during the 1980s was to require that all waste be disposed of within 35km of where it is generated to make it clearer to more citizens what the problem is.

Moving from a position where solid waste management is paid for through general revenues to one where it is paid for entirely by user charges may not be possible in the short or medium term in low- and middle-income countries. Where it is desirable or feasible, a gradual transition is sensible, particularly if the real

system costs are rising at the same time. As with other aspects of the waste service, dialogue with all stakeholders is critical.

■ Incentives

Setting up and fine-tuning financial incentives is controversial. Many cities believe that incentives at household level are essential to achieving high payment rates and high recycling rates, but the four high-recycling rate cities, San Francisco, Adelaide, Tompkins County and Quezon City, make only modest use of incentives, and Rotterdam, with a moderate recovery rate, makes none at all.

Incentives within the system, to steer the behaviour of providers, work better: the entire Dutch recycling system is basically driven by the municipal drive to reduce disposal costs by achieving high recycling and composting rates, supplemented by payments from industry to municipalities for recycling. Provider incentives can be built in contracts, fee structures or subsidies. Some of the key tools to use when setting up financial incentives include the following:

- *Cross-subsidize:* there is a demand for waste collection and a willingness to pay or spend resources. This allows, in most cases, for an affordable service at full-cost recovery, given that the level of service is set at the demanded level and is paid for through a cross-subsidized system. Lusaka is an example where payment rates and collection rates were as low as 40 per cent; but when the municipality figured out the costs

---

Box 5.17 Recovering costs of collection in Bengaluru, Karnataka State, India

In Bengaluru, India, household fees are not applicable to slums and other low-income areas. There are differential rates for commercial areas, institutions and hotels and restaurants. This collection fee ranges from 100 rupees (about US$2) to 150 rupees and most regular collection fees come from the hotels and restaurants. A large share of the budget comes from state grants and the rest from property tax collection and licence fees. In addition, the municipality also depends on additional sources of funds for covering deficits when required, such as special state grants, and sometimes on central government grants under its various schemes. The budgetary allocation is controlled by the state government, not the municipal corporation; but once it has been decided, the municipal commissioner and the standing committee jointly form the final authority on allocation of funds for different activities under solid waste management.

of the system and managed to work out a cross-subsidy scheme, the payment rates rose to 75 per cent. There are different ways of working out systems for cross-subsidies, such as setting lower fees for low-income households and higher fees for high-income households and businesses, or setting a fee per 'connection' based on average socio-economic characteristics in the different wards in the city, with the possibility of granting exceptions to the general rule based on specific characteristics.

- *Differentiate:* it is important to think through and choose financial incentives keeping in mind that recycling is a business that is profitable only due to its intrinsic value. If more recycling is targeted, incentives are needed. Thus, for example, source separation may be encouraged by having users pay less for removal of separated materials, or based on their environmental footprint, so that disposing of hazardous waste 'costs' more than delivering tree branches to a wood recycler.
- *Boost demand:* recycling may be encouraged by setting a price for a secondary material. A valuable example is San Francisco, where high recycling targets and zero waste goals are matched by market development for secondary materials and products. Buy-back systems can operate through the same principle: they set a price for end-of-life products, thereby creating a demand for them.
- *Avoid costs:* applying a high gate fee at final elimination may also increase recycling by making it a financially more attractive option to recycle. Examples of good practice are Quezon City, where municipal districts get a waste quota for landfilling; where they bring less to the landfill due to recycling, they receive a refund of their costs.

- *Fine-tune:* beware of the costs associated with interfering in the market. Markets are complex mechanisms that need to be fine-tuned in order to avoid adverse impacts, such as increased illegal dumping to avoid costs when high gate fees are applied at landfills, or when collection is paid by quantity collected in 'pay-as-you-throw' systems. The cost of implementing a system should never outweigh its benefits. In other words, if you are spending too much money on controlling illegal dumping or keeping away informal-sector scavengers from your containers or enforcing payment, then you need to examine and fine-tune your system.

■ Emerging issues: Carbon financing and extended producer responsibility (EPR)

When it comes down to it, many of the questions about sources of funds and investments rely on finding a payer with 'deep pockets' because outside of high-income countries, the users normally cannot or are not willing to pay for the full costs of disposal or other measures of environmental protection. There are two other emerging sources of these funds: producers of the products, through extended producer responsibility, and high-GDP countries and companies, through the mechanism of carbon financing.

Extended producer responsibility (EPR) is a way of subsidizing recycling by obliging polluters to either recycle or pay for recycling. Some of the best-performing cities from the point of view of collection coverage and recycling, such as Adelaide and San Francisco, tap into both of these sources for financing. Carbon financing is a good alternative for financing because it is operational financing, meaning it only will be available when the investments deliver the emission reductions that were projected.

KEY SHEET 18
SOLID WASTE, RECYCLING AND CARBON FINANCING: FACT OR FICTION?

Nadine Dulac (World Bank)

FICTION AND LEGENDS

Until now, 1834 projects have been registered by the United Nations Framework Convention on Climate Change's (UNFCCC's) executive board. The projects have been developed according to Clean Development Mechanism (CDM) rules and conditions and are grouped by sectoral scopes. There are 15 UNFCCC sectors. The solid waste management CDM projects belong to sectoral scope number 13, entitled 'Handling of waste and disposal'. Out of the 400 projects listed under this scope, 136 are projects directly related to landfill gas capture/extraction (LFG) with or without utilization (so-called gas-to-energy projects) and composting that have been registered by the UNFCCC's executive board. The remaining CDM projects (264) are about methane capture and combustion from liquid waste, such as manure or agricultural residues. In addition, there is another group of 142 projects related to LPG or composting under preparation, mainly at validation stage. Since the beginning of 2009, CDM can also be implemented through a programme of activities. As of today, only one programme has been developed for composting in Uganda and the project has been passed at validation stage. Another one will be developed in the near future in Morocco, where the World Bank has signed a partnership agreement on municipal solid waste (MSW) sector development, which includes a carbon finance/CDM component under a programmatic approach.

Clean Development Mechanism (CDM)	All projects registered	All programmes registered or under validation
Total	1834	1
Solid waste management	136	1 (at validation, composting)

Table K18.1

Numbers of projects and programmes

Table K18.1 presents an overview of the total number of projects and programmes in the UNFCCC pipeline; Table K18.2 shows the level of preparedness of the projects.

There is a belief that carbon finance can help to develop more controlled waste disposal because additional revenues will flow to the project. Carbon finance provides hard-currency annual payments for performance in emissions reduction. However, carbon revenues from CDM LFG or composting projects depend on the performance of the project – in other words, carbon payments are tied to certified performance. The portfolio analysis suggests that many of the registered and implemented projects are underperforming relative to initial estimates of methane anticipated to be captured and destroyed.

Table K18.2

Preparedness of the projects

	At validation	Request registration	Registered	Total projects
Landfill flaring	31	4	61	96
Landfill power	63	6	59	128
Combustion of MSW	12	1	4	17
Gasification of MSW	2	1	1	4
Landfill aeration	1	0	0	1
Composting	20	1	11	32
Total	129	13	136	278

Figure K18.1

Carbon market growth in 2008 (US$ millions)

Note: JI = joint implementation

Source: adapted from Capoor and Ambrosi (2009)

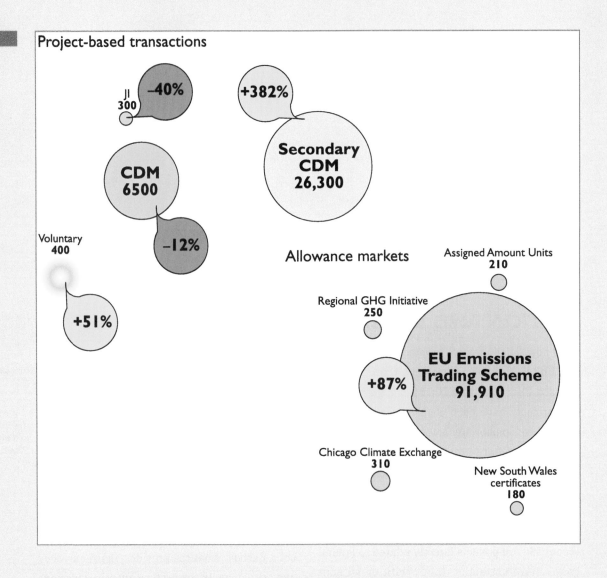

Table K18.3

Carbon market at a glance: Volumes and values, 2007–2008

Note: NA = not available.

Source: adapted from Capoor and Ambrosi (2009)

	2007		2008	
	Volume (million tonnes of CO_2e)	Value (US$ millions)	Volume (million tonnes of CO_2e)	Value (US$ millions)
Allowances markets				
EU Emissions Trading Scheme (ETS)	2060	49,065	3093	91,910
New South Wales	25	224	31	183
Chicago Climate Exchange	23	72	69	309
Regional GHG Initiative (RGGI)	NA	NA	65	246
Assigned amount units (AAUs)	NA	NA	18	211
Sub-total	2108	49,361	3276	92,859
Project-based transactions				
Primary Clean Development Mechanism (CDM)	552	7433	389	6519
Secondary CDM	240	5451	1072	26,277
Joint implementation	41	499	20	294
Voluntary market	43	263	54	397
Sub-total	876	13,646	1535	33,487
Total	2984	63,007	4811	126,345

STATE OF THE ART OF THE CARBON MARKETS

The overall carbon market continued to grow in 2008, reaching a total value transacted of about US$126 billion (86 billion Euros) at the end of the year, double its 2007 value. Approximately US$92 billion (63 billion Euros) of this overall value is accounted for by transactions of allowances and derivatives under the European Union Emissions Trading Scheme (EU ETS) for compliance, risk management, arbitrage, raising cash and profit-taking purposes. The second largest segment of the carbon market was the secondary market for certified emissions reductions (sCERs), with spot, futures and options transactions in excess of US$26 billion (18 billion Euros) representing a fivefold increase in both value and volume over 2007.

Confirmed transactions for primary CERs (i.e. CERs purchased directly from entities in developing countries, or pCERs) declined nearly 30 per cent to around 389 million CERs from 552 million CERs in 2007. The corresponding value of these pCER transactions declined 12 per cent to around US$6.5 billion (4.5 billion Euros) in 2008, compared to US$7.4 billion (5.4 billion Euros) reported in 2007. Confirmed transactions for joint implementation (JI) also declined 41 per cent in value to about US$294 million (201 million Euros) for about 20 million tonnes of carbon dioxide equivalent (CO_2e) transacted in 2008. The supply of CDM and JI in 2008 and early 2009 continued to be constrained by regulatory delays in registration and issuance, and the financial crisis made project financing extremely difficult to obtain.

To further complicate matters for CDM demand, 2008 and early 2009 also saw several pioneering transactions of about 90 million assigned amount units (AAUs) with related green investment schemes (GIS) at various stages of elaboration. The economic blues also affected the voluntary market, which saw transactions of 54 million tonnes of CO_2e in 2008 (up 26 per cent over 2007) for a value of US$397 million, or 271 million Euros (up 51 per cent), but still fell short of the exponential growth of previous years.

	Number of registered projects that deliver certified emissions reductions (CERs)	Issued CERs (total)	Issuance success in comparison to PDD estimates
Landfill gas capture/ extraction (LFG) project	40	6,968,000	34%
Methane avoidance/composting	51	5,231,000	48%

Table K18.4 presents the number of registered CDM projects that are implemented and delivering CERs.

Table K18.4

Issuance rate in 2009 for CDM landfill gas and composting projects

STATE OF THE VOLUNTARY MARKET AND CARBON OFFSET STANDARDS, INCLUDING COMPENSATION PROGRAMMES

Besides the Clean Development Mechanism and the regulatory carbon market, solid waste management projects with landfill gas extraction or avoided emissions are also developed under the voluntary market. There are more than nine standards and half of them accept such projects (i.e. the Gold Standard (GS), the Voluntary Carbon Standard (VCS), the Chicago Climate Exchange (CCX) and the Voluntary Offset Standard (VOS).

Table K18.5 presents the number of projects registered. The most relevant market is the voluntary market, characterized by the VCS and the VOS.

Table K18.5

Verified emissions reductions for landfill gas, composting and incineration projects in the voluntary carbon market in 2009

	Number of projects registered under Gold Standard (GS)	Number of projects registered under the Voluntary Carbon Standard (VCS)	Chicago Climate Exchange (CCX)	Voluntary Offset Standard (VOS)
Landfill gas capture/ extraction (LFG)	4 (Turkey and China)	4 (US and China)	31 (US)	8
Composting	1 (vermin-composting in India)		1	
Incineration		1 (China)		
Total VERs			6,121,000 (LFG) 20,200 (avoided emissions)	Number not available

POTENTIALS AND DRAWBACKS

Critical issues

Sustainability is critical to the success of any waste-to-energy project, which is expected to increase the social, environmental, economic and technological well-being of the host country and address the treatment technology adopted, location of the project, waste handling and disposal practices, pollution control measures, and type of supporting fuels used. In some instances, especially in India, such LFG projects were rejected. In South Africa, NGOs and local communities expressed concerns.

Metering and monitoring

Improper data on waste composition and the amount of non-biomass materials (such as plastic and rubber) in the waste stream pose a serious threat to the emission reduction claims by the project developer. Therefore, emission reduction verifications are frequently delayed. Another shortfall is that project developers typically fail to provide this data due to improper and/or inadequate data monitoring practices; however, adhering to the monitoring frequency of parameters specified by the CDM methodology applied would streamline the verification process.

Regulatory requirements

Obtaining the No-Objection Certificate (NOC) and having an up-to-date Consent for Operation Certificate from the appropriate local government authority is essential to prove that the project meets all environmental regulations in the region. It is also critical to obtain the host country's approval and permits for the project. Monitoring key parameters and optimizing the plant operation with the best waste management practices will reduce the difficulty in meeting local regulatory requirements most of the time. Securing the ownership of the emissions reduction credits might also be a problem, especially if the landfill site belongs to a public entity and the site is operated by a private company, who is also the project developer.

Methodology selection

Project developers tend to use small-scale and inappropriate methodologies to avoid complex monitoring procedures and the comparative ease of proving additionality of the project. Improper choice or application of a methodology (e.g. using AMS III E instead of AM0025) not only delays the project registration, but also raises doubts on the project's applicability.

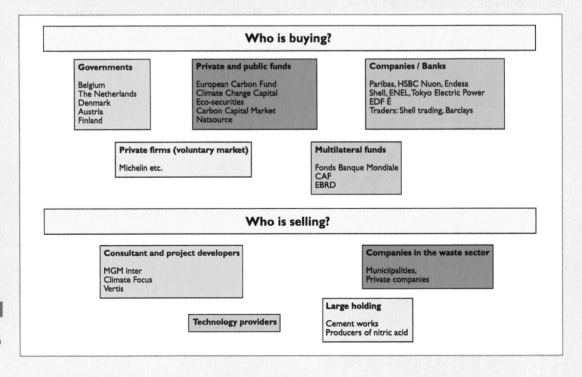

Figure K18.2

Who is buying and who is selling carbon credits?

CURRENT AND NEW METHODOLOGIES

Existing methodologies

While there are currently 15 methodologies in use under sectoral scope 13, only the ones presented in Table K18.6 are applicable to LFG, composting and solid waste management projects.

New methodology

Even though recycling of the materials in municipal solid waste has been shown to reduce the level of greenhouse gas (GHG) emissions relative to the use of virgin feedstock, a methodology that will measure and monitor the emission reductions from a recycling programme has yet to be approved by the CDM executive board.

The World Bank is working to remedy this situation with the development of a methodology for the recycling of materials from municipal waste. Except for paper, recyclables such as plastics, aluminium, ferrous metal and glass do not degrade in a landfill to form methane. The greenhouse gas benefit of recycling is a reduction in emissions from the use of fossil-fuel energy in the extraction and manufacture of products from virgin materials versus secondary materials. The methodology initiative to measure and monitor GHG emissions is focused on the recycling of products manufactured from high-density polyethylene (HDPE), such as plastic bottles, and low-density polyethylene (LDPE), such as plastic bags. Once this methodology is approved, other methodologies will be developed for paper, metals and glass. The process for development of a methodology requires a specific project to demonstrate the application. A recycling project in the area of Buenos Aires, Argentina, is the sample project being used for this purpose. The CEAMSE plastic recycling project in Buenos Aires consists of eight recycling plants where plastics will be recovered from post-consumer waste, classified, packaged and sold as raw materials for the production of HDPE and LDPE products, such as bottles and bags. Around 500 tonnes will be recovered, corresponding to

	Large scale	Small scale	Coupled with power generation
Landfill gas capture/ extraction (LFG)	ACM 001	AMS III G (landfill methane recovery)	AMS I D (grid connected renewable energy <15MW)
Composting	AM0025	AMS III F	
Other	AM0083 (avoidance of landfill gas emissions by *in-situ* aeration of landfills) AM0039 (methane emissions reduction from organic waste water and bio-organic solid waste using co-composting)	AMS III E (avoidance of methane production from decay of biomass through controlled combustion, gasification or mechanical/ thermal treatment) AMS III F (avoidance of methane emissions through controlled biological treatment of biomass)	

Table K18.6

List of current methodologies

around 914 tonnes of CO_2e per year.

The methodology is using the life-cycle approach and compares the GHG emissions from all steps involved in the project activity. The development of the methodology is also based on:

- the modalities and rules for small-scale methodologies; and
- simplified rules and monitoring protocol.

The new methodology is based on the assumption that there is a difference in energy/electricity use for the production of material from virgin inputs (i.e. from extraction of feedstock to manufacturing) versus recycled inputs (i.e. from collection to manufacturing).

In parallel, the World Bank is also working towards an urban methodology that will enable small individual activities to gain the financial benefits of incremental reductions from three urban sectors: waste management, energy efficiency and transit.

REFERENCES

www.worldbank.org

www.carbonfinance.org

http://go.worldbank.org/A5TFX56L50 (waste management)

www.cfassist.org

www.ipcc-nggip.iges.or.jp/public/2006gl/vol5.html

http://cdm.unfccc.int/index.html

SOUND INSTITUTIONS, PROACTIVE POLICIES

In most countries, local authorities are responsible by law for the 'public good' of safeguarding public health and the protection of the environment related to solid waste. But the degree of interest and diligence that a local council or city mayor brings to solid waste varies widely in the cities (although the fact that the cities agreed to be profiled for this Global Report suggests at least a modest level of ownership, so the general level of commitment in this particular sample is probably skewed to be higher than the modal).

As such, ISWM tests the full range of governance skills for 'managing' waste: priority-setting, strategic planning, consultation, decision-making, law-making, delegation, contracting, human resources management, financial management, enforcement and conflict resolution. Waste management services consume between 3 and 15 per cent of municipal budgets, not as much as the 20 to 50 per cent reported by the World Bank, but still a large proportion of the operational revenue of a city or municipality. They employ or contract or provide welfare-based work for many hundreds, if not thousands, of municipal staff. Because waste management is one of the most visible of urban services, the state of waste management tells a lot about the management capability of the local authority.

One of two regional composting facilities in the Adelaide metropolitan area

© Justin Lang, Zero Waste SA

Issues

What are the issues for institutional development and governance? For better or worse, people often judge whether or not a mayor is doing a good job by the cleanliness of the streets and the quality of the waste management service. The adequacy of services and their equity across income levels and social status also reflects how committed the city administration is to addressing urban poverty and equity, suggesting that the solid waste sector can be a useful proxy indicator of good governance.[24] Where waste management continues to work through periods of election and during and after changes of political administration and mayors, it is likely that the city has also addressed and opened up underlying organizational arrangements, management structures, contracting procedures, labour practices, accounting, cost recovery, transparent budgeting, institutional memory-building, documentation procedures, and corruption.

In this section we look at institutions and governance in the reference cities in terms of the following issues and related questions.

Policy commitment and ownership of the problem
Perhaps the most critical issue is whether there is a sustained political commitment to sustainable solid waste management and, closely related, that there are parts of the municipal administration which own the problem and consider it part of their responsibility to protect public health and the environment. Gathering and maintaining accurate information on the solid waste management system can be seen as a sign of commitment from the authorities' side and enables municipal staff to build upon accumulated knowledge and wisdom.

When existing data comes only from external donor projects and/or is old, this is a strong indication of lack of proactive management and priority for solid waste. The waste management agency Superintendência de Limpeza Urbana (SLU) in one of the reference cities, Belo Horizonte, Brazil, has an extensive internal documented institutional memory of annual activity reports, financial reports, strategic plans, performance and cost indicators, and evaluation reports that enables current staff responsible for solid waste management to draw from previous experiences. In 2000, the SLU even published a book: *A Century of Experiences in Solid Waste Management in Belo Horizonte.*

Issues about proactive policy examine the existence of clear policy frameworks established at municipal level related to sustainable waste management in both strategic municipal urban development plans and sector-specific environmental and/or waste management plans. Sustainability can be threatened when policies regarding solid waste management change every time a new mayor takes office, but are strengthened when ownership is passed on from one administration to another, building upon the existing institutional memory.

National policy context and framework

The relationship between the city authorities and the enabling policy environment has a strong influence on ISWM. Modernization changes and challenges this relationship. The larger political context is shaped by environmental ministries, inspectorates and agencies; by national and international health, economic and finance institutions; by rule of law; by commitments to stakeholder participation; and by national and global rules about the private sector and financial institutions.

A clear and transparent policy framework is critical for ISWM, and guides the city authorities in the processes related to planning an implementation. Understanding the number of different national institutions that have something to say about solid waste provides information about national institutional development. Recent activities in modernization of the legal framework indicate some level of interest and commitment. The way in which national institutions are organized may have more influence on the cleanliness of the city than the type of trucks or the location of the landfill; but they usually get far less attention.

Institutions and organizational coherence.

Municipal solid waste management in many cities is institutionally fragmented and lacks administrative coherence. For example, a small cleansing department under the city council may depend on labourers and supervisors managed on a decentralized basis in all of the individual districts. Vehicles and drivers for both the waste and highway departments may work for a central transport department, rather than for their operational divisions. Salaries for these labourers and drivers may be paid by the personnel organization or finance ministry, meaning that there are three, four or even more separate departments involved in day-to-day operations.

Fragmentation makes it difficult to assign responsibility or accountability; but consolidation is difficult because of established bureaucratic claims, traditional organizational structures or national legislation related to recruitment and assignment of *confidence* functions of middle- and higher-management staff.

In looking at institutional coherence in the reference cities, key lines of enquiry focus on the consolidation of the solid waste function and the presence of clear lines in the organizational chart. An indicator is the number of different budgets that contribute to solid waste expenditures. Another one is dedicated funds: in some cities, city authorities are allowed to retain the revenues collected from local taxes or to levy direct charges for services; but in Bulgaria these go into the general fund and have no direct relationship to solid waste budgets.

Regionalization and inter-municipal cooperation

The modernization trajectory in many cities is based on closing local dumpsites and regionalizing disposal. While engineers can calculate the economy of scale for disposal quite easily, the success of regionalization, in practice, depends more on institutional capital and skills for cooperation than on the quality of the soils.

Failures to organize inter-municipal cooperation can increase costs substantially and/or delay construction of controlled disposal for years. Overcoming biases based on political party differences between municipalities or hierarchy rivalry between municipal and regional (provincial) authorities is crucial in this cooperation. Making regionalization work depends on clear and transparent processes that build trust and facilitate cooperation.

Private sector involvement and pro-poor public–private partnerships (5-Ps)

The former recycling manager of the State of Massachusetts in the US used to talk about good practice in organizing public–private cooperation as the 'Caesar–God Principle' based on a biblical parable ascribed to Jesus Christ. When Christ was asked to choose who should get taxes, the Roman Empire or the Church, the apocryphal answer is reported as: 'Render unto Caesar what is Caesar's, render unto God what is God's.'

In solid waste practice this is taken to mean: let the private sector do what they are best at and leave the rest to the public sector to do the best they can, respecting always that final responsibility for waste management lies with the city. Or, more specifically, consider the kinds of solid waste and recycling activities that work well as private-sector activities, and separate them from those operations which the public sector, in general terms, must retain to itself. The discussion on business models in the previous section also contributes to this topic. Some key questions are:

- How much of solid waste and recycling is already in private hands? Is there a need for intervention, and how can that intervention be organized so that what works continues to work, and what doesn't work is improved?
- Is there an enabling framework for working with the private sector, and does this apply equally to MSEs, CBOs and informal enterprises?
- What are the rights and responsibilities of the city when it comes to providing services itself or managing services provided by others?

- What are the unique responsibilities of the city authorities? And how does the city ensure that it remains in control?
- If certain services are outsourced, is the city able to define what it needs, supervise and control to ensure that it will receive what it needs, and does it have the resources to pay for it?

Professional competence and networking

ISWM systems are run by people, and the quality of service is determined by the professional capacity of those making the policies, plans, contracts and operational decisions. Waste is generally seen as dirty work, even for managers and planners, so professional support from others in the field is usually not only helpful, but necessary. Formal and informal contacts, training, conferences and meetings all contribute to strengthening the skills, credibility and, above all, the spirit of municipal waste management organizations. And maintaining and increasing skills in a changing and modernizing context also requires investment and sustained commitment to capacity-building and institutional strengthening.[25] Questions include:

- What mechanisms of information exchange are available?
- How are staff in waste and recycling classified and what kinds of structures do they work in?
- What kinds of support and development are offered?
- What are the recruitment procedures for municipal staff, both higher- and middle-management level?
- Who is responsible for recruitment? How agile is the process, or do bureaucratic procedures delay timely filling of vacancies?
- Are there job descriptions for solid waste functions; and is there coherency between these function requirements, recruitment procedures, staff functioning evaluation practices, training needs identification and capacity-building?

KEY SHEET 19

WASTE MANAGEMENT AND GOVERNANCE: COLLABORATIVE APPROACHES FOR PUBLIC SERVICE DELIVERY

Ellen Gunsilius (GTZ) and Andy Whiteman (WasteAware)

BACKGROUND

Waste management is one of the most visible of municipal services such that inappropriate waste management in low- and middle-income countries is a good indicator of government's difficulties in delivering public services to their population. In developing countries, even the most basic targets of safe waste collection and disposal are not met. Unclear distribution of responsibilities, inefficiency of public waste management services, non-transparent management processes and poor financing mechanisms are only some of the common problems in waste management in developing countries. It has been shown that without focusing on governance aspects, waste management systems are less effective.

Governance describes the sum of decision-making processes and structures in the public sector, the 'rules of the game', so to speak, with which the state structures the possible interventions of society and the private sector, as well as defines the relations between government and the governed. It encompasses processes in which the classical hierarchical model of political regulation is complemented by forms of cooperation, co-production and communication between and among stakeholders. Good governance implies transparency of decision-making processes and responsibility of the relevant actors. As central governance weaknesses have to be overcome to guarantee efficient delivery of waste management services, the rate of waste collection and other waste management indicators can also serve as indicators for good urban governance.

EFFICIENT DISTRIBUTION OF RESPONSIBILITIES BETWEEN GOVERNMENT LEVELS

In governance approaches, the principle of subsidiarity, which states that political functions should be fulfilled by the political authority at the level nearest to the population, is central. In many countries, areas needing waste management competencies are many and widespread, widening the scope of action for local governments. For the multilevel governance approach to work, national waste management planning and its legal framework have to serve as strong foundations that can provide proper orientation for local authorities on waste management. The decentralization framework of governance has to be strengthened to ensure real transfer of competencies and resources from the national to the local level. In parallel, municipalities have to be supported in their institutional development, planning processes, and in the application of appropriate systems for waste collection, transport, valorization and minimization of waste in order to be able to deliver these services

Sweepings left in the street, Cairo, Egypt

© David Wilson

autonomously. As waste management infrastructures require high investments and economies of scale to operate efficiently, the creation of joint service councils (for example, in the Palestinian Territories) is one approach that uses synergies through inter-communal cooperation between municipalities in order to increase public service efficiency.

An important aspect of efficient local waste management systems is the design of efficient and transparent financing systems to support recyclers, contractors for waste collection and disposal, operators of compost plants, landfills, etc., which will require the cooperation of other stakeholders, such as the local banks and non-bank institutions, credit guarantee institutions, concerned NGOs and others.

COLLABORATION WITH CIVIL SOCIETY

The success of waste management strategies depends largely on their acceptance and the participation of the population. People have to learn to segregate their waste properly according to type of waste and to put them in separate covered containers or bags while awaiting their collection by personnel from the waste management services. They have to also understand that they have to pay for these services. For this interaction between government and people to happen and be successful, the development and use of communication strategies and involvement of the population in waste management-related decision-making actions are of utmost importance. Encouraging community associations and civil societies to participate in planning and decision-making processes through local committees or regular consultations are essential in order to take into account the priorities of the population and, in the process, ensure social balance and the acceptance of waste management interventions.

Awareness-raising campaigns and transparent information management have proved to be good approaches to convince people of the necessity of behaviour change and to minimize objections against new waste management measures. In a project in Mozambique, for example, the willingness of the population to pay for waste management services was improved through open communication between the municipality and civil society. In this project, after the population had been informed in detail about the reasons for the new waste management fees and experienced improvement of services people expressed more willingness to pay the fees.

Involvement of civil society in the design of fee systems can also contribute to ensuring social equity – for example, by imposing smaller fees on poorer households, using income or energy consumption as a parameter to determine the amount of fee, thus probably charging wealthier households more. In Mozambique, traditional-sector representatives were also involved in the monitoring of private service providers, allowing the inhabitants to obtain a stronger voice in the waste management system. With increased citizen participation, the rate of waste collection improved from 25 to 60 per cent in the poor suburban areas of Maputo, where public satisfaction with waste management services is very high.

PUBLIC–PRIVATE COLLABORATION

Participation of the private sector in waste management takes place because of two reasons. First, public institutions even in a developed country often depend on private investments to augment their limited financial resources. Second, and related to the first, private enterprises specializing in solid waste management have the necessary technical expertise and equipment that municipalities lack. They have the capacity, equipment and other resources to work more efficiently than public institutions. As a result, private enterprises are engaged widely in waste collection, as well as in the management of waste management infrastructures, becoming service providers for the public sector. The most common form of engaging the services

of private enterprises is through 'contracting', a situation where the national or local government contracts or enters into an agreement with a private company and pays for its services. To support this type of collaboration, it becomes necessary for government to generate revenues, perhaps through taxation or fee charges, in order to be able to afford all of the system costs. But it has to ensure transparent and competitive bidding procedures as well as effectively monitor and supervise private service providers. The success of public–private partnerships depends considerably on a thorough preparation of contracts and clear guidelines that form the framework upon which the public and private partners identify and combine their respective forces.

Itinerant waste buyer in Ghorahi, Nepal

© Bhushan Tuladhar

COOPERATION WITH THE INFORMAL SECTOR

Cooperation with the private sector does not relate only to cooperation with large companies, but to all types of enterprises, large-, medium- or small-scale community-based enterprises and whether these are national or multinational corporations engaged in waste management. Most of these entities belong to the formal private sector. But there are small-scale community-based enterprises that do not belong to the formal sector because they are not registered with any public entity and therefore do not have legal personality. They are referred to as the informal sector. Establishing collaboration with the informal sector is quite difficult for governments. It creates a dilemma since cooperation may be construed as accepting the actions of the informal sector that contravene existing rules and regulations while, on the other hand, it is an important aspect of a collaborative governance approach. In many cases, governments have realized that they cannot ignore or prevent the activities of informal waste collectors, resellers or recyclers, and that collaboration with these actors can actually bring about great advantages on the way to a broad-based and resource-efficient waste management system. The focus of the informal sector on the valorization of waste materials can, in addition to strengthened resource efficiency, reduce the costs of the collection and disposal systems and thus raise the service levels in waste management.

CONCLUSIONS

The active involvement of private and civil society actors in waste management demands substantial institutional changes from municipal governments. But it can result in much more efficient and publicly accepted waste management policies and thus result in efficient public services for the population.

REFERENCES

Coad, A. (2005) *Private Sector Involvement in Solid Waste Management: Avoiding Problems and Building on Successes*, Collaborative Working Group on Solid Waste Management in Low- and Middle-Income Countries, St Gallen

Wehenpohl, G. und Kolb, M. (2007) *The Economical Impact of the Informal Sector in Solid Waste Management in Developing Countries*, Eleventh International Waste Management and Landfill Symposium, GTZ, Eschborn

Whiteman, A., Smith, P. and Wilson, D. C. (2001) 'Waste management: An indicator of urban governance', Paper prepared for the UK Department for International Development (DFID) and submitted by DFID to the UN-Habitat Global Conference on Urban Development, New York, June 2001, www.davidcwilson.com/Waste_Management_An_Indicator_of_Urban_Governance.pdf

Cities and experience with good governance practices

■ Policy commitment and ownership of the problem

Solid waste management is a permanent and continuous responsibility to be assumed by elected representatives when they take office. It is part of the deal of being mayor or sitting on the city council – a responsibility not only shared with the administrative municipal staff, but one to be organized and delegated within the municipal structure. Two key issues to be examined are coherency in policy commitment and organizational coherence.

Commitment to a clean urban environment goes beyond the pamphlets of the pre-election campaign or the timely acquiring of new collection trucks by the mayor seeking re-election. It concerns having strategic vision regarding how waste issues are to be tackled and laid down in strategic plans that span periods beyond a term of office. This then provides the basis for planning the yearly activities and budgets for that department, section or specific agency to which the waste management responsibility has been assigned.

The current mayor of Quezon City is one example of a highly committed public official who knows that solid waste is a priority. Similarly, Kunming City has a very strong political commitment to earning the official title of an environmentally clean city. In Kunming, the local politicians have focused effectively on using the political initiative of China's central government Western Region Development Strategy 2000–2010 to push the modernization process in their city.

South Australia and Adelaide represent an example of state-of-the-art governance over the waste and resources sector. Key factors are that:

- urban services (including waste management and recycling) are provided under a stable high-GDP economy and mature democratic system; and

Centro de Tratamento de Resíduos Sólidos (CTRS; Solid Waste Treatment Centre) in Belo Horizonte, Brazil, includes a sanitary landfill, a small composting plant, a recycling facility for construction waste, a unit for storing used tyres, and a unit for environmental education

© SLU Belo Horizonte, Brasil

- politically, the South Australian government has taken a major step by establishing Zero Waste SA as a specialized agency focusing on supporting the movement of waste and resources management 'up the hierarchy'.

In India, it is mandatory for the municipalities to protect public health and the environment. The 74th Amendment of the Indian Constitution gives a wide range of powers, functions and responsibilities to the municipalities. But these powers, functions and responsibilities are not commensurate with the resources provided. Even a large municipal corporation such as Bengaluru has not been able to implement its plan, owing to the lack of both financial and human resources. The municipal staff have little power to establish and enforce a cost-covering service fee without sustainable support from the political parties or

Box 5.18 Urban environmental accords: Setting an example in policy commitments

San Francisco, US, has had a series of mayors who were willing to be national, even international, environmental leaders. On 5 June 2005, San Francisco hosted United Nations World Environment Day, during which Mayor Gavin Newsom presented mayors from around the world with a unique opportunity: to create a set of objectives for an urban future that would be 'ecologically sustainable, economically dynamic and socially equitable'. Based on existing best practices and applied to issues such as energy, waste reduction, urban nature, transportation and water, the urban environmental accords have since been signed by more than 100 mayors who have begun applying accord principles in their own cities across the globe. Since that time, San Francisco has been adopting three urban environmental accord actions per year.

India

The Indian Civil Service is characterized by very short terms of office and high turnover for city high commissioners, who are the city chief executives. While the professional staff of the city are charged with implementation, their hands are tied during the process of changing commissioners, the policies made seldom get implemented, and there is no official functioning institutional memory. Apart from this, the system also becomes paralysed if there is a change in the party of the elected councillors. The priorities of one political party are often not the same as the previous regime, and this discontinuity does not allow long-term plans to be fully implemented. Enforcing service is also not a priority among the elected councillors, as it is a populist measure – the party, which claims to remove such charges, is anticipated to win elections, rather than those who put forward a viable financial model. For example, during the period of 1995 to 2004, when Bengaluru was part of the Urban Waste Expertise Programme (UWEP) and UWEP Plus programmes, there were no fewer than 15 commissioners, with the longest serving in office for about two years. This made it virtually impossible to make progress on waste management.

Belo Horizonte, Brazil

In 1973, the Superintendência de Limpeza Urbana (SLU) was created. This Public Cleansing Authority has been operating at arm's length from the municipality until now. It is endowed with a corporate entity and its own patrimony, and has administrative, financial and technical autonomy. By municipal law, the prime objective of the SLU is the exclusive execution of all solid waste management services (sweeping, collection, disposition, treatment and transformation of the garbage) for the entire City of Belo Horizonte, including the sales of its services, products and by-products.

Created as a public autarchy, the SLU continued until now to enjoy considerable autonomy regarding decision-making, with some control of expenditure over a long period of its history, which might help to explain its continued drive for modernization, even through changes of numerous mayors. The SLU was also devised to have a central planning unit, but decentralized operational units, which might have contributed to this process of modernization. Investments on the qualification of technical staff have also been an important feature.

their elected officials. The existing fees, covering only 40 per cent of costs, are not strictly enforced either, owing to a belief that voters prefer lower costs. Moreover, there is a kind of fear that enforcing payment of the fee could lead to more complaints about the quality of service delivery, irregularities and absenteeism among the municipal workforce. The result is a kind of 'gentleman's agreement' for the government not to raise fees and for the providers and users to accept what the system offers without critique.

While the city information does not support quantitative comparison, there is at least an attempt to say something in a qualitative way (e.g. a score) about institutional coherence. Belo Horizonte, Curitiba and Porto Alegre in Brazil are examples of municipal governments that

have chosen a prolonged and sustained approach towards improving solid waste management without the benefit of a national solid waste policy, which is still awaiting approval in congress. In addition to having strategic waste management plans, the municipalities have a Multi-Annual Municipal Action Plan (PGAP), which establishes a detailed performance assessment of the municipality, specifying explicit policy, objectives, action, programmes and goals to be reached, defining necessary resources for their implementation.

An important feature of this four-year municipal plan is that it does not correspond to the four-year term of the local government administration. Instead, it covers two election periods: three years for the current administration and the first year of the new administration. This means that newly elected authorities cannot ignore what previous administrations have planned and budgeted; this leads to continuous building upon the institutional experience gained.

When there is no organizational coherency, creating ownership of the waste problem can be severely hindered. It is unclear who is responsible within the municipality, leading to a lack of a common vision on how to address the solid waste issues. No one individual takes a lead role in defining what needs to be done.

In Managua, Nicaragua, the municipality recognizes that institutional fragmentation is one of its key weaknesses that need to be addressed in its current strategic planning process. Solid waste-related activities are undertaken by at least eight different municipal sections, including at the central municipal level, operational solid waste activities under responsibility of the Public Cleansing Department; environmental awareness campaigns led by the Urban Environmental Department; and street sweeping, drainage and illegal dump cleaning by decentralized districts. What is missing is one entity preparing and implementing an overall strategy with a cumulated budget. Instead, waste expenditures budgeted and reported are more than half of what the municipality spends in reality because only that part designated to the Public Cleansing Department is conceptualized as solid waste costs.

Documenting information, establishing archiving protocols, monitoring and evaluating mechanisms all contribute to consolidating ownership of knowledge on solid waste issues and to keeping track of how the policy commitment is being implemented. For goals and targets set in annual plans to be consistent with budgets, the municipality needs to be sure that the required budgeting and planning skills are common knowledge for higher- and middle-management staff responsible for solid waste.

■ National or regional/state policy context and framework

When national or international policy provides a framework, it can be easier for municipalities to translate objectives and targets for recycling or treatment strategies into their yearly plans and budgets. In Adelaide, the South Australian government established a Zero Waste SA as a specialized agency focusing on supporting the movement of waste and resources management 'up the hierarchy'; it is also responsible for implementing policies established by the federal government. In compliance with the national framework, the agency develops statewide policies and programmes to achieve waste reduction and recycling targets. The state government also often funds and make decisions relating to major waste management facilities or programmes.

In Quezon, the Philippines, the national Waste Act 9003 calls for each village (*barangay*, or local political unit) to create its own solid waste management committee to supervise management of waste in the city. *Barangay* local governments have the responsibility of implementing an ecologically sound management system for recyclable and biodegradable wastes.

San Francisco, US, has the most ambitious waste diversion goals at national level. More than just lofty ambition, it can claim to be well on its way to achieving zero waste – currently diverting more than 72 per cent of its waste! Materials are diverted from landfills according to the hierarchy of source reduction, reuse, and recycling and composting. While waste prevention, composting and recycling programmes are generally proceeding well, each additional 1 per cent of diversion is more difficult to achieve than the previous 1 per cent. All programmes and activities are laid down in different waste management-related plans: a zero waste plan, a sustainability plan and an environmental plan.

Legislation and policy-making at state level contributed to this significant progress. The 1989 Integrated Waste Management Act (AB 939) mandated cities and counties in the State of California, working off the base year of 1990, to reduce the amount of waste landfilled in 1995 by 25 per cent and in 2000 by 50 per cent. The California Integrated Waste Management Board

In many Eastern European countries, complying with EU waste directive recycling norms has meant reserving public space for recycling banks, as here in Prague, the Czech Republic

© Jeroen IJgosse

With the support of the Pan-American Health Organization (PAHO) in August 1998, Peru prepared a sector analysis of solid waste management with the participation of the main stakeholders from public and private sectors. In July 1999, the Ministry of Health through the General Direction of Environmental Health prepared the General Law of Solid Waste. Following a revision process of one year, Law 27314 came into effect, with the technical norms and related decrees approved four years later. The national law, including mandatory strategic solid waste planning and municipal laws and decrees, were developed and approved within the national framework. Now, ten years after the process was initiated, most municipalities have strategic solid waste plans, while some provinces and large municipalities are into their second generation plans.

(CIWMB), the state agency that oversees solid waste, requires each jurisdiction to develop an integrated waste management plan and to submit data annually in order to demonstrate how they are meeting the mandated requirements. The components included in the plan are waste characterization; source reduction; recycling; composting; solid waste facility capacity; education and public information; funding; special waste (asbestos, sewage sludge, etc.); and household hazardous waste. This mandate carries a US$10,000 per day fine for cities and counties that do not participate. As a result of this commitment by citizens, cities, counties, solid waste management companies and other recyclers, an infrastructure costing hundreds of millions of dollars is now in place in California, collecting, sorting, processing and transporting recovered recyclables.

During the 1990s an overall and in-depth assessment of the solid waste sector at national level was conducted in more than ten countries in Latin America. These *sector-analysis studies* were initiated by the Pan-American Health Organization (PAHO) and carried out by international and national multidisciplinary teams. In a number of countries, this was an important turning point in their modernization process and a departure point for mayor changes (see the Peru case study in Box 5.20).

Whereas national policy is important for setting goals and targets for waste management strategies at the municipal level, realizing that solid waste infrastructure requires active planning is just as relevant. Preparing urban development plans means making choices for reserving urban space for different urban functions. In this process, it is just as important to reserve a land-use function for managing solid waste.

Land is reserved for healthcare through building hospitals, for employment creation through developing industrial parks, and for education through reserving space for universities and schools. Even leisure or public green areas will have a zoning code assigned to them protecting a specific area of a city from being used for another purpose.

Reserving land for solid waste management, whether a landfill, composting facility or a transfer station, is rarely done during the early stage of the modernization process, although in most cities waste is very much present within the urban physical context and can occupy plentiful of land through illegal dumping and blocking of waterways.

In Managua, Nicaragua, the land-use zones laid down in urban plans and used by its Urban Infrastructure Department for project evaluation do not include codes related to solid waste functions for transfer stations. This has made it difficult to find possible locations for a small transfer station as part of the strategic planning process. While finding reasons for not accepting different location proposals has proven easy, asking the municipality to indicate which areas it has actually reserved for this specific urban function has been left without reply.

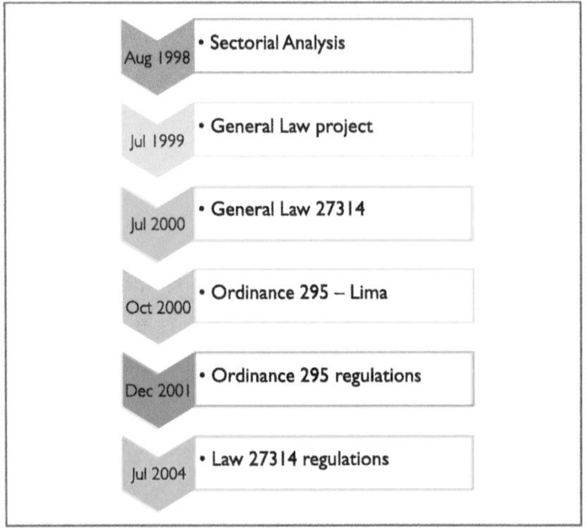

Figure 5.1

Regulation process of solid waste in Peru

Source: IPES, Lima, Peru

- Aug 1998 · Sectorial Analysis
- Jul 1999 · General Law project
- Jul 2000 · General Law 27314
- Oct 2000 · Ordinance 295 – Lima
- Dec 2001 · Ordinance 295 regulations
- Jul 2004 · Law 27314 regulations

The example of Tompkins County, New York, previously discussed, demonstrates an innovative approach to mitigating the NIMBY effect. Interestingly, Tompkins County never built the landfill that it sited because in negotiations with the private sector, contracting for export of waste to a different county appeared to be a better choice. Agreements between the local authority and community need to respond to financial and political circumstances, environmental ambitions and citizens' opinions. Agreements also need to be laid down in plans and binding documents that are respected and enforced by future municipal administrations.

During the modernization process it is important to go beyond the phase of upgrading an existing landfill and to take the institutional steps required to reserve space for a landfill, as was exemplified in the Belo Horizonte, Brazil, case study (see Box 5.21). For example, it is also necessary to avoid other urban activities encroaching into the buffer zone reserved around the area allocated to the landfill. Especially in large urban areas where pressure on housing is great, illegal neighbours can quickly invade the area adjacent to the landfill and its access roads. Usually, the landfill is seen as the culprit, even if it was already there for a long time.

■ Institutions and organizational coherence

Institutional and organizational coherence relate to the location of the solid waste function situated within the municipality, the ultimate responsibility for solid waste, and the degree of consolidation in the organogram. Where one entity is responsible for the vision, implementation and monitoring of solid waste, there is more coherence than when waste functions are spread across different departments so that no one is ultimately accountable. Some of the reference cities can say that all waste management functions fall under a single node in their organizational charts, while for the others, the situation is too fragmented, or they were not able to report their coherence (itself a likely indicator of general lack of policy support).

Box 5.21 Reserving land for a landfill to avoid being caught by surprise

From 1973 until 2007, most of the municipal solid waste from Belo Horizonte, Brazil, was disposed of in a designated disposal site in Bairro Califórnia, to the north-east of the city The site forms part of the so-called Centro de Tratamento de Resíduos Sólidos (CTRS). This 115ha treatment centre houses, besides the landfill, a number of other facilities. The landfill first operated as a controlled disposal site; but after 1997 operations were upgraded to a sanitary landfill standard in compliance with the national sanitary norms.[26]

When the landfill was allocated in 1973, no provision was made at the time for its estimated capacity and, as such, no end of life was foreseen. However, during the 1990s, with the process of upgrading, it was realized that the landfill would actually one day reach a maximum capacity and that alternatives would be needed. This maximum capacity was established as a function of the height of the landfill receiving domestic waste.

In anticipation of this future closure, the municipality started a process of selecting an alternative site. A suitable area was identified belonging to the municipality. However, its land use was not designated and no provisions were made for its future use as a sanitary landfill within the urban plans. As a result, when the time came to commence project preparation for the new landfill, the foreseen access roads and adjacent lands were occupied by dwellings and popular pressure forced the municipality to seek another solution.

In August 2007, the established maximum capacity of the landfill was reached and the landfill was officially closed for domestic waste. The Federal State Environmental Authority (FEAM) approved the closure plan for the landfill. Although the city went successfully through the modernization process of going from a controlled dumpsite to a sanitary landfill, which was institutionally and financially supported by the municipality, guaranteeing a continuous long-term option for disposal has proven more challenging and a hurdle not yet surpassed within the modernization process. To a certain degree, the municipality 'was caught by surprise' when the landfill did fill up and has had to resort to '*ad hoc* measures'. A private sanitary landfill provides a temporary solution. Major investments had to be made to construct a transfer station at short notice to reduce hauling costs. In the meantime, negotiations for an alternative more than 50km from the city have been lingering for more than three years without a foreseeable outcome.

This could, perhaps, have all been avoided had the location of the previously agreed-upon original site been respected and institutionalized.

Consider the institutional situation in Varna, Bulgaria. The waste management activities are coordinated between three directorates in the city administration: the mayors of the city administrative regions; the mayor, represented

Municipal staff assessing the different roles in waste management governance in Galati, Romania

© Mathias Schoenfeldt

Box 5.22 Roles in waste management

- *User:* direct beneficiary, also frequently paying for the services and benefits.
- *Service organizer or client:* has responsibility for providing an adequate level of service that protects public health and the environment, at an affordable cost, to all of the population.
- *Comptroller or fiscal manager:* is responsible for financial management, but not for setting policy. Typical responsibilities include collecting, receiving or raising funds, managing the process of allocating them to operations; budgeting; disbursement; and accounting.
- *Provider or operator:* has the obligation to implement physical systems and to deliver actual services for street sweeping, waste collection, transport, transfer, treatment and disposal, or some combination of these. Occasionally, recycling and organic waste management are included in operations; sometimes they are considered to be separate.
- *Regulatory body or inspectorate:* is charged with monitoring and inspecting operations; applying and enforcing environmental legislation, regulations and standards; issuing site licences or permits to treatment and disposal facilities; and inspecting/monitoring their operations to enforce the licence conditions and policing illegal dumping elsewhere.
- *Adjudicator or process manager:* is an independent authority for adjudicating conflicts, managing environmental impact assessment (EIA) and strategic environmental impact assessment (SEIA) processes; ensuring that laws and rules about inclusivity and participation are observed; and handling complaints.

---

by a deputy mayor, the chairman of the municipal council of the Municipality of Varna; and two committees. The organizational chart as related to waste management appears quite fragmented. This is, in fact, the case: Varna is a large city and different departments work on different aspects of waste management. The principal role in waste management, however, is performed by the Directorate of Engineering and Infrastructure Development.

In Bengaluru, India, all formal structures of the waste management system come under the Solid Waste Management Division or Cell, with the final authority resting with the commissioner, indicating a high degree of institutional coherence. The Municipal Solid Waste Rules 2000 comprise the guiding principle. In Nairobi, Kenya, most solid waste functions fall under Nairobi City Council; but the operational department is the Department of Environment, and there is no solid waste organization at director level.

One important aspect is whether the solid waste management (SWM) department reports directly to the political level, or whether it is part of another department. In Tanzania, for instance, SWM generally falls within the remit of the Department of Health.

Modernizing waste systems and improving performance is easier to implement and monitor when all or most waste-related functions come together under a single node in the organogram, making it possible for accountable and transparent management, budgeting and operations. Where this is not possible, relations between different departments and their specific roles and responsibilities need to be spelled out and endorsed by all parties. Prior to modernization, these functions may be combined, fragmented or difficult to identify. One element of good practice is to clarify responsibilities and differentiate among the key roles presented in Box 5.22. It is also useful to say something in a qualitative way, for example through a score, about institutional coherence and accountability.

Good practice guidelines suggest that introducing a clear division of responsibilities results in better outcomes. Often it works best if certain functions are shifted out of waste management to other municipal organizations, even if this contradicts the principle of coherent institutional structure. For example, in the modernization of municipal waste management in Hungary during the early 1990s, functions such as managing parking meters, operating the nursery for street plantings, and managing sport facilities were shifted to another municipal entity, giving the municipal waste company a clearer mission and focus.[27] When a municipality decides to stop operating its collection system, for instance, and contract it out to private operators, it may close its operational division, but the functions of supervisor, comptroller, regulator and adjudicator still remain. This is because public cleanliness and public health are 'public goods', and the local authority holds the responsibility for ensuring that there is an ISWM system, which delivers them.

■ Regionalization and inter-municipal cooperation

Waste collection is usually best provided at the lowest appropriate level of municipal administration; but waste treatment and disposal may need to be organized on a unified basis across the metropolitan area as a whole, or even between cities in a region, province, county or specially created solid waste district. Inter-municipal cooperation is thus essential.

There is a growing trend for landfills and other waste management facilities to be regionally shared, accepting waste from multiple municipalities, the city and surrounding towns.[28] The US approach in many states has been to keep responsibility for collection at the level of municipal authorities, while shifting responsibility for disposal to the county or regional level. Traditionally, in North American policy, legal responsibility sits one level higher, with the states themselves in the US and with the provinces in Canada. In The Netherlands, since modernization started in 1979, municipalities have had continuous responsibility for waste collection. But disposal responsibility has shifted from provinces to groups of municipalities and back several times. Producer organizations have financial and logistical responsibility for the end of life of their products, including processing for recycling supply chains and safe disposal; but they delegate the collection of recyclables to the municipalities. In Bulgaria and Romania, under the influence of European Union (EU) accession, old regional structures have been reinvented as the institutional home for regionalized disposal systems.

Developing the accompanying regional cooperation and fair cost-sharing arrangements represents an ongoing challenge to city authorities. When the municipalities differ significantly in population size and related amounts of waste generated, in particular, there is an unfortunate tendency to find a perfect site for the landfill in the weakest, least populated and poorest of the municipalities. The position of the host municipality that will host the landfill should not be

considered lightly. When this involves a smaller municipality, the impression of the big neighbour dumping the waste in a small backyard needs to be neutralized. Many *environmental justice* conflicts are about asymmetrical power relations between large cities and their poorer surrounding towns and villages. The injustice is especially bitter when the host community doesn't get any compensation, or when their own levels of service are not satisfactory to their citizens. This sense of injustice may improve when users of an out-of-town landfill pay a surcharge, in the US called a 'host community fee', which goes to improve infrastructure in the host community, as was seen in the case of Tompkins County, New York. Since all stakeholders get used to the concept of paying for final disposal, and the requirement for paying tipping fees to enter and

Siting of a landfill has to compete with many other land uses as here to the north of Rio de Janeiro, Brazil

© Jeroen IJgosse

Box 5.23 Owning up for the landfill in Bamako, Mali

The main challenge the City of Bamako, Mali, faces for waste disposal is the lack of a designated final disposal site. There is a site designated for a controlled landfill about 30km outside the city limits in Noumoubougou. There are a few fading orange markers that were placed in the ground 15 years ago; but the landfill has never been constructed. Two important issues still need to be resolved:

1   Which authority is responsible for paying for the operation and maintenance of the landfill?
2   Who will pay to transport the waste collected in Bamako to the landfill?

While these issues remain unresolved, the city's waste is either dumped in illegal sites or sold to farmers who will accept contaminated waste because the organic content is so high.

use a landfill is accepted, it becomes both easier and more commonplace for host communities to receive compensation for offering their open space as a *sink* to the other users.

Management structures also might need to be developed, as is the case in Bamako, Mali (see Box 5.23), especially if the landfill is to be operated by the private sector. Relevant and effective institutional frameworks are essential because many inter-municipal alliances fall outside of existing jurisdictions and boundaries. Agreements between different municipalities will need to be made that will not be affected by a future change of one or more mayors, and political differences between neighbouring municipalities will have to be set aside.

Private enterprises collecting recyclables play an increasing role in Romania

© Jeroen IJgosse

■ PPPs and 5-Ps: Public–private partnerships and pro-poor PPPs

In many of the reference cities, collection and sweeping are provided by municipal departments; in others, there is a strong private or PPP component. Kunming, as part of the Chinese central planning approach, has a tradition of direct municipal service; but in recent years, following private investment in disposal facilities, is moving towards privatization: street sweeping has been privatized in the last four years in some of the districts of the city. Moshi, Tanzania, does its own work because it cannot attract private parties for other than CBO collection. Varna, Bulgaria, moved gradually during the period of 1999 to 2009 to complete zonal privatization of collection, and the landfill is also now operated privately.

There is clear 'cherry-picking' privatization in some cities. Nairobi, Kenya, is one of several cities where the city authorities allow private companies to be active in the relatively profitable central business district and high-income neighbourhoods, while the public sector struggles – and fails – to provide services at the city borders and low-income areas, where cost recovery is much more of a challenge. Dar es Salaam, Tanzania, takes the other approach: the city and its main large contractor limit themselves to secondary collection and the central business district, and the emphasis in the lower-income areas is on collection via micro-privatization with the involvement of MSEs and CBOs who collect their own fees.[30] Public–private partnerships in service delivery are an option for improving both cost effectiveness and service quality and coverage. Tompkins County, US, when implementing its solid waste plan, went through extensive consultation with private waste collectors – but still was not able to secure a government monopoly on processing. The more mature systems, such as Rotterdam and Adelaide, depend on the private sector as an integral part of their service provision approach, something that is equally true for Nairobi, Quezon City and Bamako. Rapidly modernizing cities such as Delhi and

Bengaluru and Managua are still working out their relations to the private sector.

The principle is that a formal or informal private company, motivated to produce income and support its owner and workers, has more incentive and flexibility to deliver services efficiently and cost effectively, but needs the counterweight of a public authority to protect the public good of a clean waste-free city. Municipal authorities, as providers, can contract out their operational responsibilities while retaining the other functions. In ISWM, sound contracting practice begins with setting operational goals, defining performance standards and specifications, and producing a document that communicates these to private, semi-private, NGO, CBO or other economic actors who would like to participate as service providers.

A competitive tendering procedure is a common, but not the only, mode for engaging private actors in operations; but all instruments have to ensure agreed-upon performance levels, enforce contract conditions and introduce sanctions for non-performance. On-time payment is another provider function, and even when the arrangement calls for the contractor to collect user fees directly from the users, the provider function has to keep the system operating. As providers, municipal authorities need to ensure that the operators provide the service, and that it meets the required standards of reliability, efficiency, customer relations and environmental protection as specified in the contract.

Three standard conditions of competition, transparency and accountability[31] are associated with successful PPPs. An essential but less-frequently mentioned fourth condition is the

Door-to-door collection by community-based organizations (CBOs) using handcarts in areas of difficult accessibility in the lower-income areas of Maputo, Mozambique

© Joachim Stretz

presence of external controls and horizontal power relations that safeguard a balanced partnership. In order for PPPs in waste management to be successful, all of the four key conditions must be met.

The challenge is to find the balance between efficiency, effectiveness and fairness – the three ISWM principles. A recent publication provides many useful examples, both good and bad, to help one to avoid the problems and build on the successes presented: *Private Sector Involvement in Waste Management: Avoiding Problems and Building on Successes*.[33]

In many parts of Africa, where local governments often do not have the capacity to organize waste management services themselves, a great many waste-related services are provided through some form of private-to-private

---

Box 5.25 International Labour Organization (ILO) micro-franchising in Africa

The city of Dar es Salaam, Tanzania, organizes waste collection via more than 55 micro-, small and community-based enterprises that tender for micro-zones, some with less than 500 households. The International Labour Organization (ILO), which pioneered this model there during the late 1990s, has been working with cities all over East Africa to replicate it.[32]

Box 5.26 Safi youth group in Mombasa, Kenya

In June 1999, a group of 22 unemployed school leavers in Mombasa, Kenya, decided to organize themselves and offer a waste collection service in their neighbourhood Mikindani that at the time was not receiving waste collection services from the municipal waste department.

The group did some very basic market research by simply knocking on families' doors and offering their services and leaving leaflets. Out of 200 doors where they spread the leaflets, around 60 families accepted their offer. They started collecting household waste twice a week and charging a fee of 200 Kenyan shillings per month. The residents were so satisfied with the services that within three years the collection service area had expanded to over 1000 households. As of mid 2009, the Safi youth group is serving more than 2000 households and four companies in the area. They have expanded their services to include separate collection of PET bottles as well as street compound cleaning, car washing, and carpet and sofa cleaning.[34] School-leaver micro- and small enterprises (MSEs) of this type are receiving increasing recognition as the dominant model for micro-privatization in francophone West Africa, where the MSEs are called *Groupements d'Intérêt Économique* (GIEs).

Educating women on composting, Siddhipur, Nepal

© Bhushan Tuladhar

National field visit in India, with municipal staff from Bengaluru visiting a landfill in Navi Mumbai

© Sanjay K. Gupta

financial relationship that is mediated by local authorities.

Municipalities award zones or give micro-zone monopolies to firms, MSEs, CBOs or even to informal family enterprises in order to collect waste, sweep streets, clean out drains and gutters, maintain parks and beaches, install office paper systems, and the like. In the so-called 'ILO–Dar es Salaam' model, each micro-contractor collects fees directly from the users, itself a difficult and expensive task, within guidelines set by the municipality. Both collection zones and provider operators are classified as large, medium or small. This model, increasingly referred to as pro-poor PPP (5-Ps),[34] is receiving increasing global attention as one viable, proven variant of public–private partnerships (PPPs), which has the goal of helping small enterprises to provide services to unserved poor communities.[36]

The four conditions go a long way in ensuring that the contracting process is transparent and free from corruption; the latter has regrettably been associated with waste management services in a number of places.[37] Fair contracting and transparent legal and commercial arrangements are at the core of every functioning PPP, whether the private partner is a powerful multinational corporation or a local MSE or informal-sector co-operative operating in a situation of economic weakness.

■ **Professional competence and international networking**

Clearly, ISWM is not just an engineering discipline. The field of work attracts people from a range of disciplines and with a range of competencies. Solid waste planning and operations departments are populated by natural scientists, economists, planners, environmental activists, business managers, farmers, sociologists, lawyers, medical doctors, statisticians, information technology (IT) specialists and political scientists. As a field of specialization, ISWM offers a wide range of intellectual and practical challenges. Everybody knows something about waste but nobody knows everything!

Capacity development programmes are a popular focus for international development assistance, and solid waste and recycling have had a certain amount of direct attention in recent years. Waste and recycling training is offered by the United Nations Development Programme (UNDP) PPPs-SD (public–private partnerships for sustainable development) through the Cities and Climate Change Initiative (formerly the Sustainable Cities Programme of UN-Habitat), the International Council for Local Environmental Initiatives (ICLEI) by the World Bank Institute, and by many bilateral donors and others. The International Finance Corporation has been developing specific business training for the recycling supply chain in the Western Balkans and in Central Asia. Specific ISWM training materials are available from the organizations involved in producing this book. Many universities are now offering specific courses in

International waste professionals on a field visit to a transfer station during a CWG meeting in Ouagadougou, Burkina Faso

© Jeroen IJgosse

waste management. In some countries, there is a legal requirement for certification and accreditation of the competence of operating personnel at all levels of delivering waste management services.[38] Such initiatives work to underpin the development of the sector through training, skills transfer, and strengthening of professional competence and critical thinking.

Nine reference cities report that they recognize occupational categories in the formal sector, and two, Belo Horizonte and Quezon City, recognize informal categories. Six cities report that professional support is available for informal-sector groups. This should go hand in hand with transparent recruitment and selection procedures that test candidates on their competence and critical thinking, leading to contracting of staff (especially for top- and middle-management functions) that builds upon experience and accumulated skills rather than on *shared* political views, family or kinship ties.

Strengthening the management skills and capacities of the human resource management section can play a crucial role in ensuring that the right people are located in the right place within the organizational chart. Permanent feedback on coherence between job requirements,

function evaluations and training needs are essential, as is the possibility for staff to grow with the organization in terms of responsibility.

Networking, conferences and informal peer exchanges are also critical to strengthening the

Table 5.12

Professional recognition

Notes: NA = not available; NR = not reported; N = no; Y = yes.

City	Occupational recognition	Occupational categories, formal	Occupational recognition, informal	Professional support available
Adelaide	Y	Garbage collector, recycler	None	Union
Bamako	NR	NR	NR	Training institute, waste management committee
Belo Horizonte	Y	NR	collector of recyclables	Waste picker cooperatives, platforms
Bengaluru	Y	Sweeper, Truck Crew	None	Self Help Groups, NGOs/CBOs
Canete	N	None	None	Stakeholder consortium
Curepipe	NR	NR	NR	NR
Delhi	Y	Sweeper, collector, junkshop owner and workers, transporters	Dump pickers, Street pickers, IWBs	NGO, associations
Dhaka	NR	NR	NR	NR
Ghorahi	NR	NR	NR	NR
Kunming	NR	NR	NR	NR
Lusaka	NR	NR	NR	None
Managua	NR	NR	NR	Unions, associations
Moshi	NR	NR	NR	NR
Nairobi	NR	NR	NR	Associations, NGOs/CBOs
Quezon City	Y	Junkshop operators	Dump pickers	Cooperatives and associations
Rotterdam	NR	NR	NR	NR
San Francisco	NR	NR	NR	Associations, foundations, groups
Sousse	NR	NR	NR	None
Tompkins County	Y	NR	NR	Cooperative
Varna	NR	NR	NR	NR

human resources in the solid waste and recycling sector, especially as processes of modernization and globalization increase the benefit of exchange between and among cities of all kinds and sizes, and other stakeholders.

Many cities have benefited from establishing and maintaining networks or platforms at the regional, national and/or international level. There are networks, such as United Cities and Local Governments (UCLG) (until 2008, the International Union of Local Authorities, or IULA), City Net Asia and others, of which many cities may already be members, which are excellent platforms for exchange and sources of new information, insights and inspiration on waste management. Some networks are specific to solid waste management, such as the Association of Cities and Regions for Recycling and Sustainable Resource Management (ACR+)[39] in Europe, the National Recycling Coalition in the US, ReCaribe

in the Caribbean, and 3Rs Knowledge Hub in Asia.[40]

One global platform, the Collaborative Working Group on Solid Waste Management in Low- and Middle-Income Countries (CWG), deserves mention for its biannual workshops and capacity materials that support global networking related to facilitating cooperation between cities and their informal recyclers, informal service providers, and micro- and small enterprises in waste and recycling. The growing interdependence of cities in solid waste management is beneficial at the country level and at the regional and international levels.

Networking is also a form of capacity-building. Good practices and lessons learned, for example, have more impact and memory recall when shared directly between one mayor and another through networking.

NOTES

1 GTZ/CWG, 2007.
2 Hawley and Ward, 2008.
3 GTZ/CWG, 2007, Lusaka City Report and worksheets.
4 Sampson, 2009.
5 This refers to the not-yet-published report *Economic Aspects of the Informal Sector in Solid Waste*. Draft report is available from www.gtz.de, www.waste.nl and www.cwgnet.net.
6 GTZ/CWG, 2007.
7 Wilson et al, 2001.
8 CWG: the Collaborative Working Group on Solid Waste Management in Low- and Middle-Income Countries, www.cwgnet.net.
9 IJgosse et al, 2004b, www.wastekeysheets.net.
10 METAP, undated. Includes seven training modules on strategic planning: click on policy and planning – training. Available in English and Arabic.
11 More discussion on service-based, commodities-based and values-based enterprises is to be found in *Micro and Small Enterprises in Integrated Sustainable*

Waste Management, WASTE, Advisers on Urban Environment and Development, Gouda, The Netherlands, www.waste.nl.
12 Solid Waste Management Association of the Philippines, 2009.
13 WIEGO, 2008.
14 Dias, 2000, 2006.
15 See www.civisol.org.
16 ILO/IPEC, 2004; Rosario, 2004.
17 Anschütz et al, 2005.
18 Anschütz et al, 2005.
19 Text in this section was first prepared for the publication *Closing the Circle, Bringing Integrated Sustainable Waste Management Home*, Association of Dutch Municipalities (VNGI), The Hague, the Netherlands, 2008.
20 Keita, 2003.
21 METAP, undated. Includes regional guidelines on finance and cost recovery; available in English, French and Arabic.
22 METAP, undated.
23 See also discussion in 'Waste collection: Protecting public health in the reference cities' and 'Waste treatment

and disposal: Front lines of environmental protection' in Chapter 4.
24 Whiteman et al, 2001.
25 METAP, undated. Provides guidelines, tools and training materials, covering a wide range of aspects of an ISWM system.
26 According Brazilian norms, sanitary landfills must be fenced off; have a monitoring system, including a weigh bridge; allow no waste-pickers on the landfill; and have 'impermeabilization', daily covering of waste body with earth, drainage systems, leachate and landfill gas capture and treatment systems.
27 Scheinberg, Anne (1999) 'Worse before it gets better – Sustainable waste management in Central and Eastern Europe', *Warmer Bulletin*, no. 68, pp18–20.
28 Please see discussion entitled 'Modernization of solid waste management systems in developed countries' in Chapter 2.
29 Jacobi, 2006.

30 Ishengoma, 2003; see also Key Sheet 16 in this volume.
31 Cointreau-Levine and Coad, 2000.
32 Ishengoma and Toole, 2003.
33 Coad, 2005.
34 Case study provided by Alodia Ishengoma (ILO).
35 UNDP–PPPUE, 2003.
36 CWG, 2003.
37 Corruption is seldom talked about. An interesting exception was an interview given by the recently appointed head of Rostekhnadzor (the Russian federal environment inspectorate), Nikolay Kutjin, with the *Komsomolskaya Pravda* newspaper on 20 July 2009. He stated that the bribes taken by his staff could well reach US$3 billion per annum; see www.mnr.gov.ru/.
38 See, for example, the UK Waste Management Industry Training and Advisory Board (WAMITAB), www.wamitab.org.uk/.
39 See www.acrplus.org/index.asp?page=280.
40 See www.3rkh.net.

CHAPTER 6

REFLECTIONS AND RECOMMENDATIONS

REFLECTING BACK ON THE KEY MESSAGES OF THE BOOK

This Global Report was written in a cyclical manner. UN-Habitat required an early output, and asked the authors to prepare a prepublication version of the book, summarizing the key messages for decision-makers responsible for a city's solid waste management system. That document was prepared before the results of our research in the 20 cities were available. So, in this final chapter of the report, the editors have taken the opportunity to take a step back and to reflect on what we have learned from the cities, and how that has changed or refocused our key messages.

Leachate treatment systems form one of the key elements for upgrading landfill practices, Ghorahi, Nepal

© Bhushan Tuladhar

LESSONS FROM THE CITIES

Local solutions to local problems

We have been impressed by the progress being made on the ground by many of the reference cities, often not those which you would necessarily expect to be models of global good practice. Two of the smaller cities, in Africa and in Asia, show what can be achieved with very limited financial resources if the politicians, the city administration and the people identify that solid waste as a priority.

Moshi in Tanzania has a clear focus on the cleanliness of the city, driven by concerns over public health. A stakeholder platform on solid waste was set up in 1999, and a variety of other forums are used for two-way public communication. Pilot projects were used to test new models of service delivery, involving both the local private sector and community-based organizations (CBOs) who provide primary collection in unplanned settlements. The citizens are very supportive – the local Chaga and Pare tribes both hold cleanliness in high esteem in their culture, regardless of their income levels. As a result, Moshi has won the official title of the cleanest city in Tanzania for several years in a row. The council is committed to achieving higher levels of cleanliness and maintaining this proud status. All of this has been achieved without outside financial support from donors. User fees have been introduced, although the service is free to 36 per cent of residents, based on income. A proportion of the total cost is still met by payments from central government (i.e. from taxes collected nationally).

Ghorahi is a small and relatively remote municipality in south-western Nepal. It has shown that a well-managed state-of-the-art waste processing and disposal facility can be established if there is strong commitment from the municipality and active participation of key stakeholders. The municipality has very limited human and financial resources, but it managed to conduct scientific studies, identify a very suitable site that was accepted by the general public, and develop a well-managed facility. This includes systems for waste sorting and recycling, sanitary landfilling, leachate collection and treatment, and a buffer zone with forests, gardens and a bee farm that shields the site from the surrounding area. Key success factors included a clear vision and strong determination, which enabled them to use a small initial investment from the municipality budget to mobilize national financial support and to bring the site into operation within five years; and a strong landfill management committee involving local people and key stakeholders to ensure that the site is properly managed and monitored. It is a pleasure to walk through the landfill site as there is hardly any foul smell and you will see more bees than flies.

Taking a very different example, Adelaide and its home state of South Australia have focused on how to achieve a national target of 50 per cent reduction in waste sent to landfill. The 2004 Zero Waste SA Act established the specialist agency, Zero Waste South Australia (ZWSA) to act as a 'change-maker', catalyst and financier in order to drive waste management up the hierarchy and to promote the 3Rs: reduce, reuse, recyle. ZWSA is funded by 50 per cent of the regional revenues from the nationally mandated landfill levy and initiated the development of a state waste management strategy. ZWSA works with local government, industry, schools and households, and provides a complete range of programmes from research and education to investment grants for municipal, commercial and industrial waste reduction, reuse and recycling. A high level of political commitment, and the resulting institutional structures, financing mechanisms and organizational capacity have been instrumental in exceeding the 50 per cent waste diversion target, so that they are reporting 70 per cent recovery, the highest of any of the reference cities.

These three examples might appear to be quite different. But they do share a common

thread, in that in each case the city (that is all the stakeholders working together) *care* about the particular issue they are focusing on, and so are prepared to work together to find a solution that works in their particular local situation. They have each focused on what they have identified as the next appropriate step in developing their local solid waste management system.

Different approaches to a similar problem

One of the key messages of this Global Report is that every city is different, and that the best solution in one location may not be the most appropriate somewhere else. In preparing the detailed city presentations, which accompany this book, our aim has been to make available a range of options in order to provide you with inspiration as you seek the right option for your own city.

A good illustration of this is provided by the cases of Delhi and Bengaluru in India. They each faced a similar challenge, to implement the national 2000 Solid Waste (Management and Handling) Rules, itself the result of public inter-

est litigation brought by an individual citizen to force municipalities 'to protect India's peri-urban soil and water and the health of its urban citizens through hygienic practices for waste management, processing and disposal'. They both had a similar local driver, to show a clean city, in the case of Delhi to the outside world during the 2010 Commonwealth Games, and of Bengaluru to international investors in their high information technology (IT) industry. But the successful solutions that they have developed are quite different.

Under the jurisdiction of the Municipal Corporation of Delhi, Delhi took an approach to contract out waste collection and disposal to a large private-sector company. However, this has been resisted by the city's large community of informal-sector waste workers and recyclers – our research suggests that there are 170,000 informal-sector waste workers in Delhi, the largest number in any of our 20 cities, although fewer in percentage terms that in Dhaka. The city authorities (represented by the New Delhi Municipal Council, or NDMC) have taken a different approach and have now instituted a legally recognized door-to-door collection system, which provides livelihoods to informal-sector entrepreneurs. Residents are charged a nominal amount (up to 50 rupees per month per household) for daily collection of their solid waste. Waste-pickers organized under non-governmental organizations (NGOs) are issued with a uniform and an identity card, which establishes their right as waste collectors. They are provided a rickshaw for collection and space for segregation. The informal sector is providing the primary collection service and delivering the waste to the communal bins, from which the private collector picks it up, providing a secondary collection and disposal service.

If the authorities of Delhi would take the same approach as those in New Delhi, this would be a remarkable win–win solution. Our data shows that Delhi recycles 34 per cent of its waste, of which 27 per cent is recycled by the informal sector. So, by facilitating the work of the informal recyclers rather than displacing them with the new contractor, the city would continue to save itself a vast amount of money – without the informal recyclers, the total waste quantities that the city would have had to pay for through its contract would have leapt by more than 2000 tonnes per day. From the point of view of the informal recyclers, this simple programme has been monumental in legitimizing their activities. Getting access to the waste closer to the generation point can increase the sale value since recyclables are relatively untainted and fresh. Armed with an ID card, the pickers are spared harassment by the authorities and the police. Space for segregation means that the waste-pickers can separate recyclables into more than one category and accumulate significant quantities of each – in this way increasing the selling price.

While Delhi has built its system around the traditional communal bins – in effect, transfer points between the primary collectors and the formal city secondary collection and disposal system – Bengaluru Municipality decided that the bins were part of the problem rather than the

solution. Its bins tended to be overflowing, with waste spread around them, partly by human waste-pickers and partly by cattle, thus creating a public health risk. The new system is based, instead, on door-to-door collection, implemented through public–private partnerships involving a multiplicity of small contractors. The primary collector's handcarts are loaded directly into secondary collection vehicles, eliminating multiple manual handling of the waste and thus increasing efficiency. The formal sector recycles 13 per cent of the total waste, with the parallel informal system recycling a further 15 per cent. The transition to the new system was greatly facilitated by long-standing institutions of communication and consultation, Bengaluru's solid waste platforms involving citizens and a range of stakeholders.

Data is power: Indicators of good practice in integrated sustainable waste management

It is an old saying that 'If you don't measure it, you can't manage it.' Without proper data collection and management systems, it is difficult to be accountable and transparent, or to make sound strategies and budget for them. If knowledge is power, than a city without knowledge of its solid waste system may lack the power to make positive changes. So, the quality of waste data in a city could be viewed as a proxy measure for the quality of its overall management system, of the degree of commitment of the city, or even of the city governance system.

If this is the case, then, most of our cities perform quite poorly. Despite finding so many good things going on in the reference cities, we also found relatively little hard information that we can really point to. If a city aspires to a 'modern' waste management system, then a good data collection and management system needs to be seen as a key component.

Having said that, the information provided by the 20 reference cities studied here has been used to build a database that is probably unique, and which we believe offers a better basis for the quantitative comparison of solid waste management around the world than has been available before.

For example, in the first version of this Global Report, we quoted very old data, which has been repeated many times in the literature since it was first published, that 'Cities spend a substantial proportion of their available recurrent budget on solid waste management, perhaps as much as 20 to 50 per cent for some smaller cities.' Clearly, this figure depends critically on what other responsibilities the city has within its budget; but our suspicion was that this range was rather high. We were only able to collect this data for seven of the cities, which showed a range of 2.7 to 15 per cent. The highest figure was from Ghorahi, one of the smaller cities, and the one singled out above as an example of good disposal practice, so this commitment is not surprising. The next highest, at 10 per cent, is from Adelaide, which probably says more about the distribution of responsibilities between different levels of local government. Perhaps more helpful is data that divides the total municipal budget for solid waste management (SWM) by the population, and then expresses that as a percentage of the gross domestic product (GDP) per capita: most of the cities are in the range of 0.1 to 0.7 per cent.

PUTTING INTEGRATED SUSTAINABLE WASTE MANAGEMENT INTO PRACTICE

Consider all the dimensions of integrated sustainable waste management

Looking at all of these and other examples, we are gratified that many of the key principles developed for the prepublication version of this Global Report have survived into the final publication.

The good practice examples in the previous section appear to focus on the physical elements of integrated sustainable waste management

(ISWM), such as collection in Moshi, disposal in Ghorahi, resource recovery in Adelaide, collection and disposal in Delhi, and collection in Bengaluru. However, in each case the solution depends critically on the underpinning ISWM governance features. Inclusivity, involving both the users and the service providers, is a key feature in all the examples, as is progress in developing sound institutions and proactive policies.

Building recycling rates

A particular focus of the solid waste modernization process during the 1990s and 2000s in developed countries has been to set recycling goals and to work towards high recycling rates that achieve them as a means of diverting wastes from landfill and to stem both the spiralling costs and the difficulties in locating ever bigger landfill sites. In many countries, recycling had fallen into single figure percentages, so new systems had to be built, based on parallel collection systems for various source-separated material fractions. The driving force was *not* the commodity value of the separated materials, but rather that the market for those materials could be regarded as another destination or 'sink' for a proportion of the waste. Recycling or composting becomes attractive if the *cost* is less than that of competing landfill and waste-to-energy options, which is quite different from the case where all the costs have to be met solely from the commodity value. Many developed countries have introduced economic and other policy instruments to shift the balance in favour of recycling – examples including landfill taxes, recycling targets and extended producer responsibility.

Our research shows that recycling rates in many developing country cities are already competitive with what is being achieved by modern Western systems. Our cities average 29 per cent valorization: cities such as Moshi (18 per cent), Managua (19 per cent), Nairobi (24 per cent), Bengaluru (28 per cent) and Delhi (34 per cent) compare well with Rotterdam (30 per cent), and with the average for municipalities in England, which has slowly increased from 12 per

The implementation of EU directives in Bulgaria is leading to the introduction of formal recycling programmes

© Kossara Bozhilova-Kisheva

cent in 2000, through 19 per cent in 2003 to 37 per cent in 2008. Some cities stand out: Quezon City reports 67 per cent valorization and Bamako 85 per cent, which compares well with the best-developed country cities – 68 per cent in San Francisco and 70 per cent in Adelaide.

These existing high recycling rates in developing country cities are being achieved largely by numerous private players – individuals or micro-enterprises, often informal-sector players – offering waste collection services or picking waste from streets and dumps, and upgrading and trading it. The informal sector is clearly any city's key ally – if the city had to deal with these quantities of material as waste, then their costs would rise dramatically. These informal recyclers are relying entirely on the commodity value of the waste, with no contribution from the city in recognition that they are providing a 'sink' for the waste, which the city would otherwise have to pay for.

There is much room for developing innovative win–win solutions, such as in Quezon City,

Biogas plant at night, powered by electricity from landfill gas, Quezon City, Philippines. The project is in receipt of carbon credits

© Quezon City

where itinerant waste buyers who buy source-separated wastes directly from households contribute 16 per cent to the recycling rate – showing the potential of moving from existing recycling from mixed waste to a more 'modern' system of segregated recycling. Or consider Delhi's recent compromise, where the city works with the existing recyclers, legitimizing their business and providing space for segregation. Such measures have the potential for substantially increasing present recycling rates, as well as developing the livelihoods of the pickers, improving their living conditions, enabling them to educate their children, and bringing them into the formal economy.

A focus on waste reduction

Recycling has been the focus of the solid waste modernization process during the 1990s and 2000s in developed countries. This is beginning to move on now to a focus on waste prevention, the first and second of the 3Rs: reduction, organized reuse and management of organics, all of which sit at the top of the waste hierarchy. Adelaide, and the Zero Waste South Australia organization, provides an example of global good practice, as does Tompkins County in rural New York. Reuse is in the middle, and organized reuse is important in both US cities, as well as in Rotterdam. While centralized composting has a mixed record in low- and middle-income countries, the more traditional destination for kitchen waste (livestock feeding) could receive a great deal more attention, and could be optimized and protected. And recent attention to organized home composting in countries such as Sri Lanka and Bulgaria shows potential for 'small cycles', where nutrients are cycled and loops are closed at the level of the household.

But organized reuse and waste reduction is *not* just of interest to the most developed countries, some 40 years after they began their current round of solid waste modernization. Waste quantities are growing fast in many developing country cities – due to population growth, inward migration into the city and rising living standards. Therefore, an ISWM approach is likely to come at the problem from three directions at the same time:

1   from the 'bottom', to get onto the hierarchy in the first place by phasing out open dumps;
2   from the 'middle', ensuring that wastes are increasingly diverted from disposal to reuse, recycling, organics valorization and composting; and
3   from the 'top', to reduce waste at source and to bring waste growth under control so that a city can make real progress rather than 'running hard simply to stand still'.

Use all available sources of finance

We return later to the challenge of how to achieve financial sustainability and the role of international financial institutions. This section highlights two potential innovative sources of finance.

Carbon financing through the Clean Development Mechanism (CDM) is an important new driver for improved waste management. Some of the early landfill sites developed with donor finance during the 1990s ultimately failed because the donors only supported capital costs and the city could not afford the operating costs; thus, despite the investment, the site effectively reverted to being an uncontrolled dump. This is one reason why the World Bank and other international financial institutions (IFIs) were keen to develop CDM financing for the collection of methane from landfills – the money is paid retrospectively, providing an annual payment to the city when it can be shown that the gas has been collected, which in turn depends on the site being managed effectively. CDM is now being developed beyond landfill gas. Dhaka has met global standards to receive carbon credits from composting. Current work is focusing on developing the system to obtain carbon credits for recycling.

Extended producer responsibility (EPR) has been developed in Europe and elsewhere as a

Imported collection vehicle adapted in a municipal workshop to suit local needs in Managua, Nicaragua

© UN-Habitat
Jeroen IJgosse

means of moving the financial responsibility for disposing of products at the end of their useful life, from the municipality back up the supply chain to the retailers and 'producers' who put them on the market. The system is quite bureaucratic, but Sousse provides a developing country example where it has been successful in channelling financial support from the producers to small enterprises carrying out the separate collection. Costa Rica, in a partnership with The Netherlands, represents a global first for a middle-income country developing a stakeholder-driven, consensus-based national EPR system for e-waste management.

Another innovative example is provided by the Spirit of Youth NGO, working with the Zabbaleen community of informal waste collectors and recyclers in Cairo. They noticed that shampoo bottles and similar products sold by international companies were often fraudulently refilled and resold on the local market. So they have formed a partnership with several to build a recycling school. The boys collect the shampoo bottles, bring them to school and cut them so they cannot be reused, and then fill out a form showing how many bottles of each type they have retrieved. The multinational producers pay them for each bottle. The form is a good tool to teach reading, writing and numeracy; the boys even learn Microsoft Excel skills. The shampoo bottles are recycled into plastic granulate that is sold to recyclers in the neighbourhood, with the income paying for the salaries of the staff.

Our key conclusion

The key conclusion from the Global Report, and from our study of the reference cities in particular, is that a successful solution needs to consider all of the three physical elements of ISWM and all of the three governance features. A reliable approach is to be critical and creative; to start from the existing strengths of your city and to build upon them; to involve all the stakeholders; to design your own models; and to 'pick and mix', adopt and adapt the solutions that will work in your particular situation.

![] MOVING TOWARDS FINANCIAL SUSTAINABILITY AND THE ROLE OF DONORS

It is probably fair to say that achieving financial sustainability is still a work in progress in all of the developing country cities studied for this Global Report. Financing and investment needs are serious in waste management, especially for middle- and small-sized cities and in low-income countries.

There is a sort of intellectual deadlock in this area. The investment needs are estimated based on 'internationally recognizable' standards and environmental protection; but such solutions are not affordable for the governments and their people. The result is that many consultants and experts produce studies that present strategies, action plans and investment projects that the cities cannot afford, and so the preparation work does not convert into actions on the ground. Or sometimes when it does, the result is a landfill site that waits for the landfill to be built, or an investment in a processing facility that the city cannot afford to operate.

The organizations which could provide the necessary finances are generally just not available. Solid waste budgets largely come from national governments, but they do not have the funds necessary to invest in new infrastructure. This leaves the international financial institutions and private investors, who bring a range of conditions and prerequisites; most, if not all, require 'international' standards on which they are not allowed to compromise, and which are not affordable to the recipient.

The research in this Global Report confirms

School children educated in recycling programmes in Curepipe, Mauritius

© Municipality Curepipe

The women's group
Nogo Jugu Kala
sweeping a road in
their Banconi
neighbourhood,
Commune I,
Bamako, Mali

© CEK Kala Saba

that the operational cost for primary collection is generally affordable, even in poor communities; secondary collection already raises issues of affordability and willingness to pay in many cities. But running modern landfills to donor standards is often beyond the capacity of municipal governments: the Ghorahi site is an exception and was funded from local sources, and makes use of a natural clay 'liner', which might or might not be acceptable to some donors.

If the donor capital is a grant, two issues arise. The first is the capacity of the city to operate and maintain the equipment or facility as it was designed, whether a collection vehicle, a landfill site or a treatment plant – the world is littered with examples of donated compactor trucks or incinerators which don't work, and landfill sites which have reverted to open dumps because the city cannot afford to run them or to repair them. Even if this first challenge can be met, the second remains: how to replace the vehicle or the landfill site at the end of its life. Grant funding may be helpful in the right circumstances and if the vehicle or facility is appropriate in the local

circumstances; but it is not a long-term solution.

If the investment is a loan, then the issue is not just about operational cost, it is also about debt servicing. A city can only afford to borrow a certain amount if it is to meet the repayments, so solid waste must compete with other funding priorities, such as health and education.

These decisions often go above the level of the individual city, involving both regional and national governments. This book is thus aimed at all decision-makers, not just the city mayor and local politicians and officials, but also at national ministries of environment, and particularly finance ministries, who may take many of the decisions; it is also aimed at the international financial institutions and the governing boards who set their rules.

This discussion brings us full circle and underpins the emphasis throughout this book on the need for good governance in waste management. The issues and need for knowledge, information and a new set of social contracts are critical. While good progress is being made in some higher- and middle-income countries, this is still a fairly unknown area in the lowest-income countries. As a result, a major priority remains capacity- and knowledge-building in order to develop inclusive approaches, sound institutions and proactive policies, and to work towards financial sustainability.

But individual cities and countries cannot solve this on their own. We believe that there is a growing international consensus for the key conclusion of this Global Report as stated above: that a sustainable local solution must be acceptable, appropriate and affordable in the local circumstances.

However, neither international financial institutions, nor national governance structures are geared to this 'pick-and-mix' approach. IFIs and their governing boards need to look again at their policies, particularly at their insistence on 'international standards' as a condition for financing. It has taken 40 years of the current phase of solid waste modernization for developed countries to achieve these standards across the board, so it seems unreasonable to insist that the same standards form part of the next step in every developing country as a condition of providing financial assistance.

National governments also need to look at a range of underlying governance issues, including corruption, powers, controls, etc.

CLOSING WORDS: WHAT MAKES AN ISWM SYSTEM SUSTAINABLE?

As a city official or politician, this Global Report invites you to really understand your city waste issues, to identify and name problems, and to take the next critical steps – together with a range of stakeholders – to find solutions that are appropriate to your specific local situation in order to set off from where you are now on the journey to where you want to be.

If you are at a relatively early stage of this journey of modernizing your solid waste management system, then it is important to identify simple, appropriate and affordable solutions that can be implemented progressively, giving your constituents the best system that they can afford.

Early steps are likely to include extending collection to the whole city and phasing out open dumps. But that is not enough: an ISWM approach is likely to include a focus on building your existing recycling rates and on taking measures to bring waste growth under control. This is particularly important, as every tonne of waste reduced, reused or recycled is a tonne of waste for which the city does not have to pay for its transport and safe disposal. A key message of this book is that there are win–win solutions, where the city and the informal/micro-enterprise sectors work together to progress the 3Rs.

There is only one sure winning strategy and that is to understand and build upon the strengths of your own city – to identify, capitalize on, nurture and improve the indigenous processes that are already working well. We hope that this Global Report will inspire you to be both creative and critical, and to work with your local stakeholders to develop the appropriate next steps for your local situation.

You and your citizens and stakeholders deserve the best system for your circumstances, and nothing less. If this Global Report can contribute to that, we will have done our work well.

GLOSSARY OF TERMS

There are many different terms in use for different parts of the solid waste and recycling systems. The terms in this glossary are the ones that the project team has agreed to use. Many of the working definitions are those of the project team – where they are taken from elsewhere, an explicit reference is given at the end of the table. Wherever possible, the definitions are drawn from standard English language use in the UK and in the US.

Term	Other terms or abbreviations used, or other things this term can refer to	Working definition
Annex I countries	Industrialized nations, OECD countries	These are the industrialized countries that have carbon reduction targets to reach under the Kyoto Protocol.
Avoided cost of disposal	Diversion credit	The amount that would have been paid per kilo for disposing of materials in a controlled or sanitary landfill and paying the official tipping fee.
Beneficiation	Processing, pre-processing, upgrading	Preparation of recovered materials for transport, marketing and recycling.
Biogas	Methane	Typically refers to a gas produced by the biological breakdown or digestion of organic matter in the absence of oxygen. Biogas originates from biogenic material and is a type of biofuel.
Bio-solid	Human excreta, wastewater treatment facility solids, animal wastes, agricultural wastes	Plant and animal wastes that have value as a soil amendment with fertilizer value that can be used as an input to agriculture, horticulture and silviculture.[1]
Broker	Stockist, dealer, trader, exporter	A trader in one or more types or grades of recyclables who trades without ever being the physical owner of the materials, usually having no storage place.
Capital cost	Investment cost, capital, purchase cost	The amount that it costs to purchase new equipment, facilities, space, buildings, etc.
Capture rate	Separation rate	A percentage relationship between the amount of recoverable materials that are directed to processes of recycling or composting and the total amount collected.
CBO	Community-based organization, grassroots organization	A private group organized to provide a waste collection, recycling, composting, community clean-up, environmental management, or solid waste function or service in a community, often fully or partially staffed by volunteers.
Certified emission reduction (CER)	Carbon credit, issued carbon credit	Climate credits (or carbon credits) issued by the Clean Development Mechanism (CDM) Executive Board for emission reductions achieved by CDM projects and verified by a third party or designated operational entity (DOE) under the rules of the Kyoto Protocol.
Clean Development Mechanism (CDM)	Kyoto project financing	An international institutional mechanism that allows industrialized countries that have targets under the Kyoto Protocol to invest in emission reductions in non-Kyoto countries and count those reductions towards their own legal commitments. A CDM project is issued with certified emission reductions, which may then be traded.
Commercial waste	Business waste, shop waste, small-quantity generator waste	Waste that comes from shops, services and other generators that are neither residential nor industrial. Sometimes includes institutional or public-sector waste.
Commingled materials	Mixed, multi-material, co-collected material, combined streams, single-stream collection	Specific mixing of recoverable materials for the purposes of efficient collection. The combination is designed for post-collection separating or sorting. Commingled materials do not include mixed waste.
Communal container	Container, skip, dumpster, box	A vessel to contain waste, usually larger than 1 m^3 and used for more than one household.
Community	Barrio, Barangay, district, neighbourhood, ward	A physical or social subdivision of or within a city, it may be as small as a group of neighbours or as large as a formal sub-municipal division that may or may not have its own governance functions.
Composition	Characterization, make-up of waste, physical or chemical nature	Quantitative description of the materials that are found within a particular waste stream in the form of a list of materials and their absolute quantities per day or per year, or as a percentage of total materials.
Composting	Organic waste management, aerobic decomposition	The decomposition of materials from living organisms under controlled conditions and in the presence of oxygen.
Construction and demolition waste	Debris, C&D, rubble, contractor waste	Waste from the process of construction, demolition or repair of houses, commercial buildings, roads, bridges, etc. Generally divided into commercial construction waste from construction companies and do-it-yourself (DIY) waste from homeowners making their own repairs.
Controlled waste disposal site	Controlled dumpsite, upgraded dumpsite	An engineered method of disposing of solid wastes on land, in which, at a minimum, there is perimeter fencing, gate control and the waste is covered every day. Some form of reporting is usual, often in the form of a weighbridge (scale house), and some form of tipping fee is usually charged. A controlled waste disposal site differs from a sanitary landfill in that it is not sealed from below and does not have a leachate collection system.
Co-operative	Co-op, buyers' association, sellers' association, MSE, CBO	An enterprise organized as a co-operative with multiple owners who participate in the activities. In some Latin American countries, co-operatives have a special tax status and so are a favoured form for establishing a business.[2]

Term	Other terms or abbreviations used, or other things this term can refer to	Working definition
Coverage	Percentage service availability	The percentage of the total (household and commercial) waste-generating points (households or businesses) that have regular waste collection or removal.
Depot	Deposit, drop-off, community collection point, community container	A place where individuals can bring their own waste or recyclables, varying from a single container to a recycling centre, a reverse vending machine, or a special site or facility designed to receive waste materials, kitchen, food and yard waste, demolition debris, and/or separated recyclables directly from the generator.
Disposal – llegal	Dumping, wild dumping, littering	Disposal of waste at a site different from one officially designated by the municipal authorities, especially where it is specifically prohibited. May also refer to disposal at the wrong time or in the wrong quantities, even if all other aspects are correct.
Disposal – legal		Disposal of waste at a site designated by the municipal authorities.
Diversion	Recovery, avoided disposal	The process or result of keeping materials out of a disposal facility.
Dumpsite	Dump, open dump, uncontrolled waste disposal site	A designated or undesignated site where any kinds of wastes are deposited on land, or burned or buried without supervision and without precautions regarding human health or the environment.
Dump-picker	Scavenger, waste-picker	Woman, man, child or family who extracts recyclable materials from disposal sites.[3]
Effectiveness	Reach, performance	The extent to which the solid waste or recycling system meets its goals and does what it claims to do; the cleanliness of the city.
Efficiency	Collection efficiency	One or more measures of the performance of the collection system, usually expressed as households/vehicles per day, or tonnes per litre of fuel used, or distance travelled per litre of fuel.
Emission reduction unit (ERU)	ERU	Climate credits (or carbon credits) issued by the countries participating in joint implementation (JI) projects, or the Joint Implementation Supervisory Committee, emission reductions achieved by JI projects and verified by a third party (accredited independent entity, or AIE) under the rules of the Kyoto Protocol.
Ferrous metals	Iron, steel, magnetic metals	Metals which contain iron and which react to a magnet and are subject to rusting.[4]
Formal sector	Official, government, municipal	Encompasses all activities whose income is reported to the government and that are included within a country's gross national product; such activities are normally taxed and follow requisite rules and regulations with regards to monitoring and reporting.[5]
Formal waste sector	Solid waste system, solid waste authorities, government, materials recovery facility	Solid waste management activities planned, sponsored, financed, carried out or regulated and/or recognized by the formal local authorities or their agents, usually through contracts, licences or concessions.
Full-cost accounting (FCA)	Total cost analysis, true cost accounting	A systematic approach for identifying, summing and reporting the actual costs of solid waste management. It takes into account past and future outlays, overhead (oversight and support services) costs, and operating costs. FCA attempts to quantify environmental and social external costs.[6]
Generator	Waste producer, household, business, user	The source of the waste (i.e. the first point at which it is discarded as a useful object and is redefined by its owner as waste).
Hazardous wastes	Toxic wastes	A material that poses substantial or potential threats to public health or the environment and generally exhibits one or more of these characteristics:[7] • ignitable (i.e. flammable); • oxidant; • corrosive; • radioactive; • explosive; • toxic; • carcinogenic; • disease vector.
High-income countries	OECD countries, developed countries, the North	Countries with a gross national income per capita of US$11,905 or higher[8] or which are located in Europe, North America or Oceania.
Household container	Set-out container, garbage can, waste can, waste bin, dustbin, bin	The vessel or basket used by a household or commercial generator to store and set out the waste materials, commonly made of metal, plastic, rubber or wood.
Household waste	Municipal solid waste, domestic waste, MSW, non-dangerous waste	Discarded materials from households that are generated in the normal process of living and dying.
Incineration	Burning, combustion	Controlled process by which solid, liquid or gaseous combustible wastes are burned and changed into gases.[9]
Informal sector	Waste-pickers, rag-pickers, scavengers, junk shops, street vendors, bicycle taxis, etc.	Individuals or businesses whose economic activities are not accounted in a country's gross national product (GNP); such activities are not taxed; exchange of goods or services is on a cash basis; and the activities are not monitored by the government and often the activities operate in violation of, or in competition with, formal authorities.[10,11]
Informal waste sector	Waste-pickers, scavengers, junk shops	Individuals or enterprises who are involved in waste activities but are not sponsored, financed, recognized or allowed by the formal solid waste authorities, or who operate in violation of, or in competition with, formal authorities.
Integrated sustainable waste management (ISWM)		ISWM is a systems approach to waste management that recognizes three important dimensions of waste management: (1) stakeholders; (2) waste system elements; and (3) sustainability aspects.
Itinerant waste buyer (IWB), itinerant waste collector (IWC)	IWB, IWC, house-to-house collector	Woman, man, child, family or enterprise who purchases or barters source-separated waste materials from households, shops or institutions, usually focusing on one specific material or type of materials.[12] In the case of an IWC, there is no payment for the goods.
Joint implementation	Kyoto project financing	The CDM allows industrialized countries that have targets under the Kyoto Protocol to make emission reductions with Annex 1 Kyoto countries and count those reductions towards their own legal commitments. A JI project is issued with emission reduction units, which may then be traded.
Junk shop	Dealer, MSE recycler, trader	A business that buys, packs and trades recyclable materials, usually with limited or no upgrading.
Landfill	Engineered landfill, engineered waste disposal facility, controlled disposal facility	'An engineered method of disposing of solid wastes on land, in which, at a minimum, there is perimeter fencing, gate control and the waste is covered every day. Some form of reporting is usual, often in the form of a weighbridge (scale house), and some form of tipping fee is usually charged. A landfill differs from a sanitary landfill in that it is not necessarily sealed from below and does not necessarily have a leachate collection system.'[10]
Low-income countries	Developing countries, non-OECD countries, poor countries	Countries with a gross national income per capita of US$975 or less.[13]

Term	Other terms or abbreviations used, or other things this term can refer to	Working definition
Materials recovery facility (MRF)	Materials recovery facility, intermediate processing centre (IPC), intermediate processing facility (IPF), recycling processing centre	An industrial facility of moderate scale that is designed for post-collection sorting, processing and packing of recyclable and compostable materials. It is usually of moderate technical complexity with a combination of automated and hand sorting. The inputs are usually commingled or mixed recyclables and not mixed waste. The outputs are industrial grade materials, usually crushed or baled and separated by type, colour, etc.
Micro- and small enterprise (MSE)	Micro-enterprise, junk shop, small recycler	The smallest businesses, smaller than SMEs, usually having less than ten workers[14]
Middle-income countries	Medium-income countries, emerging economies	Countries with a gross national income per capita from US$976 to $11,905.[15]
Municipal solid waste (MSW)	Household waste, domestic waste	Wastes generated by households, and wastes *of a similar nature* generated by commercial and industrial premises, by insttutions such as schools, hospitals care homes and prisons, and from public spaces, such as streets, bus stops, parks and gardens.
Municipality	Local government, local authority, mayor's house, city hall, city council, mayoralty, city, town, village	A unit of local government with its own level of governance, responsibility and representation, combining elected and appointed officials.
Operating and maintenance (O&M) cost	Operating and maintenance cost, operating cost	Costs associated with ongoing operations, such as energy, supplies, labour, rents, etc.
Organisation for Economic Co-operation and Development (OECD)		The OECD is an international organization of 30 countries that accept the principles of representative democracy and free-market economy. Most OECD members are high-income economies with a high Human Development Index and are regarded as developed countries.[16]
Opportunity cost		The imputed or estimated loss associated with making a choice for option *a* and not choosing option *b*.
Organic waste	Bio-waste, green waste, wet waste, organics, food waste, putrescibles, compostables	The decomposable fraction of domestic and commercial wastes, including kitchen and garden wastes and, sometimes, animal products.
Organized reuse	Repair, reuse, product recycling	A commercial or livelihood activity focused on extraction, repair and sale of specific items in the waste stream. An example is the recovery of up to 20 different types of glass bottles in the Philippines.
Pig slops	Swill, food waste, swine feed, organic waste	Food wastes collected from restaurants, hotels, markets, etc. and from households, which are either sold or used as food for pigs or other livestock.
Pre-processing	Sorting, screening, sieving, compaction, densification, size reduction, washing, drying	Preparing recoverable materials from the waste stream to be used for subsequent processing without adding significant value to them.[17]
Primary collection	Pre-collection, house-to-house collection	Organized collection of domestic waste from households, taken to a small transfer station or transferred to a truck or container.
Processing	Beneficiation, upgrading	Manual or mechanical operations to preserve or reintroduce value-added into materials. Usually involves densification, size reduction, sorting, and packaging or transport.
Recovery rate	Rate of recycling, percentage recycled, diversion rate	A percentage relationship between the amount of recoverable materials that reach recycling, composting or energy recovery and the total amount generated.
Recyclables	Recoverables, materials to be valorized	Materials contained in municipal solid waste which have an intrinsic value to the industrial value chain as represented by a price.
Recycler	Scavenger, waste-picker, MRF, junk shop	Entrepreneur involved in recycling.
Recycling	Valorization, materials recovery	Extraction, processing and transformation of waste materials and their transfer to the industrial value chain, where they are used for new manufacturing. In some definitions, recycling is only considered to have occurred when materials have been sold.[18]
Recycling or composting market	End-user industry, buyer, dealer, broker	Business, individual, organization or enterprise who is prepared to accept and pay for materials recovered from the waste stream on a regular or structural basis, even when there is no payment made.
Residual waste	Rest-waste, rest-fraction, residue, rejects	The discarded materials remaining in the waste stream or on the sorting line because they are not recyclable or compostable since they are perceived to have little or no monetary value.[19]
Resource recovery	Energy recovery, materials recovery	Process of extracting economically usable materials or energy from wastes. May involve recycling. In English-speaking countries, the term is usually restricted to recovery of energy.[20]
Reuse	Second-hand use	Use of waste materials or discarded products in the same form without significant transformation; may include a system developed to repair/refurbish items.[21]
Sample	Sub-set	A representative part of a whole that allows conclusions to be made about the whole by investigating only a small part.
Sanitary landfill	Landfill, state-of-the-art landfill	An engineered method of disposing solid wastes on land in a manner that protects human health and the environment. The waste is compacted and covered every day. The landfill is sealed from below and leachate and gas are collected, and there is a gate control and a weighbridge.
Sanitation	Wastewater management, urban environment, urban cleansing	In the 'French sense', used to refer to urban environmental activities, including waste water and solid waste management.
Secondary collection	Transfer, small transfer station	The movement of wastes collected from households from their first dumping point to processing, larger-scale transfer or final disposal.
Separate collection	Segregated collection, collection of recyclables, organics collection, selective collection	Collection of specific types of materials at a designated time, in a different container or vehicle, or in another way in order to maintain the separation potential and maximize the recovery.
Shadow price	Proxy price, hedonic price, contingent valuation	A reasonable estimate for the value of something based on extrapolating the price for something similar.
Single stream	Unsorted material, commingled, blue bag	System in which some combination of all recyclable materials (e.g. paper fibres and containers) are mixed together in a collection truck, instead of being sorted into separate commodities (newspaper, cardboard, plastic, glass, etc.) by the user and handled separately throughout the collection process. In a single stream, both the collection and processing systems must be designed to handle this fully commingled mixture of recyclables.[22]
Small and medium-sized enterprise (SME)	Small- and medium-sized business, small business	Businesses usually having between 11 and 50 employees or workers.
Solid waste	Garbage, trash, waste, rubbish	Materials that are discarded or rejected when their owner considers them to be spent, useless, worthless or in excess.[23]
Sorting	Classification, high-grading, selection	Separating mixed materials into single-material components, mechanically or manually, either at the source or after the collection process. In some cases, classifying a mixed single-material stream into specific grades or types of that material.
Source	Generator, origin, waste service user	The point at which a material is defined as waste and discarded, usually either a house or a business.
Source separation	Separation at source, segregation at source	Actions taken to keep and store certain materials separately from commingled (mixed) waste at the point of generation.

Term	Other terms or abbreviations used, or other things this term can refer to	Working definition
Stakeholder	Interested party, constituent, concerned citizen, affected party	Individual or institution (public and private) interested and involved in related processes and activities associated with a modernization process, plan, project goal or desired change.[24]
Street cleaner	Street sweeper	Formal or semi-formal worker assigned by the city authority to remove litter from streets that cannot be attributed to any specific waste generator.
Street-picker	Street scavenger, waste-picker	Woman, man, child or family who removes recyclable materials from communal containers, streets and public places.[25]
Tipping fee	Gate fee, disposal fee	The amount that is charged for disposing of waste at a facility, usually per tonne, per cubic metre or per vehicle.
Transfer	Transit, collection point, depot	The movement of wastes from their first point of discharge to final disposal; it usually includes some very basic processing: compaction, pre-sorting or size reduction.
Transfer station	Transit point	A place where waste from collection vehicles is aggregated and organized before being transported to disposal sites or treatment facilities.[26]
Treatment	Decontamination, processing, incineration, anaerobic digestion, biogas production, pyrolisis, composting	Labour-based or mechanical methods to reduce the risk of exposure or to reduce the impacts upon the environment of toxic or hazardous materials associated with the waste stream; in some cases, can concurrently capture and increase the economic value of specific waste stream components' value-added.
Valorization	Recycling, recovery, conserving economic value	The entire process of extracting, storing, collecting, or processing materials from the waste stream in order to extract and divert value and direct the material to a value-added stream.
Waste dealer	Junk shop owner, scrap trader, consolidator, owner of a 'godown', waste buyer	Individual or business purchasing quantified (weighed or measured) materials for recycling or composting, storing them, upgrading or processing them, and then reselling them into the recycling value chain. A dealer usually has their own premises and some form of dedicated storage place.
Waste generator	Households, institutional, commercial wastes	The agent or point via which a purchased, collected or grown product is discarded.[27]
Waste-picker	Scavenger, rag-picker	Person or family who salvages recyclable materials from streets, public places or disposal sites.[28]
Waste prevention	Waste avoidance, waste minimization, pre-cycling	Strategies or activities undertaken by individuals, businesses or institutions to reduce the volume and toxicity of material discarded.[29]

NOTES

1 Adapted from Simpson et al, 1992.
2 Adapted from Rivas et al, 1998.
3 Adapted from Koeberlin, 2003.
4 Adapted from Tchobanoglous et al, 1993.
5 Adapted from the International Labour Organization definition, adopted by the 15th International Conference of Labour Statisticians, January 1993.
6 Adapted from Simpson et al, 1992.
7 Resource Conservation and Recovery Act (USA 1976).
8 World Bank, undated.
9 Adapted from Tchobanoglous et al, 1993.
10 Hart, 1973.
11 Adapted from the International Labour Organization definition, adopted by the 15th International Conference of Labour Statisticians, January 1993.
12 Adapted from Koeberlin, 2003.
13 World Bank, undated.
14 Adapted from Arroyo et al, 1998.
15 World Bank, undated.
16 Adapted from the Convention on the Organisation for Economic Co-operation and Development (1960).
17 Adapted from Koeberlin, 2003.
18 Adapted from Tchobanoglous et al, 1993.
19 Adapted from Koeberlin, 2003.
20 Adapted from Tchobanoglous et al, 1993.
21 Adapted from Koeberlin, 2003.
22 Recycling Today (2002) Single-Stream Recycling Generates Debate, www.recyclingtoday.com/Article.aspx?article_id=25326.
23 Adapted from Tchobanoglous et al, 1993.
24 Adapted from the United Nations Convention to Combat Desertification (1994).
25 Adapted from Koeberlin, 2003.
26 Skitt, 1992.
27 Adapted from Franklin Associates, 1992.
28 Adapted from Koeberlin, 2003.
29 Adapted from the European Commission definition (Directive, 2008).

REFERENCES

Ali, M., Cotton, A. and Westlake, K. (1999) *Down to Earth: Solid Waste Disposal for Low-Income Countries*, WEDC, Loughborough University, UK

Anschütz, J., IJgosse, J. and Scheinberg, A. (2004) *Putting ISWM to Practice*, WASTE, Gouda, The Netherlands

Anschütz, J., Keita, M., Rosario, A., Lapid, D. and Rudin, V. (2005) *UWEP City Case Studies*, WASTE, Advisers on Urban Environment and Development, Gouda, The Netherlands, www.waste.nl

Arold, H. and Koring, C. (2008) *New Vocational Ways and Qualifications for Professionalisation in the Second-Hand Sector, European Report: An Investigation and Analysis of the Second-Hand Sector in Europe*, www.qualiprosh.eu/download/European_Sector_Analysis_Report_english.pdf

Arroyo, J., Rivas, F. and Lardinois, I. (eds) (1998) *Solid Waste Management in Latin America: The Case of Small and Micro-Enterprises and Cooperatives*, WASTE, Gouda, The Netherlands

Ball, J. M. and Bredenhann, L. (1998) *Minimum Requirements for Waste Disposal by Landfill*, 2nd edition, Department of Water Affairs and Forestry, Republic of South Africa

Bartone, C., Bernstein, J. and Wright, F. (1990) *Investments in Solid Waste Management: Opportunities for Environmental Improvement*, World Bank, Washington, DC/WASTE, Gouda, The Netherlands

Bogner, J., Abdelrafie Ahmed, M., Diaz, C., Faaij, A., Gao, Q., Hashimoto, S., Mareckova, K., Pipatti, R. and Zhang, T. (2007) 'Waste management', in B. Metz, O. R. Davidson, P. R. Bosch, R. Dave and L. A. Meyer (eds) *Climate Change 2007: Mitigation. Contribution of Working Group III to the Fourth Assessment Report of the Intergovernmental Panel on Climate Change*,

Cambridge University Press, Cambridge, UK, and New York, NY (also published in *Waste Management and Research* (2008), vol 26, pp11–32)

Carson, R. (1962) *Silent Spring*, Houghton Mifflin, Boston, MA

Chambers, R. (1997) *Whose Reality Counts: Putting the First Last*, Practical Action Publishing (formerly Intermediate Technology Development Group), Rugby, UK

Chaturvedi, B. (2006) 'Privatization of solid waste collection and transportation in Delhi: The impact on the informal recycling sector', paper prepared as partial fulfilment of course on Urban Issues in Developing Countries, School for Advanced International Studies, Johns Hopkins University, Washington, DC, December

Chaturvedi, B. (2009) 'A scrap of decency', *The New York Times*, NYTimes.com, www.nytimes.com/2009/08/05/opinion/05chaturvedi.html

Coad, A. (2005) *Private Sector Involvement in Waste Management: Avoiding Problems and Building on Successes*, CWG publication series no 2, Collaborative Working Group on Solid Waste Management in Low- and Middle-Income Countries (CWG), St Gallen, Switzerland

Coad, A. (ed) (2006) *Solid Waste, Health and the Millennium Development Goals*, Report of the CWG International Workshop Kolkata, India, 1–5 February 2006, Collaborative Working Group on Solid Waste Management in Low- and Middle-Income Countries, www.cwgnet.net

Coffey, M. (2009) *Collection of Municipal Solid Waste in Developing Countries*, 2nd edition, UN-Habitat, Nairobi, Kenya

Cointreau, S. (1982) *Environmental Management of Urban Solid Wastes in Developing Countries: A Project Guide*, World Bank, Washington, DC

Cointreau, S. (2001) *Declaration of Principles of Sustainable and Integrated Solid Waste Management*, www.worldbank.org/solidwaste

Cointreau-Levine, S. and Coad, A. (2000) *Guidance Pack: Private Sector Participation in Municipal Solid Waste Management*, Skat, Switzerland, http://rru.worldbank.org/Documents/Toolkits/waste_fulltoolkit.pdf

CWG (Collaborative Working Group on Solid Waste Management in Low- and Middle-Income Countries) (2003) *Solid Waste Collection that Benefits the Urban Poor: Suggested Guidelines for Municipal Authorities under Pressure*, compiled by A. Coad from a CWG workshop, March 2003, www.skat.ch/publications/prarticle.2005-09-29.7288084326/prarticle.2005-11-25.5820482302/skatpublication.2005-12-02.0331566765

Defra (UK Department for Environment, Food and Rural Affairs) (2008) *Municipal Waste Management Statistics 2007/08*, Statistical Release 352/08 06.11.08, www.defra.gov.uk/news/2008/081106a.htm

Dias, S. M. (2000) 'Integrating waste pickers for sustainable recycling', Paper presented to the Manila Meeting of the Collaborative Working Group (CWG) on Planning for Sustainable and Integrated Solid Waste Management, Manila, 2000

Dias, S. M. (2006) 'Waste and citizenship forum – achievements and limitations', in *Solid Waste, Health and the Millennium Development Goals*, CWG–WASH Workshop Proceedings, 1–6 February 2006, Kolkata, India

Directive (2008) *Directive 2008/98/EC of the European Parliament and of the Council of 19 November 2008 on Waste and Repealing Certain Directives*, http://eur-lex.europa.eu/LexUriServ/LexUriServ.do?uri=OJ:L:2008:312:0003:0030:en:PDF

Doppenburg, T. and Oorthuys, M. (2005) *Afvalbeheer (Solid Waste Management)*, SDU Publishers, The Netherlands

Franklin Associates (1992) *Analyses of Trends in Municipal Solid Waste Generation, 1972 to 1987*, Prepared for Procter and Gamble, Browning Ferries Industries, General Mills and Sears, Franklin Associates, Praire Village, KS

Furedy, C. (1997) *Reflections on Some Dilemmas Concerning Waste Pickers and Waste Recovery*, Source

Book for UWEP Policy Meeting 1997 (revised April 1999), WASTE, Gouda, The Netherlands

Gonzenbach, B. and Coad, A. with Gupta, S. K. and Hecke, J. (2007) *Solid Waste Management and the Millennium Development Goals*, CWG Publication no 3, St Gallen, Switzerland

GTZ/CWG (Deutsche Gesellschaft für Technische Zusammenarbeit/Collaborative Working Group on Solid Waste Management in Low- and Middle-Income Countries) (2007) *Economic Aspects of the Informal Sector in Solid Waste*, Research report prepared by WASTE, Skat, and city partners (principal authors A. Scheinberg, M. Simpson, Y. Gupt and J. Anschütz), GTZ, Eschborn, Germany

GTZ/GOPA (2009) *Elaboration du Programme d'Action pour les Filieres [Developing a Programme of Action for Material Chains]*, Report for ANGED/EPP, July 2009, GTZ, Eschborn, Germany

Hart, K. (1973) 'Informal income opportunities and urban employment', *Journal of Modern African Studies*, vol 11, no 1, pp61–89

Hawley, C. and Ward, J. (2008) 'Naples trash trauma', *Der Spiegel* online, 3 July, www.spiegel.de/international/europe/0,1518,563704,00.html

Hay, J. E. and Noonan, M. (2002) *Anticipating the Environmental Effects of Technology: A Manual for Decision-Makers, Planners and other Technology Stakeholders*, UNEP, Division of Technology, Industry and Economics; Consumption and Production Unit, Paris, France, and International Environmental Technology Centre, Osaka, Japan, www.unep.or.jp/ietc/Publications/integrative/EnTA/AEET/index.asp

Hickman, D., Whiteman, A., Soos, R. and Doychinov, N. (2009) *Model for Global Development of Recycling Linkages*, IFC Advisory Services in Southern Europe, International Finance Corporation, Skopje, Republic of Macedonia

Hoornweg, D. and Thomas, L. (1999) *What a Waste: Solid Waste Management in Asia*, World Bank, Washington, DC

IJgosse, J., Anschütz, J. and Scheinberg, A. (2004a) *Putting Integrated Sustainable Waste Management into Practice: Using the ISWM Assessment Methodology – ISWM Methodology as Applied in the UWEP Plus Programme (2001–2003)*, WASTE, Gouda, The Netherlands

IJgosse, J., Olley, J., de Vreede, V. and Dulac, N. (2004b) *Municipal Waste Management Planning: Waste Keysheets*, www.wastekeysheets.net

Ikonomov, L. H. (2007) *Training Materials Prepared for the IFC Recycling Linkages Programme*, Private Enterprise Programme, South-Eastern Europe, IFC Recycling Linkages Programme, Skopje, Macedonia, and Consulting Centre for Sustainable Development Geopont-Intercom, Varna, Bulgaria.

ILO/IPEC (2004) *Addressing the Exploitation of Children in Scavenging: A Thematic Evaluation of Action on Child Labour*, Global Synthesis Report for the ILO, ILO, Geneva, Switzerland

Ishengoma, A. (2003) 'Structuring solid waste collection services to promote poverty eradication in Dar es Salaam City', paper presented at an International CWG Workshop on Solid Waste Collection that Benefits the Urban Poor, Dar es Salaam, Tanzania, 9–14 March 2003

Ishengoma A. and Toole, K. (2003) 'Jobs and services that work for the poor: Promoting decent work in municipal service enterprises in East Africa; the Dar es Salaam project and the informal economy', Paper presented at the Knowledge-Sharing Workshop organized by INTEGRATION.ITC, Turin, Italy, 28 October–1 November 2003

Jacobi, P. (2006) Gestão *Compartilhada dos Resíduos Sólidos no Brasil: Inovação com inclusão social*, Annablume, São Paulo

Johnson, J., Harper, E. M., Lifset, R. and Graedel, T. E. (2007) 'Dining at the periodic table: Metals concentrations as they relate to recycling', *Environmental Science and Technology*, vol 41, no 5, pp1759–1765

Joseph, K. (2007) 'Lessons from municipal solid waste processing initiatives in India', Paper presented to the International Symposium MBT 2007, www.swlf.ait.ac.th/upddata/international/lessons%20from%20msw.pdf

Keita, M. M. (2003) *Diagnostique de la Filiere de Recuperation de Dechets dans la Commune IV du District de Bamako*, Rapport Final, COPIDUC, Bamako, and WASTE, Gouda, The Netherlands

Key Note Publications Ltd (2007) *Global Waste Management Market Assessment 2007*, 1 March, Key Note Publications Ltd, Hampton, UK,

Koeberlin, M. (2003) 'Living from waste: Livelihoods of the actors involved in Delhi's informal waste recycling economy', *Studies in Development Geography*, Verlag für Entwicklungspolitik, Saarbrücken, Germany

Lacoste, E. and Chalmin, P. (2006) *From Waste to Resource: 2006 World Waste Survey*, Economica Editions, Paris

Lardinois, I. and van de Klundert, A. (1994) *Informal Resource Recovery: The Pros and Cons*, WASTE, Gouda, The Netherlands

LeBlanc, R. J., Matthews, P. and Richard, R. P. (eds) (2006) *Global Atlas of Excreta, Wastewater Sludge and Biosolids Management: Moving Forward the Sustainable and Welcome Uses of a Global Resource*, UN-Habitat, Nairobi, Kenya

Lifuka, R. (2007) 'City report for Lusaka', Resource document for GTZ/CWG (2007) *Economic Aspects of the Informal Sector*, Riverine Associates, Lusaka, Zambia

Marchand, R. (1998) *Marketing of Solid Waste Services in Bauan, the Philippines*, UWEP/WASTE, Gouda, The Netherlands

Medina, M. (1997) *Informal Recycling and Collection of Solid Wastes in Developing Countries: Issues and Opportunities*, UNU/IAS Working Paper No 24, United Nations University/Institute of Advanced Studies, Tokyo, Japan

Medina, M. (2007) *The World's Scavengers: Salvaging for Sustainable Consumption and Production*, AltaMira Press, Lanham, MD

Melosi, M. (1981) *Garbage in the Cities: Refuse, Reform and Environment, 1880–1980*, A&M Press, College Station, TX

METAP (Mediterranean Environmental Technical Assistance Programme) (undated) *Solid Waste Management Centre* website (Provides extensive guidelines, training materials and case studies for capacity building in solid waste management), www.metap-solidwaste.org/index.php?id=1

Mol, A. P. J. and Sonnenfeld, D. (eds) (2000) *Ecological Modernisation Around the World*, Frank Cass publishers, London and Portland, OR (first published as a special issue of *Environmental Politics*, spring 2000)

Monni, S., Pipatti, R., Lehtila, A., Savolainen, I. and Syri, S. (2006) *Global Climate Change Mitigation Scenarios for Solid Waste Management*, Technical Research Centre of Finland, VTT Publications, Espoo

OECD (Organisation for Economic Co-operation and Development) (undated) *Environmental Data: Compendium 2006–2008, Section on Waste*, OECD Working Group on Environmental Information and Outlooks, Environmental Performance and Information Division, Environmental Directorate, OECD, Paris, www.oecd.org/dataoecd/22/58/41878186.pdf

Packard, V. (1960) *The Waste Makers*, D. McKay Co, New York, NY

Price, J., Rivas, A. R. and Lardinois, I. (eds) (1998) *Micro and Small Enterprises: The Case of Latin America*, WASTE, Gouda, The Netherlands

Prüss, A., Giroult, E. and Rushbrook, P. (eds) (1999) *Safe Management of Wastes from Health-Care Activities*, WHO, Geneva

3R Knowledge Hub (undated) *Regional Knowledge Hub on 3Rs*, Asian Institute of Technology (AIT), UNEP Regional Resource Centre for Asia and the Pacific (UNEP RRC.AP) and United Nations Economic and Social Commission for Asia and the Pacific (UNESCAP), and supported by Asian Development Bank (ADB), www.3rkh.net/

Rivas, A. R., Price, J. and Lardinois, I. (1998) *Solid Waste Management in Latin America – The Case of Small and Micro-Enterprises and Cooperatives*, WASTE, Gouda, The Netherlands

Robinson, M., Simpson, M. et al (1990) Unpublished documents related to working with the informal sector to develop community composting in Jakarta, Indonesia, Harvard Institute for International Development (HIID), Cambridge, MA

Rosario, A. (2004) *Reduction of Child Labour in the Waste Picking Sector, India: Review and Findings of an Evaluative Field Study in Bangalore and Kolkata*, www.ilo.org/childlabour

Rothenberger, S., Zurbrugg, C., Enayetullah, I. and Maqsood Sinha, A. H. (2006) *Decentralised Composting for Cities of Low- and Middle-Income Countries: A User's Manual*, Waste Concern, Bangladesh and EAWAG, Switzerland, www.eawag.ch/organisation/abteilungen/sandec/publikationen/publications_swm/index_EN

Rouse, J. R. (2006) 'Seeking common ground for people: Livelihoods, governance and waste', *Habitat International*, vol 30, pp741–753

Rudin, V., Abarca, L., Roa, F. et al (2007) *Systematisation of the Methodology for Managing Electronic Waste in Costa Rica*, WASTE, Gouda, The Netherlands

Rushbrook, P. and Pugh, M. (1998) *Decision Maker's Guide to the Planning, Siting, Design and Operation of Landfills in Middle- and Lower-Income Countries*, SKAT/World Bank, Washington, DC

Sampson, M. (2009) *Refusing to be Cast Aside: Waste Pickers Organising around the World*, WIEGO, Paris and Johannesburg

Scheinberg, A. (series ed) (2001) *Integrated Sustainable Waste Management: Tools for Decision-Makers – Set of Five Tools for Decision-Makers. Experiences from the Urban Waste Expertise Programme (1995–2001)*, WASTE, Gouda, www.waste.nl

Scheinberg, A. (2003) 'The proof of the pudding: Urban recycling in North America as a process of ecological modernisation', *Environmental Politics*, vol 12, no 4, winter, pp49–75

Scheinberg, A. (ed) (2004) *City Case Studies of Bamako, La Ceiba, Bangalore and Batangas Bay, the Four ' PPS', Cities of the UWEP Plus Programme (Urban Waste Expertise Programme Plus) (2001–2004)*, WASTE, Gouda, The Netherlands, www.waste.nl

Scheinberg, A. (2008) 'A horse of a different colour: Sustainable modernisation of solid waste management and recycling in developing and transitional countries', Paper presented at the CWG workshop, Cluj-Napoca, Romania, 22–23 February

Scheinberg, A. and Anschutz, J. (2007) 'Slim pickin's: Supporting waste pickers in the ecological modernization of urban waste management systems', *International Journal of Technology Management and Sustainable Development*, vol 5, no 3, pp257–270

Scheinberg, A. and IJgosse, J. (2004) *Waste Management in The Netherlands*, Report prepared for UNITRABALHO, Recife, Brazil, WASTE, Gouda, The Netherlands

Scheinberg, A. and Mol, A. P. J. (in press) 'Mixed modernities: Transitional Bulgaria and the ecological modernisation of solid waste management', *Environment and Planning C*

Scheinberg, A., Mitrovic, A. and Post, V. (2007) *Assessment Report: Needs of Roma Collectors and Other Stakeholders in the PEP SE Region for Training, Technical Assistance, and Financial Services and Recommendations for Programmatic Response*, Prepared for the Recycling Linkages Private Enterprise Programme South East Europe (PEP SE) of the International Finance Corporation, Skopje, Macedonia

Scheinberg, A., with support from IJgosse, J., Fransen, F., Post, V. and representatives of the LOGO South municipalities (2008) *Closing the Circle: Bringing Integrated Sustainable Waste Management Home*, Vereniging van Nederlandse Gemeenten (Association of Dutch Municipalities), The Hague, The Netherlands, www.waste.nl/page/1673

Simpson et al (1992) *Advanced Disposal Fee Study*, Report prepared for the California Integrated Waste Management Board, Tellus Institute, Boston, MA

Skitt, J. (ed) (1992) *1000 Terms in Solid Waste Management*, International Solid Waste Association, Copenhagen

Solid Waste Management Association of the Philippines (2009) *National Framework Plan for the Informal Waste Sector in Solid Waste Management*, Funded by the UNEP International Environment Technology Centre, the Philippines

Spaargaren, G. and van Vliet, B. (2000) 'Lifestyles, Consumption and the Environment', Environmental Politics, vol 9, no 1, pp50–77

Spaargaren, G., Oosterveer, P., van Buuren, J. and Mol, A. P. J. (2005) *Mixed Modernities: Towards Viable Environmental Infrastructure Development in East Africa*, Position paper, Environmental Policy Department, Wageningen University and Research Centre, The Netherlands

State of New Jersey (USA) (1984) *The New Jersey Recycling Act*, Law passed by the New Jersey State Legislature, Trenton, NJ

State of New Jersey (USA) (1985–1990) *Annual Reports of the New Jersey Recycling Act*, New Jersey Department of Environment, Trenton, NJ

Strasser, S. (1999) *Waste and Want: A Social History of Trash*, Henry Holt and Company, LLC, New York, NY

Tchobanoglous, G., Theisen, H. and Vigil, S. (1993) *Integrated Solid Waste Management*, McGraw Hill, New York, NY

UNDP (United Nations Development Programme) (2007) *Human Development Report 2007/2008: Fighting Climate Change: Human Solidarity in a Divided World*, UNDP, New York, NY

UNDP (2009) *Human Development Report 2009: Overcoming Barriers: Human Mobility and Development*, UNDP, New York, NY

UNDP–PPPUE (2003) *Toolkit for Pro-poor PPPs*, UNDP–PPPUE, Johannesburg

UN-Habitat (2009) *State of the World's Cities 2008/2009: Harmonious Cities*, Earthscan Publications, London

US Environmental Protection Agency (undated) *The Consumer's Handbook for Reducing Solid Waste*, www.epa.gov/osw/wycd/catbook/index.htm

van Vliet, B., Chappels, H. and Shove, E. (2005) *Infrastructures of Consumption: Restructuring the Utility Industries*, Earthscan Publications, London

VNGI (2008) *Closing the Circle: Bringing Integrated Sustainable Waste Management Home*, Association of Dutch Municipalities (VNGI), The Hague, The Netherlands

Waste Concern (Bangladesh) and World Wide Recycling (Netherlands) (2007) *Project Design Document: Composting of Organic Waste in Dhaka*, CDM Executive Board Registry, http://cdm.unfccc.int/UserManagement/FileStorage/3QBI7J8O3ZD1UQYIIIDO71F0WAB6G0

Whiteman, A., Smith, P. and Wilson, D. C. (2001) 'Waste management: An indicator of urban governance', Paper prepared for the UK Department for International Development (DFID) and submitted by DFID to the UN-Habitat Global Conference on Urban Development, New York, June 2001, www.davidcwilson.com/Waste_Management_An_Indicator_of_Urban_Governance.pdf

WIEGO (Women in Informal Employment: Globalizing and Organizing) (2008) *Waste Pickers Without Frontiers*, First International and Third Latin American Conference of Waste Pickers, Bogotá, Colombia, March 2008, Supported by WIEGO and other organizations, www.inclusivecities.org/pdfs/WastePickers-2008.pdf

Wilson, D. C. (1993) 'The landfill stepladder: Landfill policy and practice in an historical perspective', in *Wastes Management*, CIWM, Northampton, UK, pp24–28

Wilson, D. C. (2007) 'Development drivers for waste management', *Waste Management and Research*, vol 25, pp198–207

Wilson, D. C., Whiteman, A. and Tormin, A. (2001) *Strategic Planning Guide for Municipal Solid Waste Management*, World Bank, www.worldbank.org/urban/solid_wm/erm/start_up.pdf

Wilson, D. C., Balkau, F. and Thurgood, M. (2003) *Training Resource Pack for Hazardous Waste Management in Developing Countries*, UNEP, ISWA and the Basel Convention, www.unep.fr/scp/waste/hazardous.htm

Wilson, D. C., Araba, A. O., Chinwah, K. and Cheeseman, C. R. (2009) 'Building recycling rates through the informal sector', *Waste Management*, vol 29, pp629–635

World Bank (undated) *Urban Development, Urban Solid Waste Management*,
http://web.worldbank.org/WBSITE/
EXTERNAL/TOPICS/EXTURBANDEVELOPMENT/
EXTUSWM/0,,menuPK:463847~pagePK:149018~
piPK:149093~theSitePK:463841,00.html

WRAP (Waste and Resources Action Programme)
(2009) *Waste Prevention Toolkit: Helping You Reduce Household Waste*, Interactive toolkit for local authorities and others, www.wrap.org.uk/applications/
waste_prevention_toolkit/restricted.rm

INDEX

Printed and bound by CPI Group (UK) Ltd, Croydon, CR0 4YY

23/10/2024

01777683-0001